Occupational Safety, Health, and Wellness

Fourth Edition

Lynne Murnane, Brad McLelland, and Michael Fox
Lead Senior Editors

Mark Keough
Technical Writer

Beckie Bigler and Veronica Smith
Editorial Staff

Christy Olson
Lead Instructional Developer

Missy Hannan
Senior Graphic Designer

IFSTA

Validated by the International Fire Service Training Association

Published by Fire Protection Publications • Oklahoma State University

The International Fire Service Training Association (IFSTA) was established in 1934 as a *nonprofit educational association of fire fighting personnel who are dedicated to upgrading fire fighting techniques and safety through training.* To carry out the mission of IFSTA, Fire Protection Publications was established as an entity of Oklahoma State University. Fire Protection Publications' primary function is to publish and distribute training materials as proposed, developed, and validated by IFSTA. As a secondary function, Fire Protection Publications researches, acquires, produces, and markets high-quality learning and teaching aids consistent with IFSTA's mission.

IFSTA holds two meetings each year: the Winter Meeting in January and the Annual Validation Conference in July. During these meetings, committees of technical experts review draft materials and ensure that the professional qualifications of the National Fire Protection Association standards are met. These conferences bring together individuals from several related and allied fields, such as:

- Key fire department executives, training officers, and personnel
- Educators from colleges and universities
- Representatives from governmental agencies
- Delegates of firefighter associations and industrial organizations

Committee members are not paid nor are they reimbursed for their expenses by IFSTA or Fire Protection Publications. They participate because of a commitment to the fire service and its future through training. Being on a committee is prestigious in the fire service community, and committee members are acknowledged leaders in their fields. This unique feature provides a close relationship between IFSTA and the fire service community.

IFSTA manuals have been adopted as the official teaching texts of many states and provinces of North America as well as numerous U.S. and Canadian government agencies. Besides the NFPA requirements, IFSTA manuals are also written to meet the Fire and Emergency Services Higher Education (FESHE) course requirements. A number of the manuals have been translated into other languages to provide training for fire and emergency service personnel in Canada, Mexico, and outside of North America.

ISBN 978-0-87939-638-1   Library of Congress Control Number: 2018936257

Fourth Edition, First Printing, March 2017   *Printed in the United States of America*

10 9 8 7 6 5 4 3 2 1

If you need additional information concerning the International Fire Service Training Association (IFSTA) or Fire Protection Publications, contact:

Customer Service, Fire Protection Publications, Oklahoma State University
930 North Willis, Stillwater, OK 74078-8045
800-654-4055   Fax: 405-744-8204

For assistance with training materials, to recommend material for inclusion in an IFSTA manual, or to ask questions or comment on manual content, contact:

Editorial Department, Fire Protection Publications, Oklahoma State University
930 North Willis, Stillwater, OK 74078-8045
405-744-4111   Fax: 405-744-4112   E-mail: editors@osufpp.org

Chapter Summary

Table of Contents

8 Protective Clothing and Equipment

9 Emergency Operations

10 Facility Safety

11 Safety and Health Program Administration

List of Tables

Acknowledgements

The fourth edition of **Occupational Safety, Health and Wellness** is designed to provide an overview of programs, policies, procedures, and actions that will establish and maintain a culture of safety within the fire and emergency services. The IFSTA Validation Committee for this manual wanted to provide information so that a culture based on safety, health, and wellness concerns can be instilled in generations of emergency responders. This manual addresses the objectives of NFPA 1500, *Standard on Fire Department Occupational Safety, Health, and Wellness Program, 2018 Edition*. Additional standards are listed in the introduction.

Acknowledgment and special thanks are extended to the members of the IFSTA validating committee who contributed their time and knowledge to the development of this manual.

Contract Writer

The fourth edition of **Occupational Safety, Health, and Wellness** could not have been completed without the skill and hard work of contracted technical writer Mark Keough. Mr. Keough provided the early drafts of all manual chapters to the validation committee for their review, revision, and approval. He also assisted the committee and lead editors with changes to chapter content and validation discussions.

IFSTA Occupational Safety, Health and Wellness Fourth Edition Validation Committee

Chair

Mark Pare
Director
Rhode Island Fire Academy
Plainville, Massachusetts

Vice Chair

Lance M. Loyer
Rogers Fire Department
Rogers, Arkansas

Secretary

Nathan Trauernicht
City of Davis & UC Davis Fire Departments
Davis, California

Committee Members

Robert J. Agnew
Oklahoma State University
Stillwater, Oklahoma

Russell Bainbridge
6715 Michelle Drive
Amarillo, Texas

John Barker
Champaign Fire Department
Champaign, Illinois

Kathleen Brown
Lincoln Fire & Rescue
Omaha, Nebraska

Billy Ewing
PRO-TEC Fire Services, LTD.
Green Bay, Wisconsin

Heather Goodrich
Clackamas Fire District #1
Clackamas, Oregon

Jeffrey Griffin
Oregon Fire Chiefs Association
Wilsonville, Oregon

Merlin Klassen
Brazeau County
Wetaskiwin, Alberta, Canada

IFSTA Occupational Safety, Health and Wellness Fourth Edition Validation Committee

Committee Members (cont.)

Barbara Law
Sacramento Metropolitan Fire District
Mather, California

Robert Moore
Texas A&M Engineering Extension Services
College Station, Texas

Frank Ricci
Casa Grande Fire Department
Casa Grande, Arizona

Timothy Robinson
Concord Fire Department
Concord, New Hampshire

Josh M. Stefancic
Largo Fire Rescue
Largo, Florida

Don Turno
Savannah River Nuclear Solutions
Aiken, South Carolina

James S. Williams
Loudoun County Fire and Rescue
Leesburg, Virginia

Greg Willis, Chief
Wetumpka Fire Department
Wetumpka, Alabama

David Zayonce
Camrose County
Viking, Alberta

Dedication

This manual is dedicated to the men and women who hold devotion to duty above personal risk, who count on sincerity of service above personal comfort and convenience, who strive unceasingly to find better and safer ways of protecting lives, homes, and property of their fellow citizens from the ravages of fire, medical emergencies, and other disasters

...The Firefighters of All Nations.

Last, but certainly not least, gratitude is extended to the following members of the Fire Protection Publications staff whose contributions made the final publication of this manual possible.

IFSTA Occupational Safety, Health and Wellness Fourth Edition, Project Team

Lead Senior Editors
Lynne Murnane, Senior Editor
Brad McLelland, Senior Editor
Michael Fox, Senior Editor

Contract Writer
Mark Keough, Battalion Chief
Mesa Fire and Medical Department
Mesa, Arizona

Lead Instructional Developer
Christy Olson, Curriculum Developer

Director of Fire Protection Publications
Craig Hannan

Editorial Manager
Clint Clausing

Curriculum Manager
Colby Cagle

Production Coordinator
Ann Moffat

eProducts Manager
Justin Smola

Editorial Staff
Beckie Bigler, Senior Editor
Tara Gladden, Editorial Assistant
Veronica Smith, Senior Editor

Illustrator and Layout Designer
Missy Hannan, Senior Graphic Designer

Curriculum Editor
Lindsey Dugan, Curriculum Developer

IFSTA/FPP Photographers
Jeff Fortney, Senior Editor

eProducts Staff
Janice French
Kelly Naas
Ryan Souders

Indexer
Nancy Kopper

The IFSTA Executive Board at the time of validation of the **Occupational Safety, Health, and Wellness, Fourth Edition** was as follows:

IFSTA Executive Board

Executive Board Chair

Steve Ashbrock
Fire Chief
Madeira & Indian Hill Fire Department
Cincinnati, OH

Vice Chair

Bradd Clark
Fire Chief
Ocala Fire Department
Ocala, FL

IFSTA Executive Director

Mike Wieder
Associate Director
Fire Protection Publications
Stillwater, OK

Board Members

Steve Austin
Project Manager
Cumberland Valley Volunteer Firemen's Association
Newark, DE

Mary Cameli
Assistant Chief
City of Mesa Fire Department
Mesa, AZ

Dr. Larry Collins
Associate Dean
Eastern Kentucky University
Safety, Security, & Emergency Department
Richmond, KY

Chief Dennis Compton
Mesa & Phoenix, Arizona
Chairman of the National Fallen Firefighters Foundation Board of Directors

John Hoglund
Director Emeritus
Maryland Fire & Rescue Institute
New Carrollton, MD

Tonya Hoover
State Fire Marshal
CA Department of Forestry & Fire Protection
Sacramento, CA

Dr. Scott Kerwood
Fire Chief
Hutto Fire Rescue
Hutto, TX

Wes Kitchel
Assistant Chief
Sonoma County Fire & Emergency Services Dept.
Santa Rosa, CA

Brett Lacey
Fire Marshal
Colorado Springs Fire Department
Colorado Springs, CO

Robert Moore
Division Director
Texas A&M Engineering Extension Services
College Station, TX

Dr. Lori Moore-Merrell
Assistant to the General President
International Association of Fire Fighters
Washington, DC

Jeff Morrissette
State Fire Administrator
State of Connecticut
Commission on Fire Prevention and Control
Windsor Locks, CT

Josh Stefancic
Division Chief
Largo Fire Rescue
Largo, FL

Paul Valentine
Senior Engineer
Nexus Engineering
Oakbrook Terrace, IL

Steven Westermann
Fire Chief
Central Jackson County Fire Protection District
Blue Springs, MO

Introduction

Introduction Contents

Introduction

The first edition of the International Fire Service Training Association's (IFSTA) **Fire Department Occupational Safety** manual dates back almost 40 years. And while the world is very different than it was in 1979, educating fire and emergency service professionals on safety, health and wellness **remains a top priority.** This 4th edition of **Occupational Safety, Health, and Wellness** gives those in the field an education on these important topics.

Risks come with the job, and providing a safe work environment for members should be a goal of all emergency service organizations. Because fire fighting, rescue, emergency medical activities and other work involves hazardous situations, all personnel need to prioritize safety. Some professionals operate with the goals of protecting the lives and property of the community while providing a safe work environment for emergency responders. This manual offers guidelines to help organizations and individuals establish a culture of safety in the midst of occupational risks and hazards. It also provides fire and emergency responders with recommendations for a healthier lifestyle.

To reach these goals, this manual includes significant changes from its 2010 predecessor. The information has been updated to include current research and it is delivered with easily understood, efficient writing.

On its face, this manual is about occupational safety and wellness, but **in practice** it's about you. It will point you toward a safe work environment; it will give you tips that will help you lead a healthier life; and it will help you create worthwhile safety policies and procedures. That's true for newcomers and department chiefs. Some of the policies and procedures are geared toward the following:

- **Individual:** Adhere to an organization's safety and health policies, maintain a healthy lifestyle, and behave safely. You are responsible for the actions of fellow workers and team members, and your own. Correct bad behavior, report unsafe equipment or situations, and recommend possible changes to policy and procedures that will increase personal safety for yourself and others.
- **Supervisor:** Company officers and chief officers are responsible for the survival, health, and occupational safety of the team. Set a positive example by following rules and procedures, enforcing safety and health policies, and monitoring the safety program. Stop unsafe actions and correct them immediately, fix unsafe conditions, report unsafe equipment or situations, and forward notifications of issues and correction recommendations to the proper authorities right away.

- **Manager:** As the manager of your organization, be aware of trends in accidents, injuries and line-of-duty deaths (LODD), locally and nationally. The figures may indicate breakdowns in safety and health programs. Provide leadership with any and all resources available.
- **Organization:** An organization, and its individual members, must create and maintain a culture that prioritizes safety and health issues. This culture will lead to a safer work environment for all. To create a safe and healthy workplace, fire and emergency service management, members, and union leadership (if applicable) must work together.

While many programs focus on emergency incident safety, research indicates that most LODDs come from stress-induced cardiac arrests and strokes. The risks that lead to these conditions can be controlled and reduced by taking control of these key factors:

- What you eat and how much
- Your physical fitness routine
- Your use of drugs, tobacco or alcohol
- How you deal with stress

Personal and professional behaviors can change for the better. With additional research, training, and educational tools like this updated manual, positive changes continue to improve the safety and well-being of emergency service professionals and their organizations.

NFPA Requirements

The National Fire Protection Association (NFPA) professional qualifications standards include safety requirements that apply to each job title or function. While this manual is based on NFPA 1500, *Standard on Fire Department Occupational Safety and Health Program,* the appropriate job performance requirements (JPRs) of each NFPA professional qualification will be referenced in the text. Appendix A includes a correlation between the NFPA JPRs and the chapters and pages where they are addressed.

The chapters that contain safety requirements include the following:

- NFPA 471 *Recommended Practice for Responding to Hazardous Materials Incidents*
- NFPA 472, *Standard for Competence of Responders to Hazardous Materials/Weapons of Mass Destruction Incidents*
- NFPA 473, *Standard for Competencies for EMS Personnel Responding to Hazardous Materials/Weapons of Mass Destruction Incidents*
- NFPA 1001, *Standard for Fire Fighter Professional Qualifications*
- NFPA 1002, *Standard for Fire Apparatus Driver/Operator Professional Qualifications*
- NFPA 1003, *Standard for Airport Fire Fighter Professional Qualifications*
- NFPA 1006, *Standard for Technical Rescue Personnel Professional Qualifications*
- NFPA 1021, *Standard for Fire Officer Professional Qualifications*

- NFPA 1031, *Standard for Professional Qualifications for Fire Inspector and Plan Examiner*
- NFPA 1041, *Standard for Fire Service Instructor Professional Qualifications*
- NFPA 1051, *Standard for Wildland Firefighting Personnel Professional Qualifications*
- NFPA 1071, *Standard for Emergency Vehicle Technician Professional Qualifications*
- NFPA 1081, *Standard for Industrial Fire Brigade Member Professional Qualifications*
- NFPA 1670, *Standard on Operations and Training for Technical Search and Rescue Incidents*

Nonprofessional qualification standards that have references to safety and health requirements include the following:

- NFPA 921, *Guide for Fire and Explosion Investigations*
- NFPA 1403, *Standard on Live Fire Training Evolutions*
- NFPA 1521, *Standard for Fire Department Safety Officer Professional Qualifications*
- NFPA 1561, *Standard on Emergency Services Incident Management System and Command Safety*
- NFPA 1581, *Standard on Fire Department Infection Control Program*
- NFPA 1582, *Standard on Comprehensive Occupational Medical Program for Fire Departments*
- NFPA 1583, *Standard on Health-Related Fitness Programs for Fire Department Members*
- NFPA 1584, *Standard on the Rehabilitation Process for Members During Emergency Operations and Training Exercises*

Purpose and Scope

Fire and emergency responder safety is a primary concern of all organizations and individuals engaged in providing emergency services. This 4th edition of **Occupational Safety, Health, and Wellness** intends to meet the requirements for specific ranks or positions in the fire and emergency services organization by providing the general knowledge that may not be found in other manuals and textbooks.

This manual has two purposes and is written for both associate's level college students and emergency services personnel. First, the manual is intended to meet the Fire and Emergency Services Higher Education (FESHE) learning outcomes and lesson outlines for the associate's course *Occupational Safety and Health for Emergency Responders*. Second, this manual provides guidance for meeting NFPA 1500, *Standard on Fire Department Occupational Safety and Health Program*. NFPA 1500 contains the minimum requirements and procedures for a safety and health program. It calls upon the fire and emergency services to recognize the importance of member safety and health, establish official safety and health objectives for the department or organization, and provide as safe and healthy a work environment as possible.

The scope of this manual is to provide students and emergency responders with the motivation, knowledge, skills, and abilities that they need to safely engage in emergency and nonemergency activities on a daily basis. The chapters of this manual are correlated to the FESHE Outcomes; however, NFPA 1500 is heavily referenced throughout the text as are many other NFPA safety-related standards.

Safety and Health topics in this manual include the following:

- Risks, hazards, and risk management
- Suggestions for creating a safety culture
- Physical fitness, wellness, and health considerations
- Training safety
- Apparatus safety
- Tool and equipment safety, maintenance, and operation
- Personal protective equipment selection, care, and use
- Emergency scene safety
- Facility safety
- Safety program establishment and implementation

Terminology

IFSTA has traditionally provided training materials for use throughout the U.S. and Canada. Manuals have been focused on that audience, even though the audience has grown to include personnel from around the world. That means less jargon, and more precise descriptions and definitions. The shift toward more-accessible verbiage will make translated versions more accurate and ultimately help professionals around the world.

Key Information

Some information in this book appears in shaded boxes marked by icons. Here are some examples:

Information

Information boxes offer information that complements the text discussion. It should be a point of emphasis, something that deserves special attention. Some information boxes may be a summary of points, examples, calculations or a listing of advantages and disadvantages on a topic. Some examples of information boxes include:

- Sidebar
- Information
- Key Information

Key Term

A **key term** is designed to emphasize key concepts, technical terms, or ideas. Key terms are listed at the beginning of each chapter, highlighted in bold **red** font. The definition is placed in the margin for easy reference.

Incident Commander (IC) — Person in charge of the Incident Command System and responsible for the management of all incident operations during an emergency.

Signal Words

Three key signal words are found in the book: **WARNING, CAUTION,** and **NOTE.** Definitions and examples of each are as follows:

- **WARNING** indicates information that could result in death or serious injury. See the following example:

> **WARNING**
> Ice-rescue suits are not designed to be used in moving water.

- **CAUTION** indicates important information or data that first responders need to be aware of in order to perform their duties safely. See the following example:

> **CAUTION**
> Personnel who arrive individually at an incident must always check in at the command post or staging area, if one has been implemented. Personnel must not freelance or assign themselves to a task. Freelancing is unsafe and can result in fatalities or injuries.

- Note indicates important operational information that helps explain why a particular recommendation is given or describes optional methods for certain procedures. See the following example:

NOTE: Local terminology may differ, for example warm zone and contamination reduction zone.

Metric Conversions

Throughout this manual, U.S. units of measure are converted to metric units for the convenience of our international readers. Be advised that we use the Canadian metric system. It is very similar to the International System of Units (SI), but may have some variations.

We adhere to the following guidelines for metric conversions in this manual:

- Metric conversions are approximated unless the number is used in mathematical equations.
- Centimeters are not used because they are not part of the Canadian metric standard.

- Exact conversions are used when an exact number is necessary such as in construction measurements or hydraulic calculations.
- Set values such as hose diameter, ladder length, and nozzle size use their Canadian counterpart naming conventions and are not mathematically calculated. For example, 1½ inch hose is referred to as 38 mm hose.

The following two tables provide detailed information on IFSTA's conversion conventions. The first table includes examples of our conversion factors for a number of measurements used in the fire service. The second shows examples of exact conversions beside the approximated measurements you will see in this manual.

U.S. to Canadian Measurement Conversion

Measurements	Customary (U.S.)	Metric (Canada)	Conversion Factor
Length/Distance	Inch (in) Foot (ft) [3 or less feet] Foot (ft) [3 or more feet] Mile (mi)	Millimeter (mm) Millimeter (mm) Meter (m) Kilometer (km)	1 in = 25 mm 1 ft = 300 mm 1 ft = 0.3 m 1 mi = 1.6 km
Area	Square Foot (ft²) Square Mile (mi²)	Square Meter (m²) Square Kilometer (km²)	1 ft² = 0.09 m² 1 mi² = 2.6 km²
Mass/Weight	Dry Ounce (oz) Pound (lb) Ton (T)	gram Kilogram (kg) Ton (T)	1 oz = 28 g 1 lb = 0.5 kg 1 T = 0.9 T
Volume	Cubic Foot (ft³) Fluid Ounce (fl oz) Quart (qt) Gallon (gal)	Cubic Meter (m³) Milliliter (mL) Liter (L) Liter (L)	1 ft³ = 0.03 m³ 1 fl oz = 30 mL 1 qt = 1 L 1 gal = 4 L
Flow	Gallons per Minute (gpm) Cubic Foot per Minute (ft³/min)	Liters per Minute (L/min) Cubic Meter per Minute (m³/min)	1 gpm = 4 L/min 1 ft³/min = 0.03 m³/min
Flow per Area	Gallons per Minute per Square Foot (gpm/ft²)	Liters per Square Meters Minute (L/(m².min))	1 gpm/ft² = 40 L/(m².min)
Pressure	Pounds per Square Inch (psi) Pounds per Square Foot (psf) Inches of Mercury (in Hg)	Kilopascal (kPa) Kilopascal (kPa) Kilopascal (kPa)	1 psi = 7 kPa 1 psf = .05 kPa 1 in Hg = 3.4 kPa
Speed/Velocity	Miles per Hour (mph) Feet per Second (ft/sec)	Kilometers per Hour (km/h) Meter per Second (m/s)	1 mph = 1.6 km/h 1 ft/sec = 0.3 m/s
Heat	British Thermal Unit (Btu)	Kilojoule (kJ)	1 Btu = 1 kJ
Heat Flow	British Thermal Unit per Minute (BTU/min)	watt (W)	1 Btu/min = 18 W
Density	Pound per Cubic Foot (lb/ft³)	Kilogram per Cubic Meter (kg/m³)	1 lb/ft³ = 16 kg/m³
Force	Pound-Force (lbf)	Newton (N)	1 lbf = 0.5 N
Torque	Pound-Force Foot (lbf ft)	Newton Meter (N.m)	1 lbf ft = 1.4 N.m
Dynamic Viscosity	Pound per Foot-Second (lb/ft.s)	Pascal Second (Pa.s)	1 lb/ft.s = 1.5 Pa.s
Surface Tension	Pound per Foot (lb/ft)	Newton per Meter (N/m)	1 lb/ft = 15 N/m

Conversion and Approximation Examples

Measurement	U.S. Unit	Conversion Factor	Exact SI Unit	Rounded SI Unit
Length/Distance	10 in	1 in = 25 mm	250 mm	250 mm
	25 in	1 in = 25 mm	625 mm	625 mm
	2 ft	1 in = 25 mm	600 mm	600 mm
	17 ft	1 ft = 0.3 m	5.1 m	5 m
	3 mi	1 mi = 1.6 km	4.8 km	5 km
	10 mi	1 mi = 1.6 km	16 km	16 km
Area	36 ft^2	1 ft^2 = 0.09 m^2	3.24 m^2	3 m^2
	300 ft^2	1 ft^2 = 0.09 m^2	27 m^2	30 m^2
	5 mi^2	1 mi^2 = 2.6 km^2	13 km^2	13 km^2
	14 mi^2	1 mi^2 = 2.6 km^2	36.4 km^2	35 km^2
Mass/Weight	16 oz	1 oz = 28 g	448 g	450 g
	20 oz	1 oz = 28 g	560 g	560 g
	3.75 lb	1 lb = 0.5 kg	1.875 kg	2 kg
	2,000 lb	1 lb = 0.5 kg	1 000 kg	1 000 kg
	1 T	1 T = 0.9 T	900 kg	900 kg
	2.5 T	1 T = 0.9 T	2.25 T	2 T
Volume	55 ft^3	1 ft^3 = 0.03 m^3	1.65 m^3	1.5 m^3
	2,000 ft^3	1 ft^3 = 0.03 m^3	60 m^3	60 m^3
	8 fl oz	1 fl oz = 30 mL	240 mL	240 mL
	20 fl oz	1 fl oz = 30 mL	600 mL	600 mL
	10 qt	1 qt = 1 L	10 L	10 L
	22 gal	1 gal = 4 L	88 L	90 L
	500 gal	1 gal = 4 L	2 000 L	2 000 L
Flow	100 gpm	1 gpm = 4 L/min	400 L/min	400 L/min
	500 gpm	1 gpm = 4 L/min	2 000 L/min	2 000 L/min
	16 ft^3/min	1 ft^3/min = 0.03 m^3/min	0.48 m^3/min	0.5 m^3/min
	200 ft^3/min	1 ft^3/min = 0.03 m^3/min	6 m^3/min	6 m^3/min
Flow per Area	50 gpm/ft^2	1 gpm/ft^2 = 40 L/(m^2.min)	2 000 L/(m^2.min)	2 000 L/(m^2.min)
	326 gpm/ft^2	1 gpm/ft^2 = 40 L/(m^2.min)	13 040 L/(m^2.min)	13 000 L/(m^2.min)
Pressure	100 psi	1 psi = 7 kPa	700 kPa	700 kPa
	175 psi	1 psi = 7 kPa	1225 kPa	1 200 kPa
	526 psf	1 psf = 0.05 kPa	26.3 kPa	25 kPa
	12,000 psf	1 psf = 0.05 kPa	600 kPa	600 kPa
	5 psi in Hg	1 psi = 3.4 kPa	17 kPa	17 kPa
	20 psi in Hg	1 psi = 3.4 kPa	68 kPa	70 kPa
Speed/Velocity	20 mph	1 mph = 1.6 km/h	32 km/h	30 km/h
	35 mph	1 mph = 1.6 km/h	56 km/h	55 km/h
	10 ft/sec	1 ft/sec = 0.3 m/s	3 m/s	3 m/s
	50 ft/sec	1 ft/sec = 0.3 m/s	15 m/s	15 m/s
Heat	1200 Btu	1 Btu = 1 kJ	1 200 kJ	1 200 kJ
Heat Flow	5 BTU/min	1 Btu/min = 18 W	90 W	90 W
	400 BTU/min	1 Btu/min = 18 W	7 200 W	7 200 W
Density	5 lb/ft^3	1 lb/ft^3 = 16 kg/m^3	80 kg/m^3	80 kg/m^3
	48 lb/ft^3	1 lb/ft^3 = 16 kg/m^3	768 kg/m^3	770 kg/m^3
Force	10 lbf	1 lbf = 0.5 N	5 N	5 N
	1,500 lbf	1 lbf = 0.5 N	750 N	750 N
Torque	100	1 lbf ft = 1.4 N.m	140 N.m	140 N.m
	500	1 lbf ft = 1.4 N.m	700 N.m	700 N.m
Dynamic Viscosity	20 lb/ft.s	1 lb/ft.s = 1.5 Pa.s	30 Pa.s	30 Pa.s
	35 lb/ft.s	1 lb/ft.s = 1.5 Pa.s	52.5 Pa.s	50 Pa.s
Surface Tension	6.5 lb/ft	1 lb/ft = 15 N/m	97.5 N/m	100 N/m
	10 lb/ft	1 lb/ft = 15 N/m	150 N/m	150 N/m

Risk of a High-Hazard Occupation

Chapter Contents

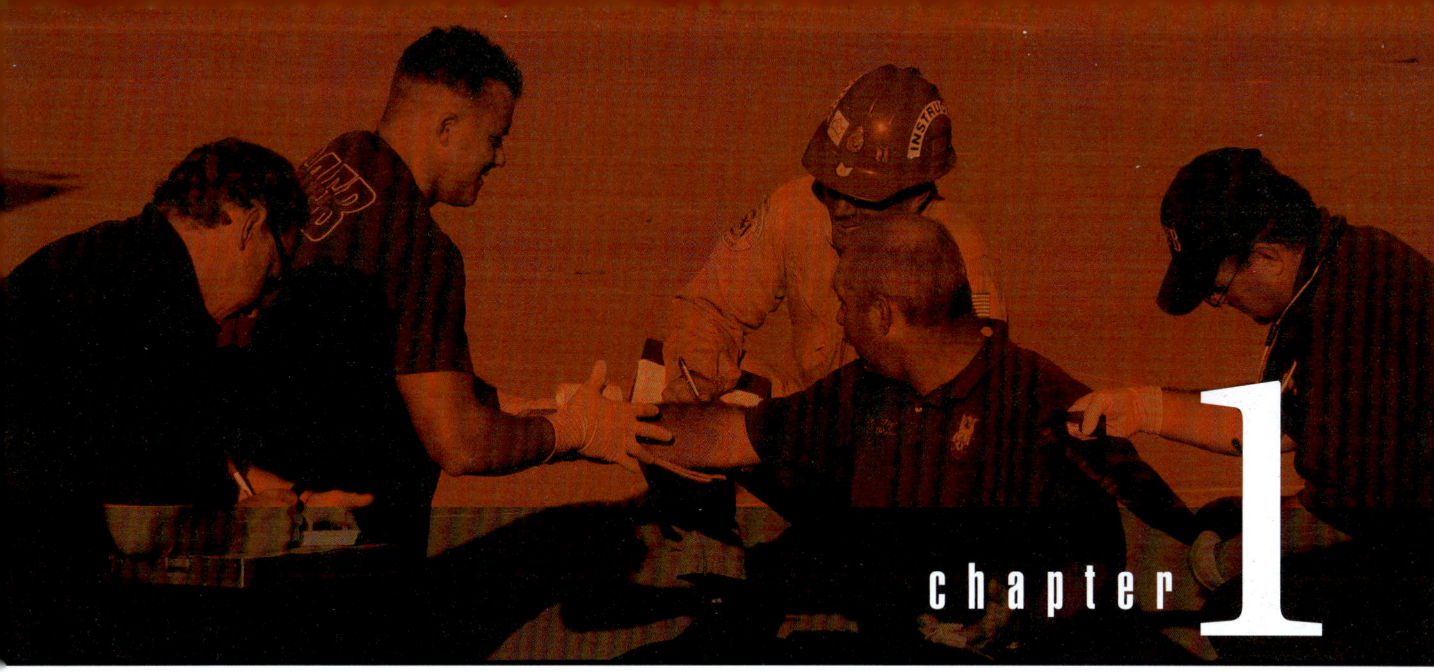

chapter 1

Key Terms

FESHE Learning Outcomes

Occupational Safety and Health for Emergency Services

4. List and describe the components of risk identification, risk evaluation, and incident management.

6. Apply the knowledge of an effective safety plan to preincident planning, response, and training activities.

10. Describe the responsibilities of individual responders, supervisors, safety officers, and incident commanders, safety program managers, safety committees and fire department managers as they relate to health and safety programs.

12. Identify and analyze the major causes involved in line-of-duty firefighter deaths related to health, wellness, fitness and vehicle operations.

Risks of a High-Hazard Occupation

Learning Objectives

After reading this chapter, students should be able to:

1. Identify hazards and risks in the fire service.
2. Describe hazard and risk management models.
3. Recognize factors that contribute to occupational risks.
4. Describe possible causes and consequences of accidents in the fire service.
5. Describe proactive approaches to accident reduction.

Chapter 1
Risks of a High-Hazard Occupation

This chapter describes fundamental concepts about hazard and risk identification and management, the all-hazards concept, differences between hazards and risks, types of hazards, hazard-control methods, and basic risk management models. Other factors contributing to occupational risks are also described. After completing this chapter, the reader should gain a greater understanding of different types of hazards, risk management, external pressures that impact risk management, and initial steps to take to reduce the likelihood of accidents.

Hazards and Risks

Firefighters encounter unpredictable environments while performing their daily activities and during emergency responses. The risk of losing health, life, or harming people is always there. Adequate preparation is necessary to avoid loss in one form or another. Emergency response organizations must be aware of types of:

- Psychological risks
- Hazard concepts
- Injury and death statistics
- Risk management models
- Hazard control techniques

Definitions

The following definitions are those used by IFSTA, which defines the terms in its **Fire Service Orientation and Terminology** manual as:

- ***Hazard*** — Condition, substance, or device that can directly cause injury or loss; the source of a risk.
- ***Risk*** — Likelihood or possibility of suffering harm from a hazard; exposure to a hazard; the potential for failure or loss.

As an example, the **hazard** may be the release of toxic gas into the atmosphere. The **risk** would be the likelihood and consequence of injury from being exposed to the toxic gas. The following should be considered when evaluating the level of risk to a toxic gas exposure:

- Toxicity
- Concentration
- Routes of exposure
- Duration
- Personal Protective Equipment (PPE)

Hazard — Condition, substance, or device that can directly cause injury or loss; the source of a risk.

Risk — (1) Likelihood of suffering harm from a hazard; exposure to a hazard. The potential for failure or loss. (2) Estimated effect that a hazard would have on people, services, facilities, and structures in a community; likelihood of a hazard event resulting in an adverse condition that causes injury or damage.

Psychological Hazards

Fire departments exist because hazards exist. Fire fighting is a high-hazard occupation, which means that firefighters engage in activities or are exposed to conditions that may result in injuries, property loss, or fatalities. Hazards come in all shapes, sizes and severities. What is less predictable is how firefighters will perform their duties under pressure, which often ends up factoring in close calls, accidents, injuries, and fatalities. Documenting these events and their causal factors is important to learn better methods to protect ourselves and those who follow us.

In addition to the complexity of the systems and environments firefighters may face regularly, three internal and external pressures face today's fire service:

- Production pressure
- Financial pressure
- Complacency

These apply to every firefighter and include all of the specialty organizational components that define a modern-day emergency services organization.

Production Pressure

All individuals and divisions face production pressure, namely expectations for delivery of services regardless of staffing levels or financial support. Production pressure may result in the following:

- Making rash decisions at an emergency scene to appease the public who may be watching
- Conducting training exercises to meet a schedule without adequate personnel or safety measures
- Causing shop mechanics to replace parts with substandard parts, or parts just outside of specification requirements, to get the apparatus back on the road as soon as possible

Financial Pressure

The U.S. recession of the late 2000s caused fire departments to become more acutely aware of financial pressures. Regular funding shortfalls have prompted staffing cuts, reduction in training, and lowered staffing of fire trucks in many communities. The subsequent psychological stress helped make fire fighting one of the most stressful occupations in the United States in 2016. Unlike production pressures, financial pressures generate various shortages at the cost of modernizing administrative practices. These shortages act to normalize deviance away from procedures meant to create safe staffing.

Complacence

A good safety record can result in pressure (real or perceived) that works against an organization. Clean safety records actually speed up work because over time, personnel may become complacent about the inherent dangers in their work. Less careful, they may lose situational awareness.

Firefighters can develop the expectation that they will continue to be successful because past efforts have worked well. They may develop a sense of being bulletproof. Those who disagree with the group may become reluctant to speak up for fear of ridicule or reprisal.

These challenges for firefighters and emergency response organizations will not go away. This manual outlines ways for individuals and organizations to improve the health, safety, and wellness of an organization and its members in a complex and ever-changing environment. Everyone is a risk manager.

All-Hazard Concept

Trying to plan for every hazard can be daunting for community fire service organizations. A systematic, all-inclusive community strategy for hazard mitigation and prevention works best in managing hazards. Using an all-hazards approach to manage risk requires the following:

- Performing risk analysis
- Prioritizing risk based on vulnerability or probability
- Planning functional strategies for managing community risk

Because the unexpected can happen, fire departments must prepare for any emergency. As firefighters train, they will develop a higher state of readiness to handle a variety of incident types. Many hazard types tend to share characteristics. With careful review and planning for similar types of hazards, fire departments can establish a baseline set of capabilities and training, known as an **all-hazard concept,** which will address anticipated risks of the community. For instance, when crews train for universal precautions when managing an ill patient, the core practices of universal precautions get reinforced for all patient encounters.

All-Hazard Concept — Provides a coordinated approach to a wide variety of incidents; all responders use a similar, coordinated approach with a common set of authorities, protections, and resources.

At its core, an all-hazards approach is based on understanding community risk and developing resources to mitigate those risks. The all-hazards approach to preparation requires fire departments to consider hazards on a broad external and internal basis. Hazards can be grouped into three general types:

- Community
- Organizational
- Individual

Assessment and hazard control are primary job functions of firefighters, company officers, and chief officers.

Community Hazards

Communities contain all types of hazards, including:

- Mechanical
- Electrical
- Chemical
- Environmental **(Figure 1.1a, p. 14)**

The term refers to any potential hazard that may affect the community. An example of a natural community hazard would be a river that passes through a city and is a possible source of flooding. An example of a human-caused community hazard would be a refrigeration plant that contains large quantities of chlorine gas. **(Figure 1.1b)**. Additional human-caused hazards include civil unrest and terrorism.

Figure 1.1a A flood is an example of a hazard caused by a natural act. *Courtesy of Chris Mickal, District Chief, New Orleans (LA) FD Photo Unit.*

Figure 1.1b When an oil spill exceeds established reporting quantities, the National Response Center (NRC) must be notified. *Courtesy of the U.S. Coast Guard.*

Organizational Hazards

A fire department places its members in harm's way whether they are well trained or not. For instance, if members of a fire department are trained at the first responder level for hazardous materials response and the community has a major chlorine gas leak, the department is likely to be underequipped and undertrained to mitigate the hazard. Many departments would attempt to mitigate the leak anyway, even if it was beyond their capabilities or if they did not have sufficient staffing. When people are in danger, the expectation is that emergency responders will act **(Figure 1.2)**.

When fire department organizations and local governments see a mismatch between community hazards and fire department training or equipment, the organization becomes part of the hazard equation if it fails to respond with funding and staffing. If government agencies, including fire prevention staff, allow for zoning variances that permit high levels of any hazard to occupy areas not adequately arranged for them, the **community risk reduction** (fire prevention) staff becomes part of the hazard equation.

Community Risk Reduction —The process of addressing the larger issue of preventable injury that is occurring in a community; involves identifying leading risks and creating mitigation strategies through the use of integrated prevention interventions.

If the fire department receives adequate funding to manage and prepare for recognized hazards, it must train staff and equip responders. Preparation needs to include procedures, plans, risk analysis, and cooperative relationships with agencies or corporations that could offer response information or assistance. As fire departments implement training for community hazards, the training and preparation phase can become highly hazardous to firefighters with inadequate training or staffing.

Figure 1.2 This decon trailer would be considered a resource at a haz mat rescue.

Individual Hazards

Firefighters come from all backgrounds, cultures, and ethnic origins. Responders bring histories, education, experience, and desire to serve on every call. The community and organizational hazards faced become individual physical, emotional, and psychological hazards.

Environmental and physical hazards are major factors in firefighter injuries. They include:

- Heat
- Cold **(Figure 1.3a and b)**
- Noise
- Uneven terrain
- Repetitive tasks such as:
  - bending
  - lifting
  - pulling work

Figure 1.3a Firefighters encounter unique hazards in cold weather, including frostbite and hypothermia. *Courtesy of Ron Jeffers, Union City, NJ.*

Figure 1.3b Facepieces can malfunction in extremely cold weather. *Courtesy of Ron Jeffers, Union City, NJ.*

Physiological stressors can contribute to the danger in physical hazards. They include:

- Sleep deprivation
- Long work periods
- Dehydration **(Figure 1.4)**
- Frequent changes in work environments

Figure 1.4 Rehabilitation breaks provide rest and rehydration.

Additional injuries and fatalities can be caused by:

- Structural fires **(Figure 1.5a)**
- Collapse
- Confined spaces
- High angles **(Figure 1.5b)**
- Contagious diseases
- Chemical releases
- Moving traffic on roadways

Figure 1.5a Proper training is needed to avoid injuries and fatalities at structure fires. *Courtesy of Mike Wieder.*

Figure 1.5b Personnel that perform high angle rescue should have a clear understanding of the safety concerns at each incident.

Firefighters get exposed to emotional hazards at many emergency scenes. Resistance or resilience to the effects of trauma and loss vary widely. Emotional hazards tend to increase as the trauma or loss impacts the families and members of the fire department. The loss of human life is especially tragic when it is unnatural in origin. The effect of trauma and loss on a firefighter's well-being should not be underestimated.

Psychological hazards result in stress, which is usually associated with some form of emotional or mental strain that results in physiological changes to the body. Stress influences changes to physical health, cognitive thought, and behaviors **(Figure 1.6)**. Psychological hazards usually result in an acute, chronic, or delayed stress response. Firefighters go from quiet periods to times of intense activity, danger, and unknown consequences in minutes on account of their shift schedules and a broadening list of hazards.

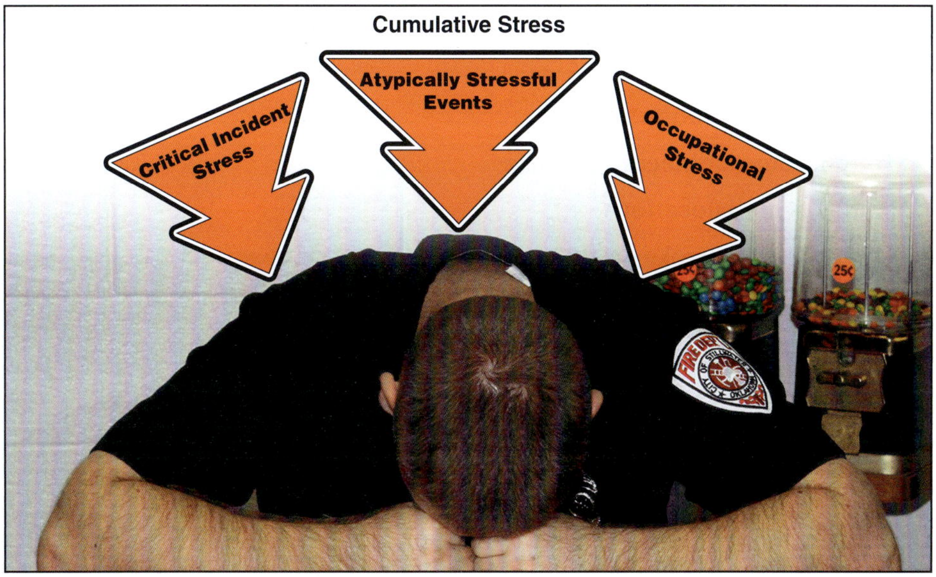

Figure 1.6 Stress is cumulative and comes from occupational duties, "routine" emergency calls, and atypically stressful events. Seeking counseling for individuals showing signs of cumulative stress is a good way to lengthen their firefighting careers and protect their health.

Hazard Categories

Government agencies, nongovernmental organizations, the insurance industry, and safety professionals use a variety of methods for categorizing hazards. Two major organizations that gather data on firefighter injuries, illnesses, and deaths are the **National Fire Protection Association (NFPA)** and the U.S. Department of Labor Bureau of Labor Statistics (BLS).

National Fire Protection Association (NFPA) — U.S. nonprofit educational and technical association devoted to protecting life and property from fire by developing fire protection standards and educating the public.

NFPA Statistics

The NFPA publishes an annual fatality and injury report: *Firefighter Injuries in the United States*. Fire department health and safety officers (HSOs) should consider adjusting accident and injury databases to match the NFPA's annual assessment of firefighter injuries. This effort would aid participating fire departments in determining if their accident and injury rates seem comparatively high or low.

The NFPA divides hazard types into five categories that include the following:

1. ***Fireground*** - Fatalities and injuries that occur during fire suppression activities
2. ***Nonfire*** - Fatalities and injuries that occur during other types of emergency incidents, including hazardous materials spills and medical, rescue, or extrication incidents
3. ***Responding /Returning*** - Fatalities and injuries that result from motor vehicle or apparatus accidents
4. ***Other On-Duty*** - Fatalities and injuries that occur during nonemergency activities such as station maintenance, inspections or physical fitness programs
5. ***Training*** - Fatalities and injuries that occur during training evolutions or courses

The type of injury further divides these categories:

- Burns
- Smoke/gas inhalation
- Wounds, cuts, or bleeding bruise
- Fractures or dislocations
- Heart attack or stroke
- Strains, sprains, or muscular pain
- Thermal stress

A further subdivision of the injuries includes:

- Overexertion **(Figure 1.7)**
- Contact with an object
- Fall or jump
- Struck by an object
- Extreme weather **(Figure 1.8)**
- Exposure to fire products
- Exposure to chemicals/radiation

Bureau of Labor Statistics

For the purpose of tracking workplace injuries and illnesses, the BLS developed a method to categorize hazards and their causes. These categories include the following:

- Type of task being performed
- Type of injury or illness
- Body part affected
- Time, date, and location of injury/illness

By establishing standardized categories of hazards, government agencies and employers can effectively gather accurate hazard, injury, fatality, illness, and accident information. Accurate record keeping permits the government and industry to recognize and respond to hazardous trends.

Figure 1.7 At some incidents, firefighters require first aid for overexertion. *Courtesy of Ron Jeffers, Union City, NJ.*

Figure 1.8 Extremely hot weather contributes to hazards at some incidents, including heat exhaustion or heat stroke. *Courtesy of Ron Jeffers, Union City, NJ.*

Hazard Control

Fire department hazard control efforts include any precautions taken to protect firefighters at work. As hazards get identified and assessed in stations, warehouses, shops, training facilities, and offices, fire department members must determine the best methods for controlling them. Developing a hazard control hierarchy or using the Three E's of Safety model work as systematic approaches to eliminate or minimize exposures to hazards.

Hazard Control Hierarchy

The common hazard control hierarchy involves moving from the most effective to the least reliable practice of eliminating or reducing hazards **(Figure 1.9, p. 20)**:

- **Elimination** is a physical removal of the hazard.
- **Substitution** involves replacing something with a less-hazardous product or process. For example: training facilities replaced burning natural products and hydrocarbons with propane props that are easier to control.
- **Engineered controls** may require installing equipment that isolates people from the hazard. An example would be diesel exhaust systems in station bays. Engineered controls are often the most expensive method of control but are highly reliable.
- **Administrative controls** require changing the way people work around hazards. These are procedures, training, warning signs, and product labels. **Safety Data Sheets (SDS)** are a form of hazard control because they inform about the hazard, offer recommendations for PPE, and list ways to decontaminate and treat individuals exposed to the hazard.
- **Personal Protective Equipment (PPE)**. PPE is least desirable because it is used around or when handling hazards. PPE can fail and it causes physiological changes to the wearer. PPE may also hinder task completion because of its cumbersome fit. All PPE has limitations that must be understood by the wearer before donning the equipment.

Safety Data Sheet (SDS) — Reference material that provides information on chemicals that are used, produced, or stored at a facility; Form is provided by chemicals manufacturers and blenders; contains information about chemical composition, physical and chemical properties, health and safety hazards, emergency response procedures, and waste disposal procedures. Also known as Material Safety Data Sheet (MSDS) or Product Safety Data Sheet (PSDS).

Personal Protective Equipment (PPE) — General term for the equipment worn by fire and emergency services responders; includes helmets, coats, trousers, boots, eye protection, hearing protection, protective gloves, protective hoods, self-contained breathing apparatus (SCBA), personal alert safety system (PASS) devices, and chemical protective clothing.

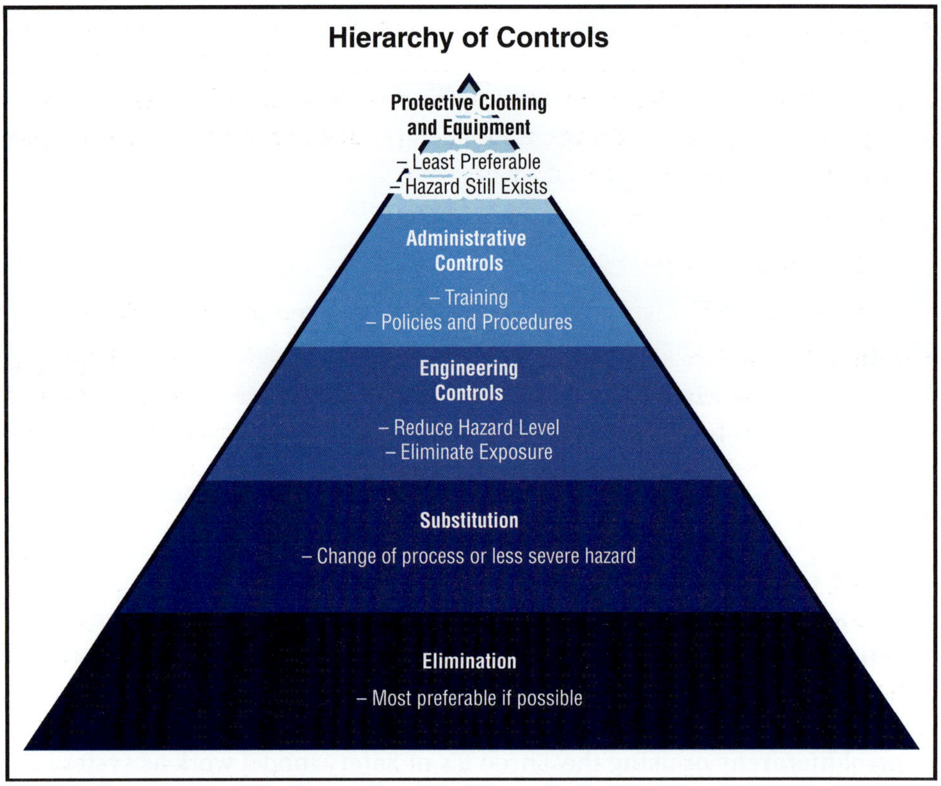

Figure 1.9 The hierarchy of controls lists risk management strategies from most effective to least effective.

Incident scene hazard control has a hierarchy. For example, fire scenes offer the opportunity to eliminate the hazard or substitute the hazard of toxic air with fresh air. Other scenes, like swift-water rescue, only allow for the application of engineered and administrative controls, along with PPE.

There are many examples of each type of hazard control method as they relate to fire department use:

- **Elimination:** Fire extinguishment, hazardous material release shutoff valves, and adding fresh air to a confined space
- **Substitution:** Removing unhealthy chemicals in stations and replacing aging equipment
- **Engineering controls**: Using seat belts or mechanical exhaust at fire scenes
- **Administrative controls**: Using **Incident Command System (ICS)** procedures and training firefighters and chief officers
- **PPE for firefighters**: Use of turnouts at fire scenes, traffic vests at roadway incidents, and Level A suits at hazardous material incidents

Incident Command System (ICS) — Standardized approach to incident management that facilitates interaction between cooperating agencies; adaptable to incidents of any size or type.

Three Es of Safety Model

The traditional approach that safety professionals use for hazard control has been the *Three Es of Safety Model:* engineering, education, and enforcement. The Three Es Model emphasizes educating the workforce and enforcing safety policies. In the fire service, the hazards vary, so the educational needs are greater. Enforcing safety policies may become safety advocacy at incidents where responders may not recognize the hazards present.

- **Engineering** — Eliminates or reduces the hazard through design or modification of an item or process. The goal behind engineering is to create a barrier between the individual and the hazard. NFPA standards were created to provide testing and design criteria to reduce fire hazard. For example, engineering applies to the following:
  - Building construction
  - Fire apparatus design
  - Personal protective equipment design, construction, and testing
- **Education** — Fire and emergency responders learn how to apply proper safety procedures when working in hazardous environments. Fire and emergency responders also teach citizens fire prevention and emergency care that result in fewer emergency responses by the organization.
- **Enforcement** — Management, supervisors, and employees are responsible for advocating safety policies and procedures. Strong safety programs result from high levels of participation from all fire department divisions and levels of rank. Management, in cooperation with labor, develops and maintains a safety program and safety culture for the organization. Management and labor must work cooperatively to involve employees to gain their acceptance of programs. The final result must be programs that provide a foundation for, and encourage development of, safety cultures within organizations.

Supervisors must set a positive example for their subordinates. They must follow the safety program. This may mean reviewing minor injuries, submitting paperwork for exposures, and altering behaviors that do not support goals of no injuries or no accidents. Supervisors must monitor the program and report its effectiveness or areas for improvement. Employees share equal responsibility for adhering to policies and following procedures. They have a duty to monitor themselves and each other and eliminate unsafe behaviors.

Classic Risk Management Model

Fire service members must understand their roles in assessing risk. For example, incident commanders will determine which strategies to employ at various incidents and decide whether the benefits outweigh the risks. Company supervisors and firefighters must maintain situational awareness and assess ongoing risks in their area of operations. Safety officers should continuously monitor conditions and forecast hazard control efforts to make sure that risks remain within the initial risk/benefit decision.

The components of the classic risk management model process help produce a community risk analysis. This model aids in development of the fire and emergency service organization's internal risk management plan required by NFPA 1500, *Standard of Fire Department Occupational Safety, Health, and Wellness Program*, 4.2.3

The NFPA adopted the classic risk management model. It has been used in general industry for decades. NFPA 1500 includes the following risk assessment components:

- Identification
- Evaluation
- Prioritization
- Control techniques implementation
- Management monitoring

Figure 1.10 Having a spotter during weight training can reduce the risk of injury.

Health and Safety Officer (HSO) — Officer authorized to manage the organization's health and safety program; performs the duties, functions, and responsibilities described in NFPA 1521, *Standard for Fire Department Safety Officer Professional Qualifications;* must meet the qualifications of this standard, or an approved equivalent.

Mitigation — To cause to become less harsh or hostile; to make less severe, intense or painful; to alleviate.

NFPA 1500 addresses risk management that should take place during emergency operations. This standard stipulates the following principles:

- Significant risk only when lives can be saved.
- Minimal risk for protection of property.
- No risk when lives or property cannot be saved.
- Excessive risks shall limit activities to defensive operations only.

In this section, the risk assessment model gets applied to safety and health of employees. Because many injuries and some fatalities occur during training evolutions or physical fitness sessions, always apply the risk assessment and analysis to both nonemergency and emergency activities **(Figure 1.10)**.

NOTE: Other risk management models are described in the following sections.

The chief of the department or organization must develop agency risk management models. Implementation of assessments and analysis often falls to the **Health and Safety Officer (HSO)** or a health and safety committee. However, risk assessment and management is every member's responsibility, especially officers. In any case, the final responsibility lies with the officer in charge of the organization.

According to the ICS glossary of terms, risk **mitigation** pertains to the activities designed to reduce or eliminate risks to persons or property or to lessen the actual or potential effects or consequences of an incident. Mitigation measures may be implemented prior to, during or after an incident. Mitigation measures often develop from lessons learned at prior incidents.

Understanding risk management as a two-phase process is important:

- The preincident phase includes mitigation efforts that help in understanding and reducing risk exposure. An example would be the purchase, education, and training related to firefighter turnout gear.
- The postincident phase includes mitigation efforts that reduce the impact of the event on people and property. An example would be the use of peer support after a traumatic event.

Risk Identification

Initiating a classic risk assessment begins with identifying the potential threat. The HSO or health and safety committee compiles a list of the organization's emergency and nonemergency operations and duties **(Figure 1.11)**. Ideally, these operations and duties include a range of possible conditions or potential events, including disasters and concurrent events.

Many sources may assist with this identification process. The first, and possibly most effective, is on the organization's loss prevention and response historical data. This data consists of annual fire-loss reports that include:

- Occupancy type
- Loss value
- Incident frequency
- Addresses
- Call volume

Figure 1.11 Fire service organizations and the events that they sponsor offer opportunities to learn about the latest fire service technology and fire fighting techniques.

Most organizations are too small to rely on their databases for statistically valid trends, but NFPA and the United States Fire Administration make national averages and trends available. National data is not always complete or accurate due to collection inconsistencies and a time lag of one to two years required to collect, analyze, and publish it.

The HSO or health and safety committee should seek input and ideas to identify potential risks from:

- Organization personnel
- Trade journals
- Training publications
- Professional associations
- Local response partners

They can also look to other emergency service providers such as hospital emergency room personnel and law enforcement to point out risks. When using information that other area fire and emergency services organizations provide, the HSO or health and safety committee should consider circumstances that might present a different set of emergency and nonemergency operations and duties. Other risk identification sources include the following:

- Risk management plans that local industry and companies developed that produce, process, transport, or store hazardous materials
- Vulnerability analyses
- U.S. Environmental Protection Agency and local, state, and regional emergency management plans
- Local Emergency Planning Committees

Risk Evaluation

Once the HSO or health and safety committee identifies the risks, they can be evaluated based on frequency and severity. Frequency, which the **U.S. Occupational Safety and Health Administration (OSHA)** refers to as incidence rate, addresses the likelihood of occurrence.

Occupational Safety and Health Administration (OSHA) — U.S. federal agency that develops and enforces standards and regulations for occupational safety in the workplace.

Injury records reveal whether particular types of incidents, such as injuries related to lifting, have occurred repeatedly. These types of injuries will continue to occur until a job hazard or task analysis has been performed to identify root causes and implement effective control measures **(Figure 1.12)**. In this example, the HSO or health and safety committee must develop and implement injury reduction measures that outline proper lifting techniques, physical fitness for injury prevention or provide mechanical aids for lifting.

Severity addresses the degree of seriousness of the incident and can be measured in a variety of ways:

- Lost work time
- Cost of damage
- Cost and time for repair or replacement of equipment
- Disruption of service
- Insurance costs
- Legal costs

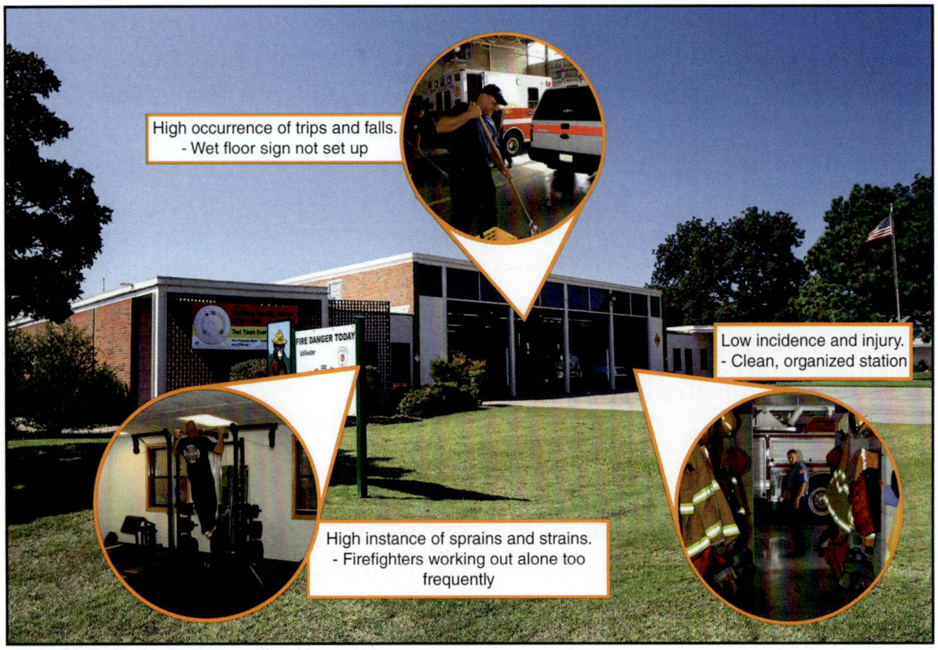

Figure 1.12 Injuries can occur in any part of a station facility. In each case, a job hazard or task analysis should be performed to identify the root cause.

Prioritization

Taken in combination, the results of frequency and severity assessments help prioritize actions. Any risk with high probability of occurrence and serious consequences deserves immediate action and is considered a high-priority item. Minor accidents with a low likelihood of occurrence become lower priorities and may be placed near the bottom of an action-required list **(Figure 1.13)**.

Hazard Control Techniques Implementation

Once the HSO or health and safety committee prioritizes risks, it becomes time to apply risk control measures. Several approaches can be taken in risk control. Two common methods are the Hazard Control Hierarchy or the Three Es method.

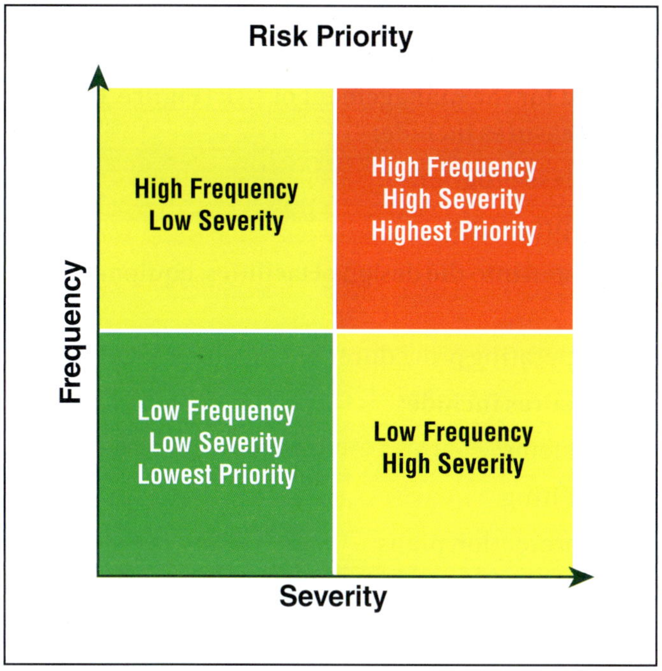

Figure 1.13 Risks should be prioritized based upon frequency and severity. Severe risks that occur frequently should have the highest priority.

There are considerations to be made before deciding on a risk control method. Some are:

- **Risk Elimination** — Remove the hazard. Some hazards can be removed through local laws or ordinances, such as mandating that condemned or structurally unsound buildings be demolished within a certain time. This action removes the risk that a fire may occur, requiring a response.
- **Risk Spread** — When the hazard cannot be eliminated, spreading the risk around can minimize its effect. Requiring that a minimum of three people lift any victim or position a ladder should reduce the potential for injury for an individual.
- **Risk Avoidance** — Avoid the activity that creates the risk (which is the best risk-control choice). Including a policy prohibiting smoking by new department hires would reduce the potential for lung cancer and other smoking-related illnesses among members.
- **Risk Transfer** — Transfer the risk to someone else or purchase insurance. Transfer of risk may be difficult, if not impossible, for a fire or emergency services organization. Contracting the cleanup and disposal of hazardous waste would transfer risks to a private contractor who accepts the liability of the risks associated with those activities. Purchasing insurance for the fire department is a way of transferring financial risk. While insurance protects organizational assets, it does nothing to prevent the station from burning. Insurance does not work as a substitute for control measures, such as installing an automatic sprinkler system.

- **Control Measures** — Designed to reduce the likelihood of accidents or to mitigate severity. Risk reduction through control measures serves as the most common method used for the management of risk **(Figure 1.14)**. Effective general control measures include:
  - — Safety, health, and wellness programs
  - — Training and education programs
  - — Safety features incorporated into the design of facilities, equipment, and apparatus
  - — Well-defined standard operating procedures or guidelines (SOPs/SOGs)

  Incident scene control measures include:
  - — Use of accountability systems
  - — Use of full protective clothing
  - — Mandatory respiratory protection plans
  - — Training and education sessions
  - — Health and wellness SOPs or SOGs
  - — Implementation of the **National Incident Management System-Incident Command System (NIMS-ICS)**
  - — Implementation of a preincident planning program for high-hazard occupancies

National Incident Management System — Incident Command System (NIMS-ICS) — The U.S. incident management system that defines the roles, responsibilities, and standard operating procedures used to manage emergency operations; creates a unified incident response structure for federal, state, and local governments.

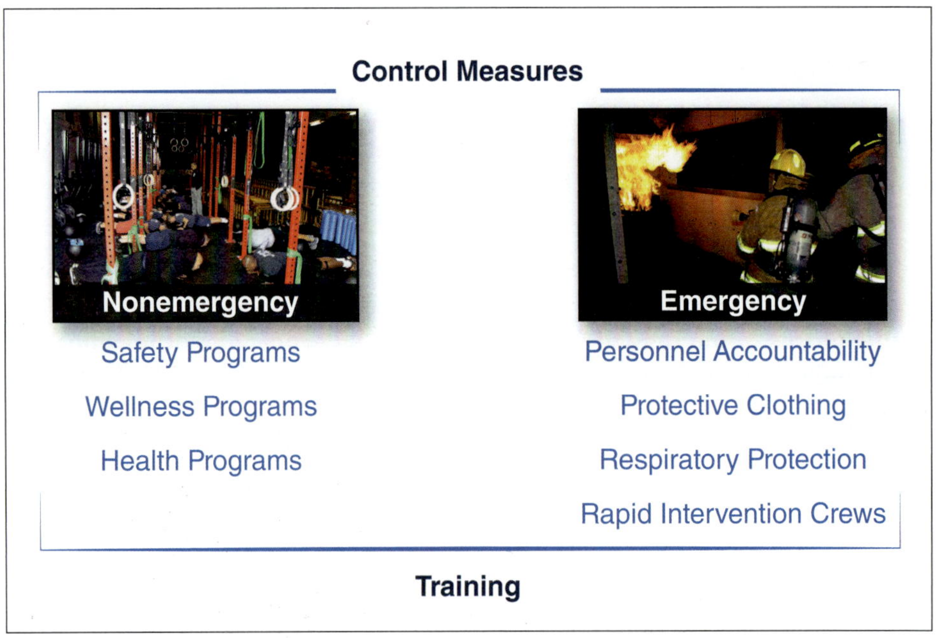

Figure 1.14 Control measures should be identified and used for both emergency and nonemergency operations. Training programs should emphasize these same control measures.

Control measures need evaluation to measure their effectiveness. For example, the risks associated with backing apparatus into station bays have been well documented. An expensive solution would be to replace single-door stations with drive-through stations. Simple solutions include:

- Improved driver/operator training **(Figure 1.15)**

- Stripes painted on the apparatus bay floors to guide the rearward movement of vehicles
- Policy that requires a second person standing at the rear of the apparatus to guide a backing operation

Figure 1.15 The use of spotters whenever an apparatus is backing should be part of all risk management plans. *Courtesy of Ron Jeffers, Union City, NJ.*

Management Monitoring

The last component in the model process is risk management monitoring. This component ensures that the system is dynamic and facilitates periodic reviews of the entire program or plan. This component is used to identify any problems that occur in the process and revise or modify the program. The risk management model intends to develop a strategy for reducing the inherent dangers associated with fire and emergency services activities.

Occupational Comparison

Bureau of Labor Statistics (BLS) reports have shown that fire fighting in recent years has not been among the 20 most dangerous occupations. Based on BLS statistics for 2014, the most dangerous professions are:

- Logging
- Commercial fishing
- Aircraft piloting
- Roofing
- Refuse collection and recycling

Although fatality rates for the fire service seem low in comparison with other dangerous professions, it is more difficult to determine how exposures to hazards over a career affect retired firefighters. Exposures to chemicals show correlations between fire fighting and specific cancers. Firefighters experience hearing loss and respiratory health problems. Fire fighting places cardiovascular stress on the body, which is why vascular events are a major contributor to firefighter deaths. Firefighters can improve their physical fitness for work at emergency scenes. Providing SCBA fit testing, annual fitness evaluations, and medical exams can help departments reduce risks.

Factors Contributing to Occupational Risk

Many factors cause or contribute to occupational risks that hazards create. Eliminating the risk or reducing its effect can control most factors. The risk management process mentioned previously can control some of these factors. The *Three Es of Safety Model* may also be applied to them. Contributing factors include the following:

- Human
- Environmental
- Mechanical
- Organizational Management
- Political
- Uncontrollable

Human Factors

Although firefighters are exposed to emergency scene risks, they can also contribute to occupational risks. Most of these factors result from personal or organizational decisions.

Many fire departments are complex organizations, with numerous internal and external relationships that are tightly coupled. Tight coupling means that relationships directly impact the safety of firefighters on an apparatus. Firefighters need to comprehend how their actions, or inaction, can influence relationships in other areas of a fire department.

Personal decisions are factors that only the individual controls. Some of these factors may contribute to accidents and will be addressed in the Accidents section that follows. Organizational factors include external and internal factors that influence the fire department. Financial, production, and safety concerns tend to influence management.

Behavior

Whether the issue is internal relations or an emergency response, individual behavior is a factor in risk. Many behaviors that contribute to risk are known to firefighters; however, some are not evident.

Examples of behaviors that contribute to occupational risk include failing to clean a hazard on an apparatus bay floor. Almost all fire departments have policies and procedures that direct and influence behaviors to make environments safer. Intentional failure to follow these policies and procedures will influence the actions of supervisors and coworkers.

At times, firefighters and fire department members are unaware their behaviors contribute to risk. When accidents or injuries occur, firefighters often say, "We've always done it that way," or "I had no idea that would happen." These situations can result from fire department members incrementally deviating from procedures in attempts to make a process more efficient or faster. Financial or production pressures outside the local crew level can drive these actions. Crews may even take small safety risks to try to return to quarters sooner. These risks are not always seen as negative, but rather in the interests of the department because they may save time and money. Safety staff should learn how to monitor behavior for trends of deviation to determine whether these actions should be seen and understood as contributions to more efficient operations or contributions to safety degradation.

Success

Fire department success can influence response behaviors. Successes, if not properly reviewed, can slowly sway members into believing any future actions will also succeed. Some see this as an invincible attitude. Fire department members can adopt the attitude that they will be successful without using their experience and education to objectively evaluate each circumstance. They can get caught off guard with some small unexpected change.

This success-engendered attitude can influence decisions at the divisional levels of a fire department and contribute to accidents. After several years of successful live fire training, training staff may choose to cut live fire training staff. Subsequently, they could overlook crew fatigue levels, and a firefighter could suffer a heat stroke injury. Fire department safety staff members need to understand how success can actually erode procedural guidelines. Overcoming this attitude about success will be discussed later in this book.

Physical and Medical Fitness

NFPA statistics make it clear that poor physical conditioning poses risks for firefighters. A lack of physical and medical fitness results from underestimating the demands of the occupation and overestimating actual fitness levels. Poor fitness could also come from laziness or complacency, as firefighters are satisfied with themselves and believe they are immune to diseases and injuries. This attitude may also be related to acquiring a success-engendered attitude about personal health.

Sometimes firefighters do not listen to their bodies. They ignore symptoms until a problem becomes incapacitating. Some firefighters believe that the way they treat their bodies is a private matter and not the function of the organization or government. Overall, fire departments must change attitudes and help firefighters believe that personal decisions impact coworkers' lives as well. Making changes may include the following:

- Participate actively in a regulated exercise program **(Figure 1.16)**
- Eliminate the abuse of tobacco, alcohol, or drugs
- Follow accepted nutritional standards
- Have regularly scheduled medical examinations **(Figure 1.17)**
- Report symptoms or changes in normal physical conditions to a doctor immediately
- Learn to identify stressful activities and cope with their effects

Figure 1.16 Physical fitness equipment is only effective if used properly and with good form. Improper use can lead to injury rather than improved fitness.

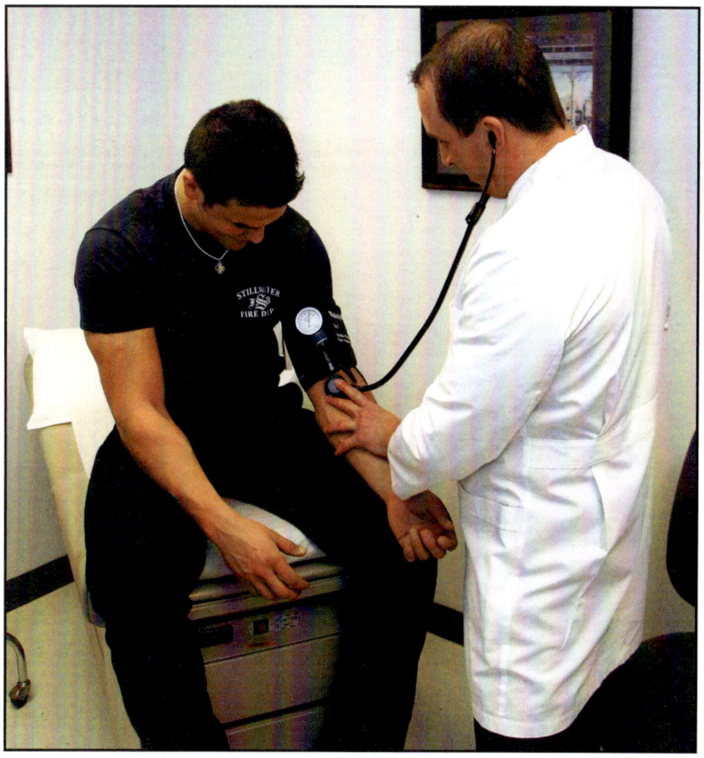

Figure 1.17 Annual medical evaluations are a key component of any fitness-for-duty evaluation program.

Figure 1.18 A Deputy Chief Officer issues orders face-to-face at an incident. *Courtesy of Ron Jeffers, Union City, NJ.*

Maintaining proper physical fitness is each individual's responsibility. Knowing physical and medical limitations is essential. Each firefighter must adhere to the organization's physical fitness program, maintain a good body composition, and follow an age-appropriate medical evaluation schedule.

Communications

Good communications are essential to effective emergency scene operations and effective relationships **(Figure 1.18)**. Poor communications contribute to risk, but may not be evident.

Inaccurate status reports can cause poor decisions during an emergency. Even nonemergency internal communications can contribute to firefighter risk. Words such as "excessive," "fast," or "abusive" mean different things to different people and may not communicate the department's intended message. Communications must be clear and concise, and the sender must confirm the message was received as intended. When providing instructions, use specific criteria, such as "within 10 minutes."

In a nonemergency, a committee chair may not want to consider all opinions on a topic. As a result, committee decision-making may result in lower service delivery or ineffectiveness because few viewpoints were considered.

NOTE: Chapter 2 describes communication between sender and receiver more fully.

Organizational Factors

Financial constraints from elected officials and town or city managers, as well as directives to do more with less, can contribute to occupational risk. Organizations with positive attitudes and high morale can see negative consequences just like groups with low morale and do-it-yourself attitudes.

Research shows shrinking resources result in lowered productivity, higher absenteeism, and increased general dissatisfaction. Lowered productivity may mean that units stay on calls longer or go out of service longer at hospitals or incident scenes. So when calls come into their primary areas, second- or third-due companies must respond. Risks to responders increase because longer response times impact fire behavior, hazardous material releases, and incoming relief for initial responders. Absenteeism places members out of rank or position who may lack training or experience to adequately respond to incidents. Member dissatisfaction creates organizational risk when members decline to participate in committees or are unwilling to take on additional responsibilities like SCBA maintenance or department health assessment administration. Dissatisfaction drains energy from members otherwise willing to engage elected officials for needed changes.

Even organizations with good morale can contribute to accidents when staffing levels are too low to ensure safe responses. Technical rescue, hazardous materials, and structure fire calls are often understaffed according to NFPA and OSHA recommendations.

In organizations with high morale, leaders often do not recognize that there can still be underlying dangers. The effects of success can raise members' belief about themselves and encourage them to continue to do more with less. A safety officer may find it difficult to step in and warn these agencies that they can be riding close to a major accident. Agencies often do not want to examine the reasons for small accidents, injuries, or customer complaints.

Efficient organizations usually have healthy morale and the ability to self-critique. 45Chief officers must continually look for indications that members have gotten away from adherence to policies and procedures in an effort to increase overall response effectiveness.

Training

Providing appropriate training for members of the organization is the responsibility of management. The supervisors provide the training, and the individual is responsible for applying the training and remembering to recertify **(Figure 1.19)**.

Figure 1.19 Volunteer and combination fire departments provide weekly training for their members to maintain efficiency and learn new skills.

Untrained personnel attempting to perform tasks can contribute to the risks and hazards. For instance, personnel trained in wildland fire suppression, but not in structural fire suppression, are not qualified to perform an interior structural attack. Officers must never order an untrained responder or unit to attempt a hazardous task.

Managers should remember that if they fail to provide sufficient information and training, firefighters may seek information through social networking sites. Unguided and unapproved practices add to potential hazards.

Supervision

Responsibility for crew readiness falls to the unit leader or company officer. That officer must ensure that all unit members get trained and qualified to perform assigned tasks. A lack of readiness on the personal or crew level may result in inefficient and ineffective activities that can increase hazards and risks. Readiness is also the responsibility of the overall organization.

Supervisors in today's fire service are asked to conduct more and more of the training activities for crews. Supervisors must not underestimate this responsibility. Being a competent supervisor requires full knowledge of training factors, including task competency and safety issues related to training.

Environment

The environment can consist of weather and earthquakes, traffic, roads, and incident site conditions, along with terrain features. The environment may be a primary hazard, such as during a hurricane, or it can contribute to a hazard, such as the effect that terrain can have on a wildland fire.

While environmental factors may contribute to a hazard, a lack of infrastructure can also increase a hazard. An absence of river crossings that can support the weight of fire apparatus can increase response times for emergency vehicles, allowing an incident to worsen.

Emergency planning and early warnings can reduce the effects of weather and earthquakes as primary hazards and contributing factors. Planning control measures for traffic and road conditions can lessen risks.

Figure 1.20 Natural fires result from phenomena outside of human causality. Source: NOAA-Public Domain.

Weather and Earthquakes

Adverse weather may be the primary hazard that requires an emergency response. These conditions may result in broken natural gas lines, downed electric power lines, fires, structural collapse, or trapped victims. Weather conditions may prevent or slow emergency responses to fires, medical situations, or rescues. It may destroy infrastructure, such as roadways, water supply systems, and radio communications systems. Weather restricts visibility during responses or at emergency incident scenes.

Examples of weather-related incidents include:

- Floods
- Ice, dust, or electrical storms
- Tornadoes, hurricanes, or high winds
- Land or mudslides
- Heavy rain or fog
- Lightning **(Figure 1.20)**
- Wildfires
- Temperature extremes

Earthquakes can cause hazards similar to severe weather. Near ocean shorelines, earthquakes can cause tsunamis.

Traffic

Traffic conditions may result from poor infrastructure or urban planning that has not kept up with population growth. Traffic can slow or prevent an emergency response, and increase the risk to emergency responders working at a road or highway incident. Planning the best routes and maintaining situational awareness about the response district can cut response delays.

Roads

The organization and community administration can ease road condition factors through planning. The company officer and crew members are responsible for keeping track of conditions in their response area and reporting deficiencies. Road conditions generally refer to the following factors:

- Width of roadways
- Condition of road surfaces
- Lack of traffic-control preemptive devices
- Lack of emergency response lanes
- Lack of water supplies along roadways
- Limited weight-carrying capacity of bridges and roads
- Obstructions such as railroad crossings or physical barriers
- Road construction **(Figure 1.21)**
- Travel against one-way traffic

Figure 1.21 Road construction can cause obstructions for apparatus responding to emergencies.

Terrain Features

Terrain — The physical features of a tract of land.

The **terrain** or the surface characteristics of the land in the immediate area of the incident can contribute to the hazard. The terrain can affect the direction and spread of a wildland or forest fire. It can also drive the spread of a liquid or gas release into low-lying areas. Terrain may also restrict access to aircraft or railroad crash sites or some sides of structures.

Incident Site Features

Besides weather, road, traffic conditions, and terrain features in the immediate area, other factors that contribute to hazards include the following:

- **Immediate physical area** — Barriers and obstructions in the immediate area around the incident can prevent or delay access. Natural barriers include trees, creeks, and terrain features. Manmade barriers include setbacks, blast barriers, fences, and walls. Some of the most common barriers that slow access to residences include security grilles and bars on windows and doors. Not only do these barriers prevent access, they also prevent emergency exits of citizens and firefighters.

Figure 1.22 Structural conditions should be considered for potential hazards. *Courtesy of Chris Mickal, District Chief, New Orleans (LA) FD Photo Unit.*

- **Structural conditions** — These conditions may result from the incident, such as a structural collapse due to an explosion. They may also result from the design or construction of the structure **(Figure 1.22)**, such as lightweight construction, firewall penetrations, large open rooms, complex room arrangements, and unprotected floor openings. A structure that has previous fire damage may lack structural integrity and be susceptible to early collapse during an ongoing fire.

- Structural contents — Structural contents may be the primary cause of the incident or contribute to the growth of the hazard **(Figure 1.23)**. The contents of a warehouse that stores flammable liquids may catch fire, exposing the structure to extreme temperatures and rapid burning. A structure containing a printing process may have a fire in the bulk paper storage area that spreads to the ignitable liquids storage area, causing the liquids to intensify the fire. Content storage may restrict firefighter access or overload upper floors, making them susceptible to early collapse. Many fires behave differently than they did decades ago. Modern furnishings cause house fires to burn faster, produce large amounts of toxic gases, and use up available oxygen faster.

Figure 1.23 Flammable contents may be either the cause or contributor to structure fires. *Courtesy of Chris Mickal, District Chief, New Orleans FD (LA) Photo Unit.*

- Adjacent exposures — Adjacent exposures may add to a hazard, such as flammable or combustible liquid storage tanks adjacent to a burning building. They may require additional resources for protection, such as actions to protect houses from a spreading wildland fire. Modern neighborhoods place residential structures closer together, which makes containing the fire to its original residence more difficult.
- Actions of others — Probably the most difficult aspect to prepare for is the actions of victims and others at an incident scene. Responding to a domestic violence call places responders in a potentially dangerous situation. Victims who are under the influence of alcohol or illegal drugs may turn violent while responders are attempting to provide assistance. Be aware that firefighters may become targets of mentally ill persons, civil unrest, and terrorist organizations.

Infrastructure Conditions

The lack of infrastructure such as roadways, water supplies, and utilities can contribute to a hazard. Without roads or bridges, it is very difficult to access wildland or forest fires in remote areas with fire apparatus. This can contribute to the growth of a wildfire, increasing the risk to firefighters.

A lack of piped supply systems means that water tenders must transport the water. Then, they are transferred to pumpers or tanks. Transporting water has often led to accidents due to poor road conditions, lack of operator training, or improper apparatus design.

Rapid fire growth and degradation of structural elements are associated with delays in securing a steady water supply. Even a shortage of radio repeaters or transmission towers creates barriers to radio communications at emergency incident scenes.

Mechanical

Mechanical failures involving apparatus or equipment can create a life safety hazard for firefighters or delay an incident response. Failure of apparatus brakes, SCBA systems, and radio communications have contributed to firefighter deaths. Hydraulic failures in aerial platforms have delayed or prohibited the deployment of elevated fire streams, which then increased the length of time it took to control a fire.

Some mechanical failures can be traced to a lack of preventive maintenance programs, a lack of daily inspections, and failing to meet industry standards. While the lack of maintenance can contribute to mechanical failure, it is also a management issue **(Figure 1.24)**.

Figure 1.24 Proper apparatus maintenance is critical in avoiding safety hazards to firefighters.

Mechanical failure can also occur when apparatus and equipment are improperly used. Using the wrong type of tool for a job can result in damage or excessive wear to the tool. Overloading an aerial apparatus often leads to ladder collapse, and oversteering has led to vehicle rollover.

Improper use may also be considered both a management issue and a training issue. Management is responsible for establishing procedures and providing training for the proper use of apparatus and equipment. Individuals must adhere to those procedures and use them within the limitations of the apparatus or equipment design. Competency training on new equipment often gets overlooked.

Organizational Management

The organization's management or top official is ultimately responsible for risk management and providing a safe workplace for its employees. Therefore, a management failure can cause a hazard or contribute to it. Management factors include the following:

- Safety culture — Management should encourage and monitor organizational culture. Providing incentives and recognition are research-supported methods of encouraging safety culture. If firefighters and fire department staff go unrecognized for behaviors that demonstrate a commitment to safety, cultural development will slow or stop. Organizations experience drift when management and supervisors do not understand how financial, production, and safety pressures play a role in incremental degradation of procedures. Also, proficient or expert operators often develop more efficient practices with unexpected negative consequences. A dispatcher may make shortcuts to algorithms and reduce dispatch times to meet alarm center production goals. This efficiency could mean that crews may not get all the information they need to most effectively respond to a call.

Figure 1.25 Health safety officers should promote safe working environments.

- Management discipline — Strong leadership that is committed to safe working environments and following procedures sends a message to employees that safety matters **(Figure 1.25)**. When members are inconsistently held accountable for actions, cultural development momentum is lost as coworkers' expectations go unmet.
- Clear goals and objectives — Mission and value statements should clearly express a commitment to occupational safety, health, and wellness. Management should develop a strategic plan that allows for measurable safety objectives. When members achieve objectives, management should recognize the successes. Failure to set and monitor measurable safety objectives allows the organization to drift.
- Planning — Planning contributes to the effectiveness of the organization by increasing opportunities for all members to be involved in developing goals, reinforcing labor management relations, and increasing communication between organizational levels.
- Required programs — Modern emergency services agencies are responsible for developing, implementing, evaluating, and monitoring safety-related programs. Some of these are mandated, such as respiratory protection, training, and hazardous materials awareness programs. Other programs are good industry practice, such as physical fitness, health and wellness, and apparatus preventive-maintenance programs.

- Resources —Without sufficient funds, apparatus, equipment, materials, personnel, and training, emergency services will not be available when needed. Based on a community risk assessment, an organization must possess or have a means of obtaining resources needed for each hazard. For instance, a storage facility might require 500 gallons (1 900 L) of foam concentrate to control a fire. If the organization only maintains 50 gallons (190 L), it may not be able to control the hazard. When a risk outweighs available resources, management must make it clear that operations must be limited to those that can be safely conducted with the available resources.
- Policies and procedures — An organization can contribute to hazards if it does not establish and enforce policies and procedures. The use of seatbelts can save lives, so failing to establish or enforce a seatbelt policy puts all personnel at risk **(Figure 1.26)**.

Figure 1.26 While an apparatus is in motion, seat belts must be worn properly.

Political

Political factors that contribute to occupational risk may be internal or external to the organization. Depending on how the person perceives a situation, political considerations can create either action or inaction.

Internal politics may consist of a lack of administrative support or a perception that an individual will face trouble if a certain decision gets made. This type of situation may be the result of any of the management factors mentioned previously. Other internal political factors may include pressures from labor groups or the wording found in the labor/management agreement. An individual may feel that making a decision could conflict with the agreement or create conflict with the labor leadership.

External political factors include influence from elected/selected officials, outside government agencies, and the public. Elected officials have agendas and priorities that the political base that supports them usually defines. Safety issues may not be a major concern for them, prompting them to support projects like repairing potholes and building recreation centers instead of maintaining fire apparatus or training programs.

Appointed officials such as city managers may make decisions based on keeping their jobs. Therefore, pleasing some politicians may prevent them from addressing or supporting safety or risk management issues. Political influence can result in inadequate or unenforceable building, zoning or fire codes that permit hazards to exist or go uncorrected.

State/provincial and federal agencies can contribute to hazards local emergency responders face. An unfunded federal mandate can be one problem. By placing requirements on the local organization, the government can force an organization to decide which services to cut to meet the new mandate.

Finally, public interference can cause or contribute to a hazard. Direct interference may exist in the form of civil unrest, which includes acts of arson, destruction of property, or obstruction of firefighter duties **(Figure 1.27, p. 38)**. Public interference may include tax protests and changes in local ordinances that reduce or eliminate fire and emergency services programs.

Figure 1.27 The remains of this incendiary device point to signs of arson, including pieces of a timer (far left).

Uncontrollable Events

An argument could be made that there are no uncontrollable factors if effective intelligence gathering and planning are applied to all situations. One could also argue that uncontrollable factors are part of the stated risk factors:

- Human
- Environmental
- Mechanical
- Organizational management
- Political

A realistic view is that uncontrollable situations exist, but they may be mitigated with planning and preparation. Uncontrollable situations may include the following:

- **Criminal negligence** — Civilians stealing fire apparatus, use of stolen fire department uniforms to impersonate firefighters or firefighters being held hostage are examples of uncontrollable events that likely have no plans for response.
- **Terrorism and acts of war** — Definitions:
  - *Terrorism*: Premeditated, politically motivated violence perpetrated against noncombatant targets, usually intended to influence an audience (according to the Central Intelligence Agency). Terrorists contribute to occupational risk. As was discovered with the Sept. 11, 2001, attacks, terrorists can be difficult to identify, locate, and control.
  - *Act of war*: One nation-state attacking another. Acts of war can be difficult to anticipate, especially those done without a formal declaration of war, such as the attack on the U.S. Pacific Fleet at Pearl Harbor on December 7, 1941.

Accidents

There has often been controversy among members of the safety profession over the use of the term *accident*. Some say the term properly describes unplanned events that result in injuries, fatalities, or property damage. Others have argued that accidents do not exist and that the proper term should be *incident* or *event*.

NFPA 921, 2014 defines an accident as follows:

... an unplanned event that interrupts an activity and sometimes causes injury or damage; a chance occurrence arising from unknown causes; an unexpected happening due to carelessness, ignorance, and the like.

The American Society of Safety Engineers (ASSE) defines the term as follows:

. . . an unplanned and sometimes injurious or damaging event which interrupts the normal progress of an activity and is invariably preceded by an unsafe act or unsafe condition or some combination therefore. An accident may be seen as resulting from a failure to identify a hazard or from some inadequacy in an existing system of hazard controls.

An accident must be unplanned, unexpected, unwanted, and may result in injury, loss of life, damage or unforeseen consequences. Accidents may result in a loss of public confidence and firefighter morale. The terms "incident" or "event" will be used to describe narratives or stories that may have resulted in an accident.

Causes

According to the National Center for Health Statistics in its National Vital Statistics Report (2014), accidents remain the fourth-leading cause of death nationally. NFPA reports 63,350 injuries to firefighters in 2014, and 27,015 of those occurred at the fireground, 41 percent of the total. Strain, sprain, and muscle pains accounted for 53 percent of the major types of injuries received during fireground operations. An estimated 10,700 injuries or 16.9 percent of 2014 firefighter injuries resulted in lost time. Although the number of fireground injuries has decreased in recent decades, the number of fireground injuries per 1,000 fires has remained relatively stable. Since NFPA began collecting injury statistics in 1981, injuries have decreased from 103,340 to 2014's total (U.S. Firefighter Injuries - 2014).

Not all accidents result in firefighter injuries. Many accidents damage:

- Apparatus **(Figure 1.28)**
- Equipment
- Members of the public
- Public confidence

Figure 1.28 Not all apparatus accidents result in fatalities, although property damage may occur. *Courtesy of Chief Mike Mallory, Tulsa (OK) Fire Department.*

Table 1.1
Categories of Firefighter Injuries

Fireground activities	39 percent
Nonfire incidents	21 percent
Other on-duty activities	18 percent
Training	14 percent
Responding to or returning from an incident	8 percent

Source: U.S. Firefighter Injuries—2016.

Before looking more closely at ASSE causes, know that firefighters may be motivated to deviate from policies and procedures in efforts to improve efficiency, reduce costs, and increase safety. Deviations can be small, happen incrementally, and may be useful at the crew level. Often supervisors look the other way, do not notice or find the deviations to be useful. Over time, deviations can be normalized.

Normalization of Deviance

Normalization of deviance happens in most if not all organizations. Over time, the deviation can become extensive and seen as normal **(Figure 1.29)**. Two considerations merit discussion in normalization of deviance. First, deviance from SOP/Gs or policies is rarely, if ever, done with criminal or malicious intent. Second, firefighters will tend to justify deviations as necessary, as an improvement, or as more efficient or safer. Either way, deviations evolve slowly or incrementally.

Normalization of Deviance — State of a safety culture in which acting against SOP/Gs becomes normal behavior rather than an exception.

Deviations become highly problematic when newcomers learn them as common procedure or when they contribute to an accident resulting in loss of health, property, public confidence, or even death.

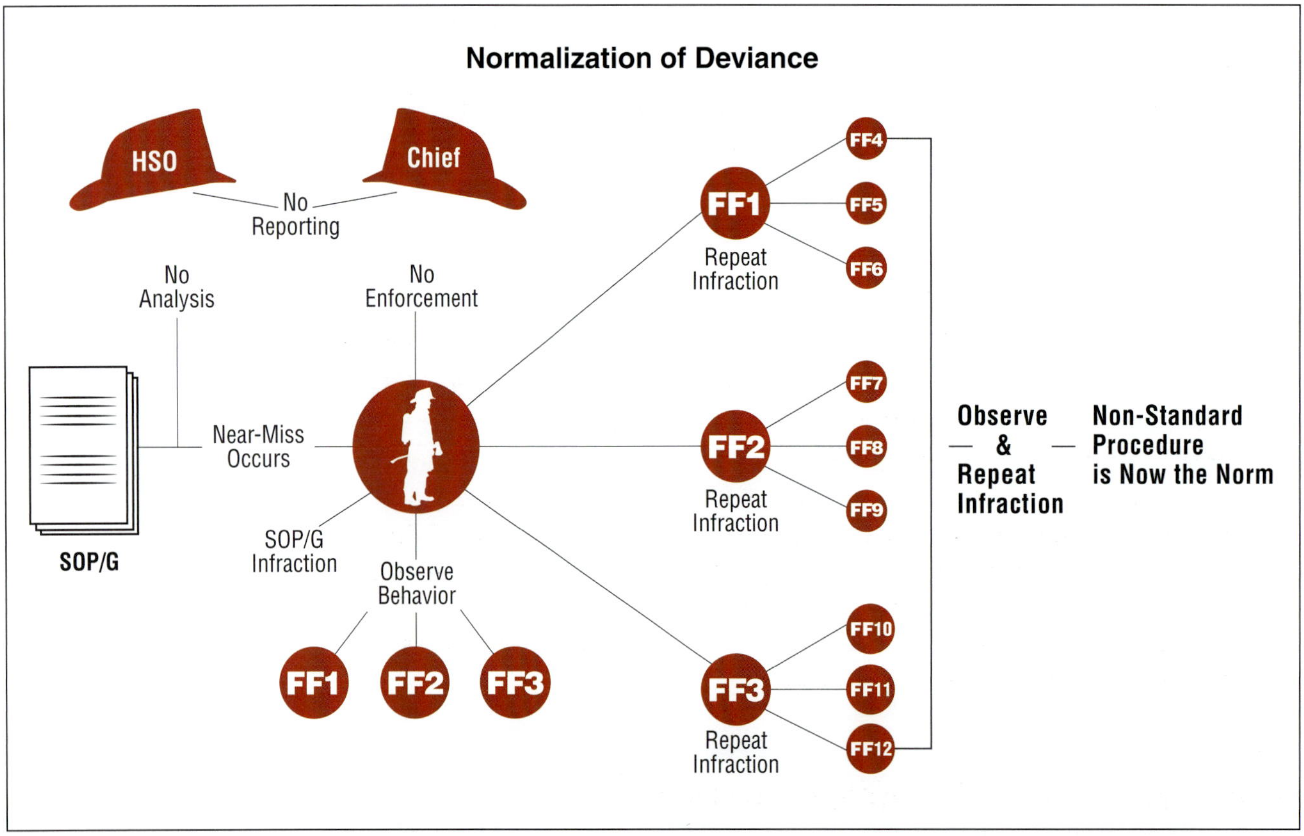

Figure 1.29 When breaking or ignoring a rule becomes common practice rather than an exception, deviation from standard operation procedures is normalized and the department culture changes. HSOs should make sure that infractions are investigated and documented, and that SOPs are revised if needed.

Similar examples of deviance exist in fire departments. Many of them get examined under investigative circumstances but are not causes for the resulting accident. Instead, human error is the primary factor, and firefighters learn not to cooperate with subsequent investigations because of the embarrassment or ridicule when they are blamed.

All members of the fire department must see normalization of deviation as a threat to the health of the organization. Once deviation is uncovered, the department can provide education about its negative effects. Alternative methods for changing policies and procedures under the continual pressure of efficiency, finances, and safety can be designed and utilized.

Unsafe Acts

The ASSE *Dictionary of Terms Used in the Safety Profession,* 4th edition, defines an *unsafe act* as:

. . . A behavioral departure from an accepted, normal, or correct procedure or practice which, in the past, has produced injury or property damage, or which has the potential for doing so in the future; an unnecessary exposure to a hazard; conduct that reduces the degree of safety normally present in an activity.

An example of an unsafe act is standing on the tailboard of an apparatus while it is in motion. Another is to enter an atmosphere that is known or suspected to be immediately dangerous to life or health without wearing the proper respiratory protection equipment.

Unsafe acts are considered to be human-caused accidents because they are the result of human behavior. The behavior may be unintentional due to inadequate training or knowledge. The unsafe behavior may also be intentional.

Human-caused accidents may also result from lack of management policies, programs, procedures or training. Management-related problems are responsible for a high percentage of workplace accidents.

Unsafe Conditions

The ASSE definition for an *unsafe condition* is as follows:

. . . any physical state which deviates from that which is acceptable, normal, or correct in terms of its past production or potential future production of personal injury and/or damage to property; any physical state which results in a reduction in the degree of safety normally present.

For example, an unsafe condition may be a lack of adequate wheel chocks to hold engine apparatus on inclines while pumping. Failing to maintain a mechanical system can result in an accident. Other examples could include the loss of engine power during pumping operations, a break in a hoseline or the loss of hydraulic pressure on an aerial platform. All accidents may result in unforeseen consequences.

Unsafe Personal Factors

An *unsafe personal factor* is defined by ASSE as follows:

. . . a mental or bodily characteristic that permitted or occasioned an unsafe act which contributed to an accident or which has the potential of doing so.

For example, being physically unable to control a pressurized hoseline or raise a ladder may result in an accident. Being mentally distracted and ignoring proper safety procedures may be considered an unsafe personal factor.

Unsafe Physical Factors

Finally, an *unsafe physical factor* is defined by ASSE as follows:

. . . an environmental or physical state or condition, of a physical hazard, that caused an accident or could do so.

Accidents that the environment causes include:

- Weather-related incidents
- Flash floods
- Lightning
- Storms
- Mudslides
- Topographic-related incidents
- Earthquakes
- Other natural occurrences

Even with sophisticated weather forecasting and warning systems, environmental accidents remain unpredictable and can have a variety of consequences.

Combination Causes

While a single cause may be identified for an accident, multiple causes usually contribute to it. For example, a firefighter is driving a baffleless water tender while responding to a reported house fire during a rainstorm. The driver exceeds a posted speed limit, enters a curve, loses control of the vehicle, and the truck rolls over. Multiple causes could include:

- ***Unsafe acts*** — Operating the vehicle at a speed greater than conditions allowed contributed to the accident.
- ***Unsafe conditions*** — Without baffles in the water tender's tank, the water sloshed, causing the apparatus to lose its center of balance in the curve, resulting in a rollover.
- ***Unsafe personal factors*** — The driver's focus on arriving quickly at the reported house fire contributed to lack of attention to conditions and exceeding the speed limit.
- ***Unsafe physical factors*** — The rainstorm reduced visibility and created the slick surface on the roadway, which contributed to the accident.

Investigations

For many organizations in the fire service, guidelines designate the immediate supervisor as a primary investigator for most minor-to-moderate workplace accidents. A general assumption in accident investigation literature from the mid-1900s to the 1990s was that human error caused more than 80 percent of accidents. As a result, research in major industrial sectors shows a supervisor bias that favors determining employee unsafe acts as a primary root cause. This is primarily related to avoiding the work of listing corrective actions and having to do the necessary follow-up implementation of those actions (Manuele, 2013). The primary determinant for supervisors doing incomplete or poor-quality investigations is the nature of the report itself. If reports are done thoroughly, they may highlight an unflattering picture of the supervision at the time of the accident.

In the fire service, the HSO or ISO is often responsible for performing or documenting accident investigations and near-miss reports. Additionally, while many fire department SOP/Gs name the HSO or the ISO as the accident investigators at incidents, municipal management policies may instruct immediate supervisors to complete accident reports **(Figure 1.30)**. Within the fire service, a gap can exist between what specific accident investigation procedures should be and what is actually practiced.

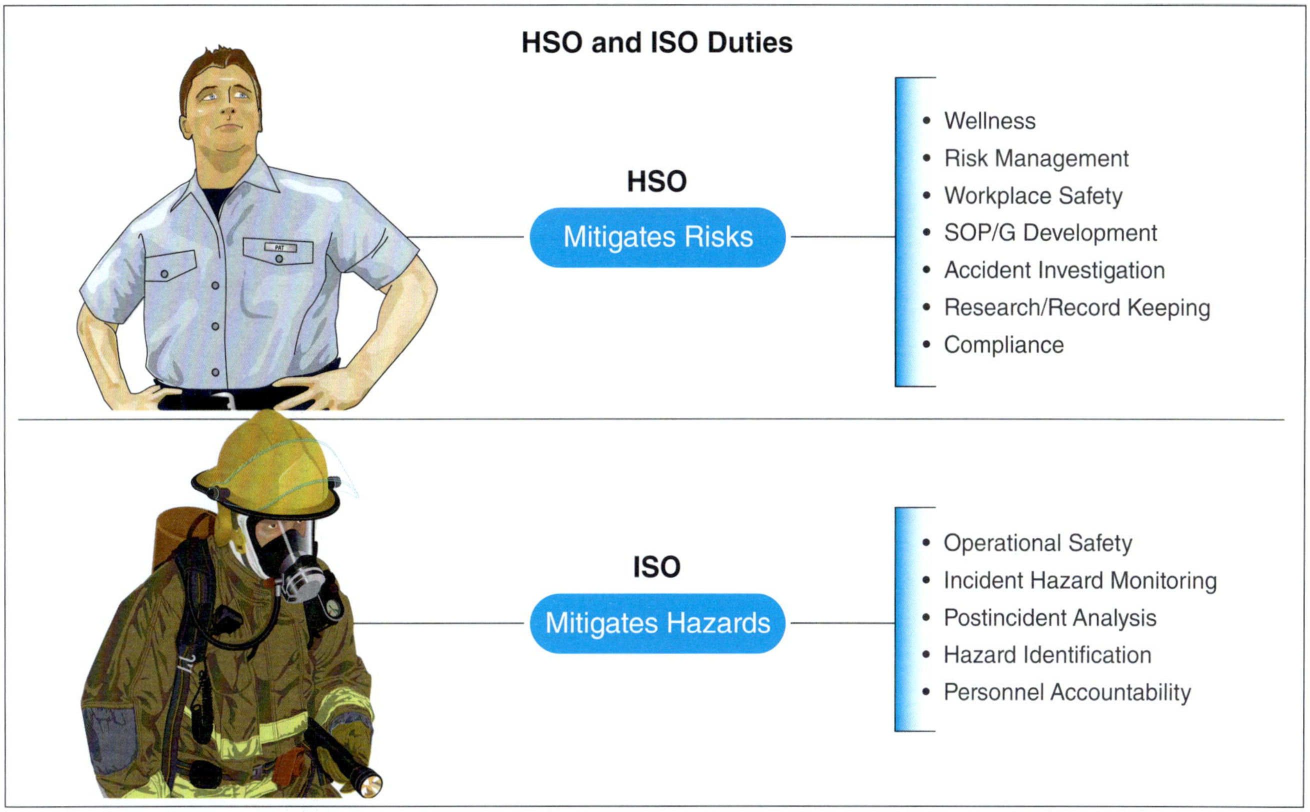

Figure 1.30 Both health safety officers (HSO) and incident safety officers (ISO) have safety functions in a fire and emergency services organization. HSOs more frequently deal with risk factors, while ISOs mitigate hazards at emergency scenes.

Accident investigation work can be susceptible to bias. Individuals who conduct investigations view failure in unique ways. Some can see accidents as a sequence or chain of events, which usually suggests that each sequence triggered the next until the accident occurred. Others may view accidents as more of a disease process in an organization, where the virus hides in management or worker decisions until it finds a way to injure or produce loss. In complex organizations, accidents often result from a myriad of unrelated systems that happen to interact and cause a failure (Dekker, 2006). *The Field Guide to Understanding Human Error* (2006), Chapter 3, The Hindsight Bias, p. 21-22. Ashgate Publishing Limited.

Although all models of accident investigation have merit, what lacks in the fire service is investigator training. Commonly the only investigation education is for arson investigators, yet NFPA documents do not place the responsibility for accident investigations upon the arson investigator class. Health and safety officer qualifications do not list arson investigation as a job requisite.

What is important for fire service safety progress is the development and training of health and safety officers in accident investigations as a means of understanding how the organization is performing and what training is necessary. This area of improvement in the fire service is needed to bring about honest appraisals for fire operations in general.

Accident investigations done competently, with the intent to identify all contributing factors without penalty or ridicule to individual members, would be a critical cultural improvement for most organizations. Although this approach does not guarantee lower accident rates initially, it does enable members to offer in-depth details about why something happened instead of assigning blame. Knowing why something happened is of far greater value to an organization that desires a thorough understanding of its members' actions and how to prevent recurrence of unsafe acts.

When an accident occurs in a volunteer department, the fire chief or an assistant fire chief can respond and begin the investigation process. For career or blended fire departments, the shift supervisor and/or the department's safety officer respond. Each agency should have procedures for including loss control professionals or insurance representatives. Roadway accidents should include law enforcement officials.

OSHA 7505

OSHA offers a training program on accident investigation. OSHA 7505, "Introduction to Incident (Accident) Investigation," is a two-day class that serves as an introductory trainer for conducting accident investigations. According to the course description on the OSHA website, course topics range from employer responsibilities related to workplace accident investigations to reasons for conducting accident investigations, as well as a six-step investigation procedure. The course targets:

- Employers
- Managers
- Employees
- Employee representatives involved in conducting accident and/or near-miss investigations

OSHA 7505 is recommended for personnel who may be responsible for conducting accident investigations in their departments.

Consequences

The individual may suffer an injury that is minor or life-threatening. Many injuries are so minor that they go unreported. Others require immediate medical care or hospitalization. Some result in permanent disability and loss of employment.

Individual

Most injuries could be prevented and almost all injuries offer information about how to improve work and training processes.

Fatalities may also result from the accident. In the fire and emergency services, approximately 80-120 firefighters have died annually in the line of duty over the past decade. Most of these deaths could have been prevented.

Exposures that can harm overall health and reduce a firefighter's lifespan include **(Figure 1.31)**:

- Chemical
- Biological
- Radiological materials
- Heatstroke
- Frostbite
- Asbestos
- Carbon monoxide
- Hepatitis

Figure 1.31 If responders make mistakes in the initial stages of the incident, the consequences can be severe. *Courtesy of the U.S. Air Force.*

Finally, accidents can result in psychological problems such as:

- Increased stress
- Emotional impairment
- Lowered morale
- Chronic depression
- Loss of confidence

These problems can affect close family members, coworkers, and friends.

Organization

Accidents can have a variety of direct and indirect effects on an organization. These effects can be financial, human, and psychological.

Financial effects include personnel costs if the accident results in physical injury to an employee. The National Safety Council (NSC) estimates that the cost per medically consulted injury is about $37,000. An accident that results in a disabling injury can cost approximately $53,000. A workplace fatality can cost as much as $1.4 million. Included in personnel costs are overtime pay, workers' compensation, medical and injury insurance, and staffing costs. Costs of training replacement personnel may result from the accident. The organization suffers from the loss of a trained employee, either temporarily or permanently.

If a government agency cites the organization, such as OSHA, then fines may occur. In addition, changes in mandated safety programs, policies, and procedures will also add costs to the organization.

If the accident involves citizens and the organization is considered liable, then other costs are incurred. Increases in insurance premiums, medical costs, and legal fees can also occur.

Apparatus and equipment repairs or replacement may also result from an accident. If an apparatus is destroyed, an organization may not be able to replace it immediately or at all. Some fire protection agencies are self-insured and will not receive an insurance payment to purchase a new vehicle. They must secure funding from another source, which could result in some services being reduced or eliminated.

The organization will also suffer psychological damage from an accident. If fatalities result, either members or citizens, the organization may suffer from a decline in morale. At the same time, the image of the organization may suffer if the accident results from organizational failure.

Community

Communities may have to adjust fiscal priorities to pay for the replacement of equipment, increased liability, or worker's compensation insurance premiums. Although financial assistance may be available from other levels of government, this aid may not cover the total loss and may be delayed. Communities may see response times increase due to lost apparatus or fewer stations.

Lost confidence or credibility in the eyes of the public is a serious negative outcome of poorly managed incidents and accidents. Consequences may be especially severe if the accident results in civilian fatalities due to firefighter behavior outside policy or procedure guidelines.

Reduction

A proactive approach to safety can reduce the accidents and their consequences. The responsibility for a proactive response rests with the community leaders and businesses, and the leaders and members of fire department organizations.

Community businesses and fire departments must be willing to expend time, resources, and funds for community risk assessment. Once assessments are complete, the community must establish programs and implement recommendations. Certainly environmental accidents cannot be prevented. However, preparing for them through land use planning, emergency response training, and public alert systems can reduce their effect.

Each fire service organization is responsible for having a safety program **(Figure 1.32)**. Establishing a safety program such as the one outlined in NFPA 1500 is less expensive than paying the costs associated with an accident. This topic is addressed in Chapter 11, Safety and Health Program Administration.

Finally, each fire department member is responsible for conducting activities in a safe manner. Some unsafe acts to be avoided are:

- Ignoring safety policies or procedures
- Modifying safety equipment
- Modifying safety procedures

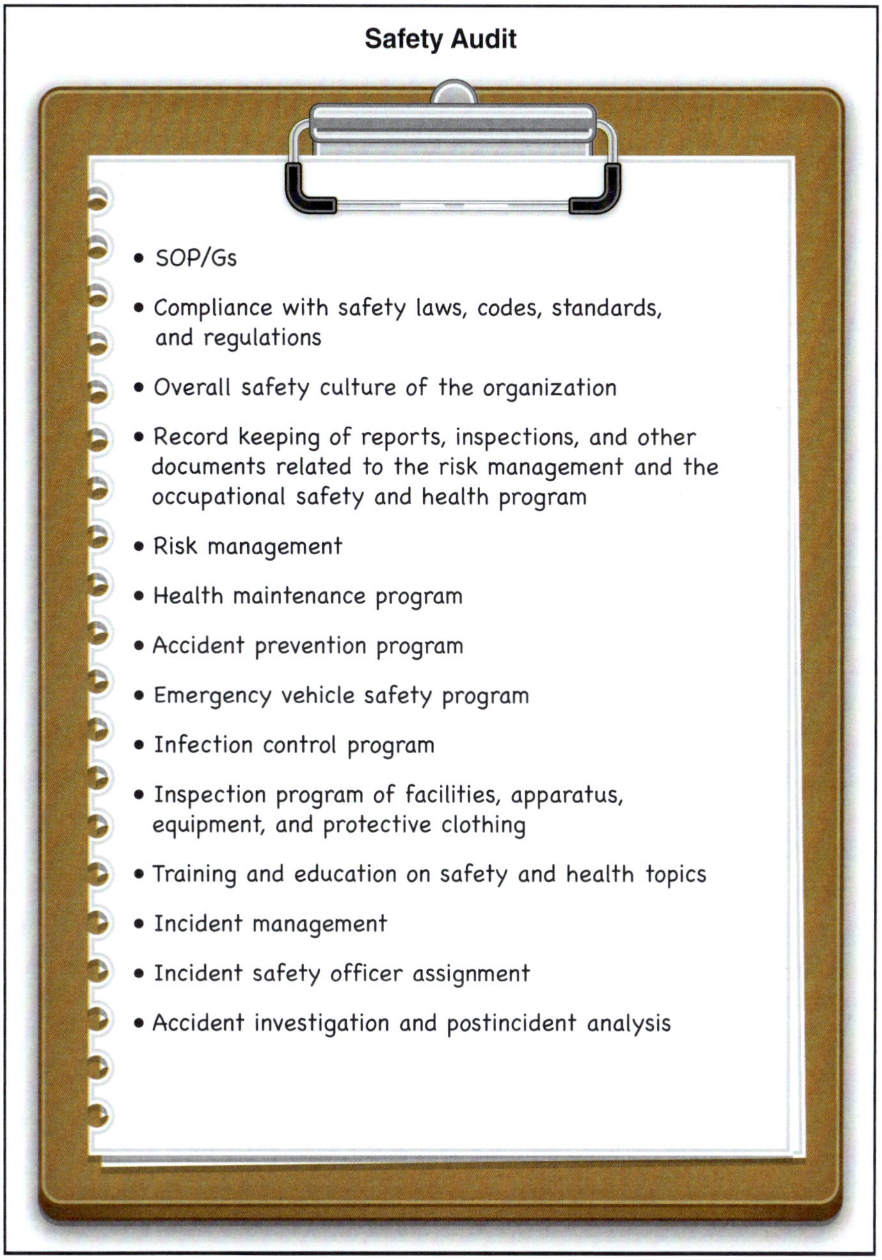

Figure 1.32 Safety audits should be comprehensive and cover all aspects of an organization's safety culture.

- Attempting tasks that you are not trained or equipped to perform
- Taking shortcuts
- Allowing a sense of urgency to replace common sense
- Ignoring warning signs or indicators of a hazard
- Being overconfident
- Failing to recognize and report near-misses
- Participating in unsafe horseplay
- Working while under the influence of drugs, narcotics, or alcohol
- Ignoring signs or symptoms of illness

If members recognize unsafe actions or behaviors, the time to do something is now. Interacting one on one works as the most effective and professional way to correct unsafe behaviors. All fire department members have the authority to stop or alter unsafe behaviors, anytime, anyplace.

Chapter Review

1. What are the three general types of hazards that the all-hazard concept examines?
2. What are three resources the HSO or the health and safety committee can use to seek input and ideas when identifying potential risks?
3. What are the differences between internal and external political factors?

Discussion Questions

1. Discuss a situation caused by bad communication. How could this situation have been avoided though the use of good communication habits?
2. Discuss an accident that was the direct result of an unsafe act.

Cultural Change in the Fire Service

Chapter Contents

chapter 2

Key Terms

FESHE Learning Outcomes

Occupational Safety and Health for Emergency Services

1. Describe the history of occupational health and safety.
2. Identify occupational health and safety programs for industry and emergency services today.
12. Identify and analyze the major causes involved in line-of-duty firefighter deaths related to health, wellness, fitness and vehicle operations.

Cultural Change in the Fire Service

Learning Objectives

After reading this chapter, students will be able to:

1. Identify why cultural changes are needed in the fire service.
2. Describe the cultural evolution of the fire service.
3. Identify barriers to change.
4. Describe common change models.
5. Describe the benefits of cultural change.
6. Identify safety initiatives aimed at fire and emergency services.

Chapter References

Rita F. Fahy, Paul R. LeBlanc, and Joseph L. Molis. 2017. "Firefighter Fatalities in the United States -2016". National Fire Protection Association. Accessed online at nfpa.org.

Hylton J.G. Haynes and Joseph L. Molis. 2017. "United States Firefighter Injuries – 2016". National Fire Protection Association. Accessed online at nfpa.org.

Chapter 2
Cultural Change in the Fire Service

If the fire and emergency services are to reduce line-of-duty fatalities, injuries, illnesses, and property loss, the culture must change. The traditional culture's emphasis on heroism needs to give way to one that prioritizes safety. The change must be complete and permanent to ensure that everyone goes home after a call, and not a return to a time when casualties were an accepted part of the profession.

Individual change takes work. Examples of personal behavioral changes, which take time and resources, include losing weight through better eating habits and stopping smoking by eliminating social or personal behavioral triggers. Cultural change within an organization or profession may present even more difficulties, because it can mean going against long-held values and beliefs.

Members of the fire service take pride in the fact that they face danger when protecting a community. Residents hold firefighters and emergency responders in high regard **(Figure 2.1)**.

Figure 2.1 A citizen meets with a firefighter at a local station.

Aggressive actions may drive the change to create an organizational and professional safety culture. They include:

- Understanding the current organizational culture by identifying values, beliefs and artifacts
- Recognizing the need for change and the barriers that may inhibit it
- Understanding the models for change that can help make progress
- Implementing change through top-down or bottom-up initiatives
- Recognizing the results of the change and knowing the benefits
- Fostering awareness of programs and initiatives that support the change to a safety culture

This chapter provides a brief look at what brought about current firefighter culture; the cultural changes taking place in the fire service; and sample models that can make change easier. The final section describes safety-related initiatives intended to reinforce the need for changing the culture.

Current Fire and Emergency Services Culture

The culture of most fire and emergency services organizations is rooted in fire service history. From the paramilitary organizational and rank structure to the can-do attitude, the culture of fire and emergency services has strong roots in valor. Changes to these long-held values can be difficult and take years, if not decades, to become established.

To understand the current culture and how it affects safety issues, one must understand the past's influence on the fire service, be able to define organizational culture and safety climate, and recognize the influence individual culture may play in organizational culture. This section:

- Describes the general history of the fire and emergency services and how a cultural image emerged
- Introduces the changes in service delivery and their effect on the culture
- Describes the organizational culture and safety climate
- Shows how demographic changes, including a culturally diverse and aging population, have created a need for change
- Shows how the culture results in unsafe actions and creates unnecessary costs to the community, organization, and individual

History of the Fire and Emergency Services

Organized fire protection dates from the Roman Empire around 24 B.C.E. The crowded population of Rome lived in shanties made of combustible materials and residents cooked over an open hearth, which resulted in more destructive fires. This led Emperor Caesar Augustus to form the *familia publica*, a group of 600 servants stationed by the gates of the city to fight fires.

In 6 C.E., following another disastrous fire, Augustus instituted the *vigiles* who would protect Rome from fire for 500 years. They were a night patrol who were distributed and quartered throughout Rome. The vigiles, equipped with buckets and axes, fought fires and enforced fire prevention measures. The vigiles had an organizational ranking system of officers and ordinary firefighters, similar to one used today by fire and emergency services organizations.

As civilization spread through Northern Europe and people congregated in cities, fires continued to cause devastation. Small cooking fires could grow to destroy villages, towns, and cities. Following the fires, residents would rebuild and start over. Unlike in Rome, these communities passed laws requiring the use of certain types of building materials, such as slate roof tiles instead of thatched straw. But they made no real attempts to form fire fighting organizations. Some communities enacted laws that required able-bodied men to respond to an alarm and gave them authority to pull down burning buildings or use gunpowder to destroy buildings to form a fire break.

The first modern attempt at forming official fire fighting units occurred following the Great Fire of London in 1666. City authorities recommended many proposals to improve fire fighting methods, but their suggestions went nowhere. The theory of the *hand of God* in fire disasters had such a hold on citizens they never considered skilled fire brigades. When citizens did think of skilled fire brigades, they considered the cost too high.

The first fire insurance company formed following the London fire. The creation of the fire insurance company had a major effect on the fire service of Britain. In turn, those traditions and ideas then spread to North America.

To protect their interests, British insurance companies formed fire brigades that responded to fires in structures that the company insured. A fire brigade would not extinguish a fire in a building unless it was insured by their employers. However, they would protect nearby buildings if they exposed insured buildings to fire. In 1717, Boston established America's first public fire department. The city experienced devastating fires, and it needed to establish organized fire protection to ensure its existence. The year after Boston formed its municipal fire department, community members formed the first mutual aid fire society. These societies only responded to fires affecting members' properties. This began the era of the volunteer firefighter in American history. The mutual aid fire societies operated independently and volunteered assistance to Boston's municipal firefighters.

In 1736, Benjamin Franklin established the first fire organization in Philadelphia, the Union Volunteer Fire Company. Franklin patterned this company after the Boston mutual aid fire societies. Unlike the Boston mutual aid fire societies, the Union Volunteer Fire Company responded to all fire calls in its vicinity. During this same period, American insurance companies formed fire brigades to protect their members' homes and businesses.

Because towns had no permanent firefighters and volunteers received no pay, insurance companies or town treasuries would authorize pay to the first-arriving firefighters who put water on a fire. This incentive started fierce rivalries that developed among volunteers in later years. The desire to be first on the scene or be the first to get water on the fire became a tradition that firefighters still strive to achieve.

NOTE: For more information on the history of fire service, see the IFSTA **Fire and Emergency Services Orientation and Terminology** manual.

Cultural Image Emerges

The creation, development, and competitive spirit of the American volunteer firefighter system did several things for the fire service. The zeal to beat rival companies led manufacturers and firefighters to work at developing new and better apparatus, equipment, and methods of fire fighting. Firefighters practiced making faster responses, drilled to save time in putting engines to work, and experimented with sleeping at the station.

Firefighters soon emerged as heroes **(Figure 2.2)**. Their daring rescues and heroic stands before fires were captured in song and verse, as well as in the drawings of Currier and Ives. Boys across America wanted to be firefighters when they grew up. The firefighter's exciting and colorful job provoked interest from the townspeople who would gather to watch them in action. Water-pumping contests would bring cheers and encouragement for local favorites.

Figure 2.2 The symbolism of the heroic firefighter dates to the earliest traditions of the fire service.

Early volunteers underwent thorough character investigations and trial periods before a company accepted them. Most were men of means — merchants, manufacturers, and professionals — who kept their fire companies as socially exclusive as a private country club. To be a firefighter was to be somebody, and anybody who was anybody usually joined.

Volunteers were expected to endure extremes of weather; the hazards of fire, smoke, and falling debris; and physical exhaustion. Fires were challenging and the volunteers reveled in fighting them. The status of belonging to a fire company, the teamwork, and the pride in the company's beautiful machinery made company membership desirable.

The heroic image of firefighters carried over into times of national crisis. At the beginning of the American Civil War, members of many volunteer fire companies in New York City fought for the Union, such as the 11th New York Volunteer Infantry, Fire Zouaves, and the 2nd New York Fire Zouaves. Firefighters from Lancaster and Philadelphia, Pennsylvania, also formed regiments to support the Union.

The military influence does not end with the American Civil War. The relationship between the fire service and the military may be found in fire fighting terms: We fight fires, wear helmets, attack the seat of the fire, and take casualties. Firefighters place a great deal of importance on taking care of our own. The rank system and unit structure terms also parallel those of the military. Even the uniforms look like military dress uniforms.

Based on the tradition and history of the fire service, certain desirable characteristics such as pride can have undesirable effects. For example, pride in accomplishments is desirable. However, that same pride may prevent a firefighter from asking for help.

Figure 2.3 Good housekeeping practices, complete with warning signs for temporary hazards, serve as preventative measures against injuries.

Fire Service Culture and Safety Climate

To understand the challenges of changing fire service culture, start by defining an organizational culture and know what the term "safety climate" means in relation to this culture. This understanding will help show what needs to develop or change.

Organizational culture can be defined as a set of shared values and beliefs around cultural touchstones.

The following characteristics should help demonstrate and measure culture:

- **Values:** Organizational values could include making a difference in others' lives, being a team member, loyalty, and honesty.
- **Behaviors:** Behaviors include actions and outcomes on response history, training, and station activities **(Figure 2.3)**. They may include participating in a charitable activity, such as donating time or money to an injured co-worker. These can be measured by safety reviews of injuries, accidents, close calls, customer complaints, and observable behaviors by safety officers or other officers of the agency or labor group.
- **Artifacts (often considered traditions):** Artifacts include shared stories about past notable events, fire department successes, ceremonies for new recruits and retirees, apparatus colors, flags, and department emblems. Participation levels demonstrate the relative importance of artifacts in sustaining the culture.

Despite adages such as "100 years of tradition unimpeded by progress," culture does change over time. Organizational members learn about, interpret, and adapt to external and internal pressures which, in turn, causes adjustments to values, behaviors and artifacts. Additionally, all members of a fire service culture bring their own values and experiences to the organization.

The safety climate describes how firefighters feel about safety and health at the local levels within their organizations. Climate is represented by individual firefighter perceptions and how firefighter actions correspond to the organization's operational or administrative policies and procedures. Firefighters at a busy station who feel safe but do not report minor accidents or close calls have a low safety climate; their behaviors show they follow only convenient safety practices. An organization may claim to have a robust safety, health, and wellness culture. However, if it does not provide time for physical activity or follow NFPA standards and other industry best practices, the safety climate it demonstrates is lacking.

Diversity of Service Delivery

Hazards and their associated risks have changed over the past half century. Modern fire behavior creates more dangerous and toxic environments in structural fires. The All-Hazards approach of most community fire departments means firefighters must be prepared to face multiple types of hazards. And those hazards require far different approaches than the ones used to fight fires.

Figure 2.4 A firefighter applying water to extinguish a fire. *Courtesy of Chris Mickal, District Chief, New Orleans (LA) FD Photo Unit.*

While fire fighting cannot be considered simple or easy, centuries of fire fighting have established fairly consistent approaches: applying water or another extinguishing agent to a fire, wearing protective clothing, and using tools to gain entry or perform rescues. Firefighters apply water or some type of extinguishing agent to a fire until it is out **(Figure 2.4)**. They wear protective equipment that repels heat, smoldering debris, and water to prevent burns. For generations, they have used tools such as pike poles to pull down walls and ceilings to expose burning structural members.

With the advent of the all-hazards concept, firefighters must take time to perform the following actions under all kinds of circumstances:

- Determine the type of hazard and the risk it poses **(Figure 2.5)**. For instance, has fire damaged the structural integrity of the building, making it too hazardous to enter the structure?
- Select the best strategy. Select offensive strategies to move crews toward the seat of the fire or the source of the problem **(Figure 2.6, p. 58)**. Select defensive strategies to keep crews at a safe distance and hold the source of the problem to its current scale.
- Select the appropriate tactics and tasks required to control or to mitigate the hazard after conducting a risk/benefit analysis. For instance, on a brush fire, should a firefighter use aircraft, drop water or chemicals; construct a fire line; or set back fires?

Figure 2.5 Firefighters discussing the hazard and risk of a fire. *Courtesy of Ron Jeffers, Union City, NJ.*

The days of rushing into the electrical vault to extinguish a fire (no matter the risk of electrocution) or diving into a swollen river to make a rescue without proper equipment and

Figure 2.6 A firefighter uses a Halligan tool to pry open a door.

Figure 2.7 Depending on the cargo it carries, a semi truck on fire could produce hazardous vapors. *Courtesy of Chris Mickal, District Chief, New Orleans (LA) FD Photo Unit.*

training are long over. Taking time to evaluate the hazard and conduct a risk/benefit analysis ensures that crews do not expose themselves to unnecessary risks.

In the last half of the 20th century, the manufacture, storage, transportation, and use of hazardous materials continued to escalate. Fire and emergency responders, depending on local fire codes, regularly responded to chemical spills, hazardous vapor leaks, flammable metal fires, and other materials that were difficult to extinguish and produced hazardous vapors **(Figure 2.7)**. The increasing risks caused by these materials changed response strategies, improved personal protective clothing and respiratory protection, and resulted in more federally mandated training in hazardous materials response. Additionally, more knowledge and certification led to the creation, training, and equipping of special units to respond to these situations.

In many jurisdictions, local laws and adoption of codes directed at hazardous materials manufacturers require corporate emergency response teams for small spill cleanups. When larger releases occur, local fire departments respond and interact with responders. This adds an element of personnel diversity that fire departments must learn to integrate into their operations. Risk/benefit analysis needs to consider these civilian factors, which can be difficult if Incident Commanders are unfamiliar with corporate capabilities.

Another major change in the late 20th century was the advent of emergency medical response. A variety of factors led to this, including:

- Increasing costs for prehospital care and growing uninsured or underinsured populations
- An aging population with life-threatening medical conditions that require immediate attention in the prehospital environment
- The need for early identification of and response to various forms of infectious or communicable diseases
- Technological and medical improvements and innovations that increase survivability
- The reliance on emergency room care

As firefighters expanded their roles as medical first responders, they needed certified personal protective equipment (PPE) that would guard against bloodborne and airborne pathogens **(Figure 2.8)**. They also required additional technical training for new medical equipment and higher levels of medical certification.

In addition to increased medical responsibilities in the late 20th century, the fire service assumed a primary role in rescue from buildings severely damaged by natural disasters. This role gave rise to the Urban Search and Rescue (USAR) programs. These programs created a genre of equipment, training, procedures, and operational safety issues. Federal government grants supported these capabilities, leading to regionalization of resources and training. Requirements for grant funding and equipment accountability have heavily impacted administrative positions in fire departments that receive grant funding. Water rescue has evolved from an incidental fire service function to a full specialty with its own equipment, training, and safety issues.

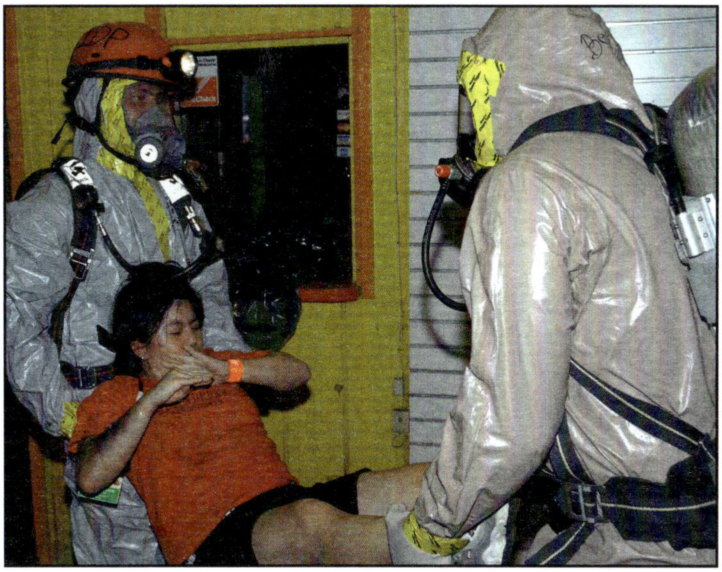

Figure 2.8 It is acceptable to assume a higher level of risk in order to save a life. *Courtesy of the U.S. Marine Corps, photo by Sgt. Christopher D. Reed.*

While firefighters have always responded to natural disasters, responses have increased with the expansion of urban and suburban communities into areas prone to natural occurrences. **Wildland urban interfaces (WUIs)** are becoming more prevalent, as are fires in those areas. Urban fire resources deployed to manage the interface areas create additional financial, training, and resource needs. Additionally, fire prevention efforts must address residential prevention efforts to strengthen interface areas against fire spread.

Wildland Urban Interfaces (WUIs) —Line, area, or zone where an undeveloped wildland area meets a human development area.

Finally, the 21st century has seen even more global and homegrown terrorism. First responders, including hospital emergency room personnel, now view **weapons of mass destruction (WMDs)** as hazards. Firefighters trained in recognizing and responding to such devices are equipped to handle them, and to mitigate the results of their use. The risks include the direct effects of WMDs on their victims and the psychological effects on emergency responders.

Weapons of Mass Destruction (WMDs) — Any weapon or device that is intended or has the capability to cause death or serious bodily injury to a significant number of people through the release, dissemination, or impact of toxic or poisonous chemicals or their precursors, a disease organism, or radiation or radioactivity; may include chemical, biological, radiological, nuclear, or explosive (CBRNE) type weapons.

Cultural Differences

As hazards have changed over the past half century, so have members of the fire service. In 1950, the majority of fire departments were composed of white males who were born in the 1920s and 1930s. Their ethnicity, gender, and generational values limited and influenced their worldviews. Today, the fire and emergency services reflect the demographics of North America: multiethnic, multigender, and multigenerational.

Ethnicity

The fraternal origins of the fire service traditionally attracted men who had similar ethnic backgrounds. For example, the New York City Fire Department was populated by men of Irish descent, while departments in Pennsylvania attracted Germans.

Shifts in populations that resulted from the Great Depression (1929-1940), the Dust Bowl (1933-1940), and World War II (1941-1945) changed the demographics of the U.S. and Canada. Populations mixed and neighborhoods became ethnically diverse as people moved to find better employment and improve their lives. In addition, a new wave of immigration following World War II brought an influx of people from East Asia and Central Europe.

Figure 2.9 The International Association of Black Professional Fire Fighters is dedicated to supporting black career firefighters. *Courtesy of Ron Jeffers, Union City, NJ.*

The gradual end of legal racial segregation that began in the mid-1950s also changed the racial composition of the fire service **(Figure 2.9)**. African-Americans have become a larger part of many fire and emergency services organizations in the last 60 years.

Instead of fraternal brotherhoods with similar cultural backgrounds, fire service organizations include ethnically diverse members with a variety of backgrounds. That diversityprovides members with new perspectives. Views on authority, work ethics, and loyalty to an organization expand as new members join organizations and gain rank and influence.

Figure 2.10 More women are serving in the fire service.

Gender

One of the major changes in the fire and emergency services is the addition of women into career organizations. Women have long played an active role in volunteer fire departments, but career organizations have only begun to hire women since the 1970s. The U.S. Bureau of Labor Statistics said approximately 8,800 women were employed as firefighters in the U.S. in 2016. **(Figure 2.10)**

By including female members, volunteer and career organizations better reflect the communities they serve. Women make up more than half of the population, according to the U.S. Census Bureau, and they have started to make up a larger part of career and volunteer fire organizations **(Figure 2.11)**. Each woman contributes her background and outlook to an organization. As with all other members entering the fire service, the values and priorities of female department members impact overall cultural values. For example, some female members of the fire service may begin changes in the safety climate; studies have shown that female workers are more safety conscious than their male counterparts.

Generational

The composition of fire departments is always changing **(Figure 2.12)**. However, with the advent of pension plans that permit members to retire after 20 years of full-time service, the average age of retirees has decreased. Members no longer need to work until the age of 65 in order to draw retirement benefits.

The differences in generations, or age groups, are determined by personal life experiences. The life experiences of a person born in 1930 are probably vastly different from the life experiences of someone born in 1990. These life experiences could lead members of different generations to have different value systems **(Figure 2.13)**.

Figure 2.11 A group of women firefighters prepare for training evolution.

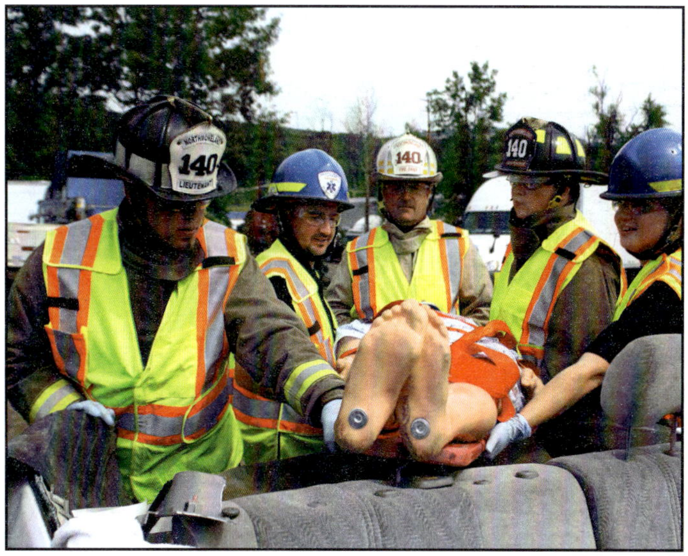

Figure 2.12 The fire service includes personnel from all age groups.

Figure 2.13 Fire officers and senior firefighters can effectively mentor personnel as a way to improve skills and knowledge and provide direction for career advancement.

Generational Diversity

North America's population can be divided, with very few exceptions, into the following four age groups:

- Traditionalists
- Baby Boomers
- Gen Xers
- Dot.coms, Millennials, or Nexters

These categories are useful to group people, not to define them. Members of each group are not identical. The influence of family, history, tradition, and media create individuals within one group that can have strong ties to the characteristics of the previous generation.

(continued)

Generational Diversity *(continued)*

Traditionalist

Traditionalists were born between 1920 and 1944. The oldest among them lived through the economic boom and bust of the 1920s, the Great Depression and the Dust Bowl of the 1930s, and World War II. Their values are considered to be loyalty, patriotism, hard work, and education. They have a reputation for being fiscally conservative, and they are known to place a high value on institutions such as universities, corporations, and religion.

Baby Boomer

Born between 1946 and 1964, baby boomers are considered idealists and place a high value on fairness, equality, hard work, and competition. However, they have a reputation for questioning authority. The leadership and upper ranks of fire and emergency services organizations are composed of members of this group. Many have experienced war, social turmoil, economic recession, and affluence. Baby boomers have been known to place a high value on education, family, and leisure time.

Gen Xer

These children of baby boomers were born between 1961 and 1980. Many require personal flexibility and thrive on feedback from supervisors. However, they have been known to dislike constant supervision and often prefer to work independently. Gen Xers are also considered entrepreneurs. They try to balance work and leisure time in their lives. Because of the mobility of their parents, the high divorce rate in the U.S., and the rapid expansion of urban America, they are accustomed to change. Members of this group make up most of fire and emergency services organizations.

Dot.com, Millennial, or Nexter

Members of this age group were born after 1980 and share the following characteristics:

- They were raised with access to technology.
- They enjoy visual stimulation (computer games and television).
- They are optimistic.
- They appreciate diversity.
- They are ready to work and learn but usually want instant gratification in the form of tangible results for their efforts.
- They have a broad worldview.
- They accept nontraditional families and lifestyles.
- They have high expectations.

In this century, they have seen terrorism and war. They are experiencing what traditionalists did in World War II and the baby boomers did during the Vietnam War. In addition, active-duty military personnel have faced drastic changes in their personal and family lives. This generation will continue to fill the ranks of the fire and emergency services and bring with them these experiences and values.

As the members of each generation advance through the ranks, their worldviews influence every personal and professional decision. For instance, a fire chief in 1940 would not have been concerned with firefighters smoking cigarettes in the station. Today, tobacco products are banned in most fire stations and facilities. Members of each generation may have different opinions about the authority of an organization to control what some may argue is a personal choice.

As people join the fire service, they accept the current level of safety as normal or expected. Therefore, new personnel will adhere to the safety policies and use safety equipment and procedures more readily than longtime veterans. Younger members **(Figure 2.14)** also see the effects that lower or nonexistent safety standards or lax enforcement have had on older members, and they want to avoid those outcomes.

Figure 2.14 A younger fire fighter learns how to avoid hazards from more experienced fire fighters.

Costs Resulting from the Current Culture

Any attempt to quantify the exact cost of fatalities, injuries, and job-related illnesses in fire and emergency services is challenging. The fire service, each level of government, and the insurance industry have all been inconsistent in collecting data related to the types and cost of injuries and illnesses. Even injuries that are not life-threatening can permanently affect continued employment and lower quality of life.

Federal Emergency Management Agency (FEMA) statistics showed an estimated 66,200 firefighter injuries annually from 2012-2014. Overexertion and strain accounted for 27 percent of all injuries; hazard exposure accounted for 18 percent of injuries; contact with objects led to 14 percent of injuries, and slips and trips prompted 12 percent of injuries. Most injuries were addressed without lost work time, FEMA reported, and 95 percent of the injuries were sustained by men. The top group by percentage for injuries was ages 40-44 for career firefighters and 20-24 for volunteers.

Firefighter injuries and job-related illnesses result in direct and indirect costs. Direct costs usually include all expenses related to treating and compensating for an injury or illness (National Institute of Standards and Technology GCR 05-874). Components of direct costs, according to the NIST report, include:

- Lost wages to the firefighter not covered by disability payments
- Overtime wages, above the cost of the injured firefighter's wages, to cover wages for the substitute for the firefighter
- Medical expenses
- Psychological counseling for any pain and anguish that the firefighter, the firefighter's family, and the firefighter's coworkers may experience
- Time spent by the firefighter, supervisors, and others in investigating the incident and writing injury reports
- External investigations
- Disability/retirement income when firefighters cannot return to work for years or retire early due to disability
- Litigation expenses

Indirect costs are those associated with preventing or mitigating injury and include:

- Additional staff sent to emergency incident scenes to prevent and deal with firefighter injuries (staff to meet the two-in/two-out rule for rapid intervention teams)
- Firefighter protective equipment, including respiratory protection
- Maintenance of protective equipment (cleaning and inspecting protective equipment after each use)
- Training for firefighter safety; for example, how to escape in extreme situations or use buddy breathing
- Administrative costs of insurance
- New safety technology such as sensors that locate firefighters
- Health and safety officers
- Union time and fire-management time spent on negotiations related to safety
- Safety aspects of firefighter personnel record keeping

The NIST report determined that no single all-encompassing method determines the annual injury costs for firefighters. Estimates range between $5,000 and $34,000 per claim. The results of the study indicate that the total cost of firefighter injuries ranges from $2.8 billion to $7.8 billion annually.

The NFPA report, *United States Firefighter Injuries - 2016* contains useful injury and motor vehicle accident statistics. Although the NFPA does not estimate the costs of firefighter injuries, it does offer recommendations for reducing injuries. NFPA's data suggests injury rates have been gradually declining over the last three decades.

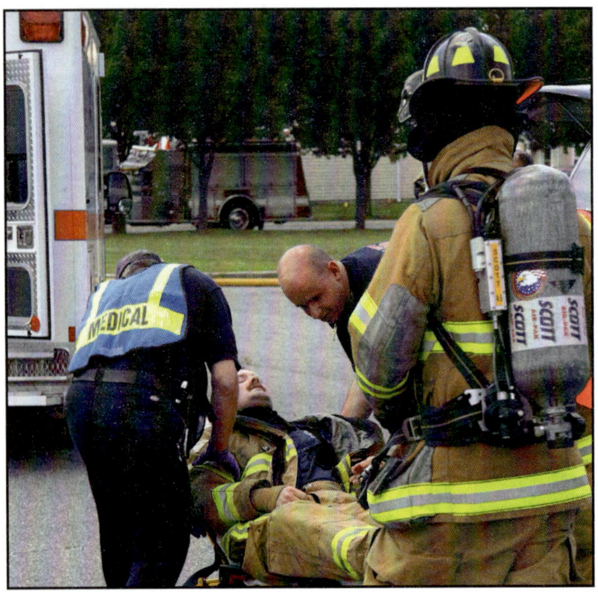

Figure 2.15 Thousands of firefighters are injured at fire scenes each year.

Recognition and Acceptance of Cultural Change

NFPA and U.S. Fire Administration (USFA) statistics make it clear that the fire service's most deadly risks come from lack of physical conditioning, which results in fatalities caused by cardiac arrest. Fire apparatus accidents, often due to poor driving or faulty equipment, result in the second highest number of firefighter fatalities.

Fire and emergency agencies must change in order to create an effective and efficient safety culture. Almost half of firefighter injuries occur on fire scenes. Year after year, strains, sprains, and muscular pain account for over 50 percent of fireground injuries **(Figure 2.15)**. The other half of firefighter injuries largely result from a combination of exposures to infectious diseases and hazardous conditions (NFPA, 2015).

Even though firefighter injuries have dropped consistently in recent decades, the rate of injury per fire remains unchanged. Although there have been significant reductions in cardiac-related deaths they still account for up to 50 percent of annual firefighter deaths (NFPA).

The fire service must not continue to accept the current level of fatalities, injuries, illnesses, and their associated financial burdens. To create change, firefighters must progress in the following ways:

- Recognize the unsafe culture and learn to recognize the disparity between cultural or organizationally motivated actions and acceptable risk
- Accept the need for change
- Recognize the barriers to change

These tasks can change at an individual or organizational level. Once accomplished, we can select a method for making the change and begin to implement it.

Recognition of Counterproductive Cultures

The current culture, based on values, behaviors, and artifacts, has good and bad influences that create a *counterproductive culture.* Adhering to negative cultural influences allows job-related fatalities, injuries, and illnesses to continue. **Table 2.1** shows some local-level characteristics and built-in activities of an unsafe and counter productive culture.

As an initial step toward positive change, agencies need to recognize the disparity between the organizational values and the actions of chief officers and elected officials. Disparity of values and actions in the fire service can be thought of as an inequality between what is valued and what is performed or offered.

Disparity at the organizational level often results from unintentional or uncontrollable budget allocations or reductions.

Table 2.1
Counterproductive Culture Symptoms and Built-in Activities

Unsafe Characteristics or Symptoms	Statistically Based Symptoms	Built-in Counterproductive Activities
• The attitude that *we have always done it this way.* • Safety is for wimps. • Self-sacrifice is a job requirement. • You get paid to take risks.	• Increase in job-related injuries • Increase in workers' compensation claims • Increase in near-miss incidents • Increase in the use of sick days • Increase in equipment and apparatus repairs • Occurrence of a job-related fatality	• Fail to enforce minimum physical fitness requirements and testing. • Hold the employee solely responsible for an accident. • Use fault-finding techniques rather than fact-finding ones when trying to determine the cause of an accident. • Use threats or discipline to correct unsafe behaviors. • Reward unsafe behavior by underreporting accidents, injuries, and near-misses. • Fail to reward supervisors for enforcing safety policies, procedures, and behaviors. • Fail to hold personnel accountable for knowingly making unsafe choices. • Lack management leadership in creating and enforcing a safety culture. • Fail to make safety an important part of every training experience

For instance, an organization may state that it values a well-trained workforce, but steady reductions to the training budget result in shortfalls. Changes that would reduce disparity may mean reducing certain services the agency offers to ensure that training keeps firefighters proficient in skills for the remaining services.

At the local level, groups can examine sick rates, injuries, accident rates, close calls, and customer complaints of a station or battalion. They can also look for disparity between organizational values, cultural influences, and the actions of individuals and crews. Once disparity is recognized, it is necessary to describe specifically how it is present and link symptoms to the potential negative results they can cause. For instance, the cost of failing to enforce physical fitness requirements can include the following:

- Excess sick leave
- Special-sized protective clothing
- Worker's compensation claims for cardiac-related medical conditions
- Damage to station furniture
- Increase in medical insurance premiums
- Need to shift tactical focus from incident control to providing care to a firefighter who needs assistance at an incident, thereby delaying control of the incident and increasing damage

Agencies could document the number of obese personnel, based on national standards, and use this documentation to track the physical health of their members. Personnel can use proper nutrition and exercise to control obesity, which can contribute to other diseases such as diabetes and high blood pressure.

Once they quantify the cost of a counterproductive culture, agencies should calculate the cost of prevention. As data has shown, fostering a safe culture can prove cost-effective and even reduce costs over time.

Acceptance of the Need for Change

Change must start at the individual level. Chapter 1 provided statistics that underscore the need for change. NFPA reports 69 on-duty deaths and more than 62,000 injuries to firefighters in 2015.

Typical firefighters tend to ignore statistics, believing that they will not become part of these numbers. This optimistic attitude, sometimes referred to as being bulletproof or success-engendered optimism, justifies actions as unsafe behavior continues. To keep from becoming a casualty, firefighters must identify organizational disparities between what firefighters value and what firefighters do. Then we firefighters must work to change negative cultural influences.

Barriers to Change

There will always be barriers to change. Some of these barriers are psychological (internal) while others are generated by the organizational culture (external). An understanding of the various types of barriers will help firefighters recognize and overcome them before they derail the change process. Change can also fail if the organization uses inappropriate methods to initiate and promote change.

Psychological (Internal) Barriers

Internal barriers result from a person's psychological conflicts and beliefs. These barriers can prevent the individual from making changes in lifestyle or behavior or accepting the need for organizational change **(Figure 2.16)**.

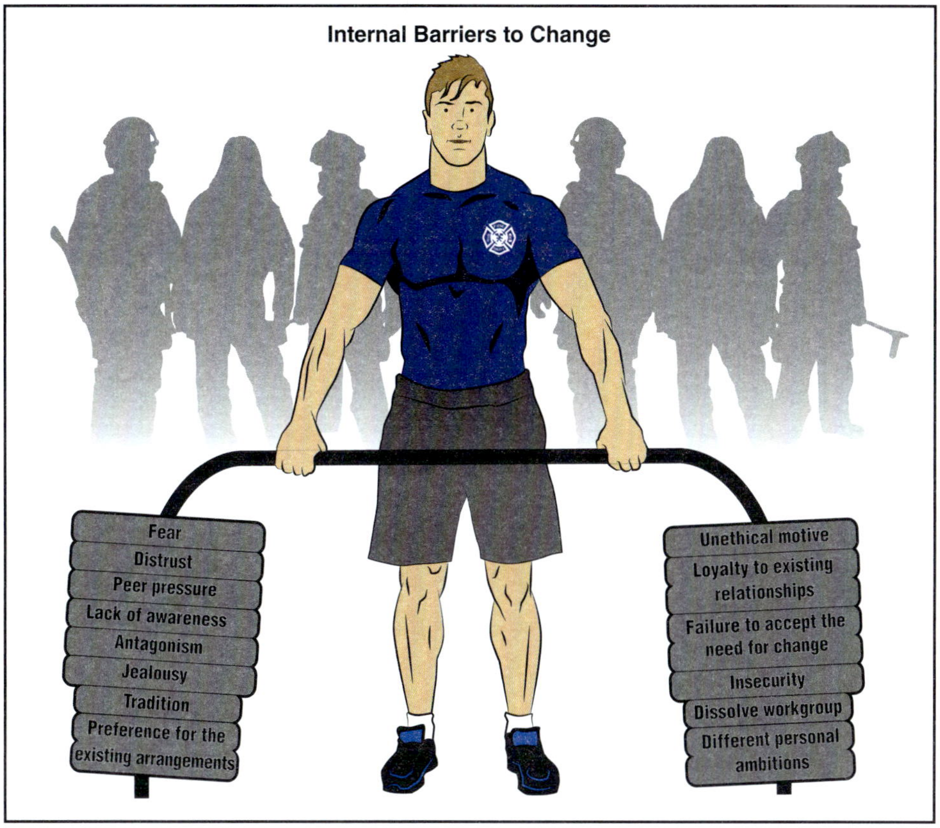

Figure 2.16 Internal barriers to change include more than a dozen attitudes and issues.

They include:

- **Fear** — Remains one of the main personal barriers to change. It may take the form of anxiety about the following:
  - — Loss of power
  - — Loss of skills
  - — Loss of income
  - — The unknown
  - — Supervisors may resist change because of a fear of losing control. They may worry that change will result in an overload of tasks and responsibilities.
- **Peer pressure** — Causes an individual to behave in a manner that others in the group will accept. The influence of a social group on an individual can create a major barrier to change. At the same time, peer pressure can have a positive effect on someone who shows inappropriate behavior. Although peer pressure is an external influence, the reaction by the individual is internal.
- **Lack of awareness** — Causes an individual to misunderstand the need for change. It may also cause a person to misinterpret information that would justify a change.

- **Distrust** — Causes lack of trust in one's own ability to implement a change and in the ability of others to provide accurate information and support to implement the change. Often the result of low self-esteem, it can cause paralysis and keep change from occurring.
- **Antagonism** — Active opposition to change by others can cause the individual promoting the change to compromise, even though the compromise might not be the best solution to the problem. Antagonism may be the result of ego, personality differences, or jealousy.
- **Jealousy** — Causes an individual to act irrationally and to block the suggestions of others. The decision-making ability of a committee can be crippled if one or more members are jealous and respond in a negative manner.
- **Unethical motive** — Causes an individual to seek personal gain, enhanced self-image, and personal protection at the expense of others. Eliminating this barrier, as well as the bad decisions it can generate, requires an insistence that individuals are motivated to base their decisions on an ethical foundation.
- **Tradition** — Bases lack of change on the history of the organization and individual loyalty to that history. Finds change in direct opposition to the way things are done, fearing it may devalue the old ways and call them into question.
- **Loyalty to existing relationships** — Forms a strong bond that creates a barrier to change, especially in cases when one is loyal to the views of a respected superior who does not agree with the change or to a supervisor who may see change as a threat to the unit composition.
- **Failure to accept the need for change** — Fails to see a need for change and rejects any argument for change. This individual believes that things are fine the way they are and does not want to be confused by facts.
- **Insecurity** — Causes an individual to feel incapable or afraid of handling a situation; generally founded on a lack of self-esteem.
- **Preference for the existing arrangements** — Prefers the comfort of the status quo or how things are arranged at present.
- **Dissolve workgroup** — Fears that change will require the unit to be dissolved; a desire to keep the status quo and to stay with the current supervisor.
- **Different personal ambition** — Conflicts with the individual's personal agenda. For instance, a proposal that officer candidates hold a college degree may present a conflict for someone who wants a promotion but does not have the educational requirements.

Organizational (External) Barriers

External barriers may be created by the organizational culture or structure and include **(Figure 2.17):**

- **Structural inertia** —Tradition, status quo, and the organizational structure can all create inertia, preventing not only change but any progressive movement.
- **Existing power structure** —A labor union, a specific division, or officers of a certain rank may interpret the proposed change as a threat to the group's influence and power.

- **Resistance from work group** —Change may be perceived as a threat to a work group's status quo. For example, fire suppression personnel may oppose performing fire and life safety inspections because of the time required to perform them.
- **Failure of previous change initiative** — Previous attempts to institute change may have failed for a number of reasons. Agencies must analyze and correct these reasons to ensure that the current initiative succeeds.
- **Lack of urgency** — Without a sense of urgency, members of the organization will not see the value of the proposed changes and will lose interest in adopting a proposal.
- **Lack of data** —A lack of accurate, sufficient, or timely information can result in making no change, making the wrong change, or accepting an inadequate compromise.
- **Lack of accurate analysis** — Wrong analysis can lead to an unacceptable decision, even when the data is accurate, sufficient, and timely. One must look at the problem as a whole, based on all available information, to determine the correct relationship of the parts.

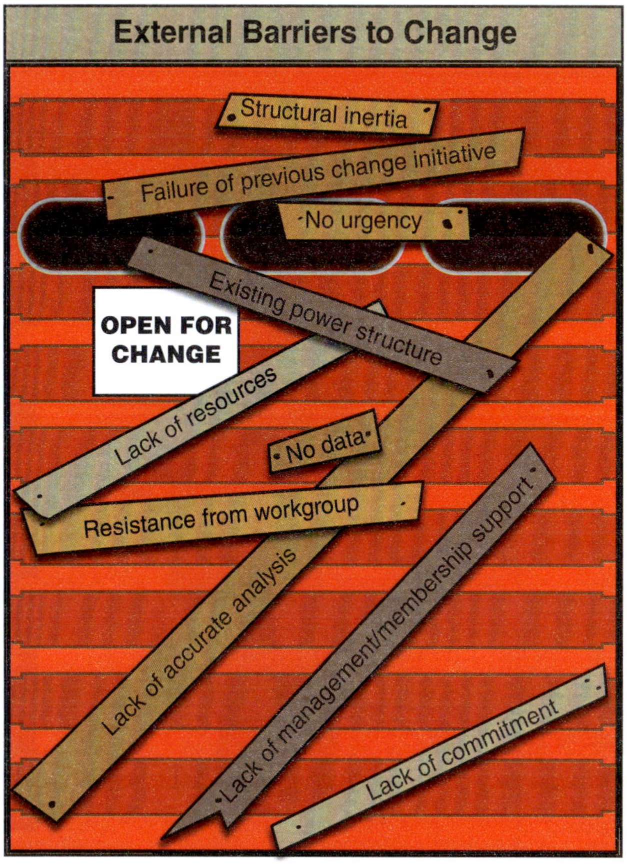

Figure 2.17 External barriers to change can stem from cultural problems in an organization.

- **Lack of resources** — Change is often delayed if necessary resources are not available. A lack of finances, personnel, time, skills, and equipment may make change impossible. Require that change be implemented in phases or postpone a change until the required resources become available.
- **Lack of management/membership support** —Without collaboration, good intentions are doomed to fail. Support grows through team building, open communication, and the empowerment of the membership. Building support can start from the bottom up or the top down.
- **Lack of commitment** —Without commitment, change is doomed to fail. Oral and written communication skills and public statements of commitment express the dedication necessary to ensure acceptance of the decision to change.

Inappropriate Change Management

Employees often resist change because the organization or change agent fails to present it in a positive manner. Inappropriate change management includes the following **(Figure 2.18, p. 70)**:

- **Failure to explain the need for change** — Failure to communicate adequate justification for the change to the leadership, membership, or stakeholders prevents personnel from understanding the need for change. Change must be based on facts and supported by accurate statistics.
- **Failure to provide information** — Lack of accurate information destroys the credibility of the change agent and prevents the leadership or membership from making an informed decision.

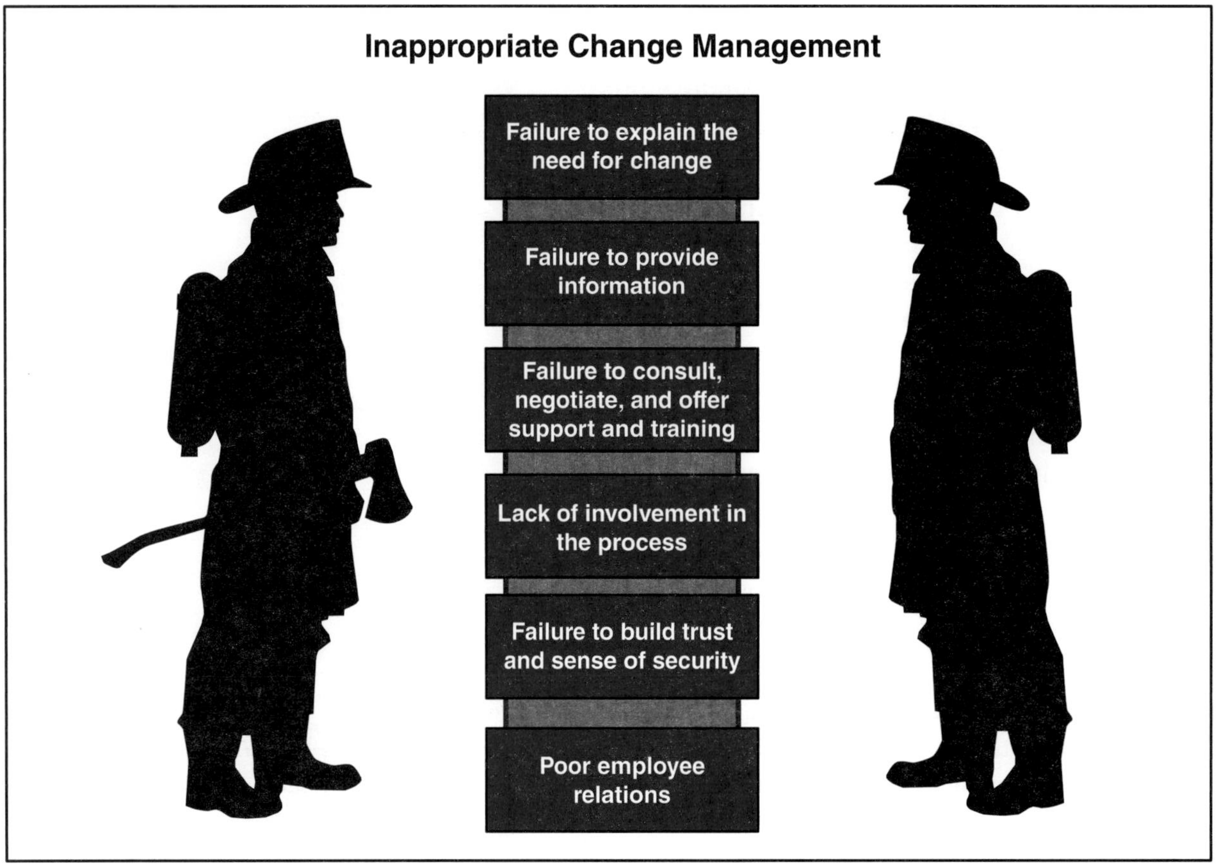

Figure 2.18 Inappropriate change management often stems from a failure to act.

- **Failure to consult, negotiate, and offer support and training** — Failure to consult with the membership and their labor representatives (stakeholders) prevents participants from knowing what changes are needed and why. When necessary, the organization must negotiate to overcome resistance and provide adequate resources to support the change. Organizations must also provide training, both in the change process and in the elements of the change.
- **Lack of involvement in the process** — Failure to involve the membership and stakeholders in the change process will alienate the affected people and those who can influence its acceptance.
- **Failure to build trust and a sense of security** — Lack of trust by the membership and stakeholders, as well as a feeling of insecurity, can prevent change. If members believe that the organization is taking advantage of them or that their jobs are in jeopardy, they will resist any attempts at change.
- **Poor employee relations** — Paternal attitudes toward employees — treating them or their concerns disrespectfully or ignoring requests and suggestions — can doom any attempt at change.

Overcoming Barriers to Change

Some simple steps that you can take to overcome change barriers, both as an individual and as an organization, include the following:

- **Communicate** — Explain the need, process, results, and benefits of the proposed change. Stress a sense of urgency to the membership.
- **Involve** — Keep members and stakeholders in the process.
- **Inform** — Keep the members and stakeholders informed of the process, progress, and results of the change.
- **Support** — Provide analysis, resources, leadership, and moral support to the change process.
- **Commit** — Provide strong commitment to the change process and the actual change.
- **Relate** — Explain how the change relates to the organization, membership, and stakeholders, as well as to the tradition and history of the organization.
- **Create** — Establish an atmosphere that is receptive to change in the organization. It must be based on mutual trust, ethical conduct, and job security.

Methods for Changing the Culture

Changing the organization's culture, and specifically its safety culture, requires planning based on a proven model. Business, industry, and the public sector have used many change models. The models can also be adapted to making personal changes. This section describes the following models:

- Five-step change
- Social-change
- Kotter's eight-step change

Five-Step Change Model

This five-step planning model, or any other similar program-planning model, provides planners with a systematic approach to decision-making.

Agencies can apply this model to either emergency or nonemergency situations including, as in this case, to facilitate organizational change. The change model can be applied to the individual as well as to the organization level. This model includes the following five steps:

1. **Identify** — Identify a problem that requires a response. For example, this manual considers the high fatality and injury rate in the fire service in general. Locally, the problem may be high lost-time injury rates or near-miss incidents.
2. **Select** — Choose the appropriate response to the problem, which requires establishing the goals, outcomes, and objectives required to meet them.

3. **Design** — Determine the steps required to meet the goals, outcomes, and objectives previously selected. To meet their goals, the National Fallen Firefighters Foundation developed 16 Firefighter Life Safety Initiatives to support organizations that share their desire to reduce the number of preventable firefighter fatalities. Locally, steps may include developing courses in proper lifting techniques or the implementation of a physical fitness program.
4. **Implement** — Perform the selected activity to mitigate the problem. This step places the program into effect. Nationally, the implementation of the Everyone Goes Home® initiative began with a campaign to convince local fire departments to adopt the 16 objectives. The local example would require personnel to attend a training course in proper lifting techniques and to participate in the fitness program.
5. **Evaluate** — Determine the effectiveness of the activities in meeting the goals or outcomes. This step includes monitoring the various elements of the program and evaluating them to determine how effective they are. Organizations may need to make some alterations in the elements to ensure that the goals are met. For instance, in the local example given above, the fitness program may result in an increase in injuries due to improper implementation or the lack of stretching exercises before exercising.

This five-step change model can be represented as a circular path or as a flowchart **(Figure 2.19)**. The results of the evaluation step may indicate the need for some changes in the activities. When organizations identify changes, a new change cycle begins.

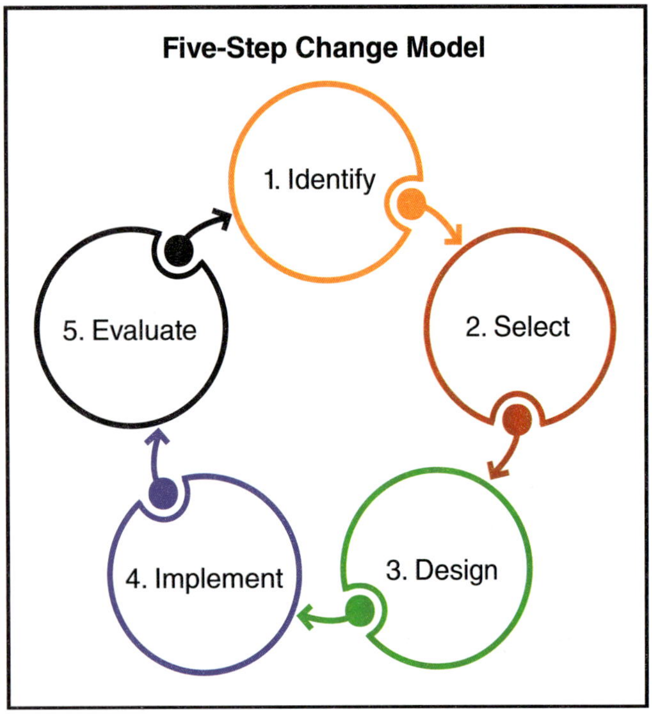

Figure 2.19 A five-step change model includes easily identifiable steps.

Social-Change Model

This value-based model of leadership places service at the core of social change.

The model's purpose is to effect change for the betterment of others through leadership. The goals of the social-change model include:

- Promote in the individual self-knowledge and an understanding of one's own interests, talents, and values.
- Increase leadership competence in order to cause positive cultural change in an institution, community, or society.

Corresponding to and interacting with these goals are seven critical values, subdivisions of the model's three areas:

- **Individual Values:**
  — *Consciousness of self and others:* Awareness of values, emotions, attitudes, and beliefs that motivate people to action

- *Congruence:* Similarity of thoughts, feelings, and actions toward others
- *Commitment:* Personal investment of time and energy for the duration of the project

- **Group Values:**
  - *Common purpose:* Shared goals and values as defined by the active participation of members of the group
  - *Collaboration:* Cooperation based on the application of mutual trust as a means of empowering others and self
  - *Controversy with civility:* Acknowledgment that group members will inevitably hold different views and that members should address differences in a civil values-based, respectful manner
- **Community/Societal Values:**
  - *Citizenship:* Acknowledgment that the members of the group have both individual rights and responsibilities to the community values
  - *Consciousness of self and others:* Awareness of values, emotions, attitudes, and beliefs that motivate people to action
  - *Congruence:* Similarity of thoughts, feelings, and actions toward others
  - *Commitment:* Personal investment of time and energy for the duration of the project

Kotter's Eight-Step Change Model

John Kotter, a recognized authority on managing change, created an eight-step model for implementing and handling change. In each step, he acknowledges a key principle that relates to people's response and approach to change. Because change is personal, people must see, feel, and then act on it. The model's eight steps include:

1. **Create a sense of urgency** — Motivate personnel and create real, relevant, and attainable objectives.
2. **Build a guiding team** — Select team members who have the appropriate knowledge, skills, and abilities and who are committed to the change process.
3. **Establish the vision** — Allow the team to establish a simple vision and strategy. Establish attainable and measurable objectives that support the vision's ultimate goal.
4. **Communicate the vision to everyone who will be affected** — Involve as many people as possible, including internal and external customers and stakeholders. Communicate the essential information in a simple and concise manner. Appeal and respond to needs of the participants.
5. **Empower the team** — Remove obstacles, permit constructive feedback, and provide support from leaders. Reward and recognize progress and achievements of the team.
6. **Create short-term objectives** — Set easily-achieved objectives and establish manageable numbers of initiatives. Complete each objective before attempting the next.

7. **Don't declare victory too soon** — Encourage determination and persistence in the team. View change as a constant process, not one that ends at the completion of the project.
8. **Institutionalize the change** — Reinforce the value of successful change. Make the change process part of the culture.

Results of a Change in Culture

Implementing the desired change and institutionalizing it may pose difficulties, but it will provide worthwhile results. There are studies from academia, government and industry that say training is the most effective way to institutionalize change in organizations. Chief officers who manage change in fire departments should evaluate the organization's need for ongoing training to normalize change. The results will help at the personal, organizational, community, and professional levels

Personal Benefits

Adopting a culture of continual safety, health, and wellness will provide firefighters with a safer work environment. Members will know they work for an organization with their best interests at heart, and one that grounds policies and procedures in safety, health, and wellness values **(Figure 2.20)**. Emergency responders will know their supervisors make decisions fully aware of the hazards they face and the risks those hazards can create.

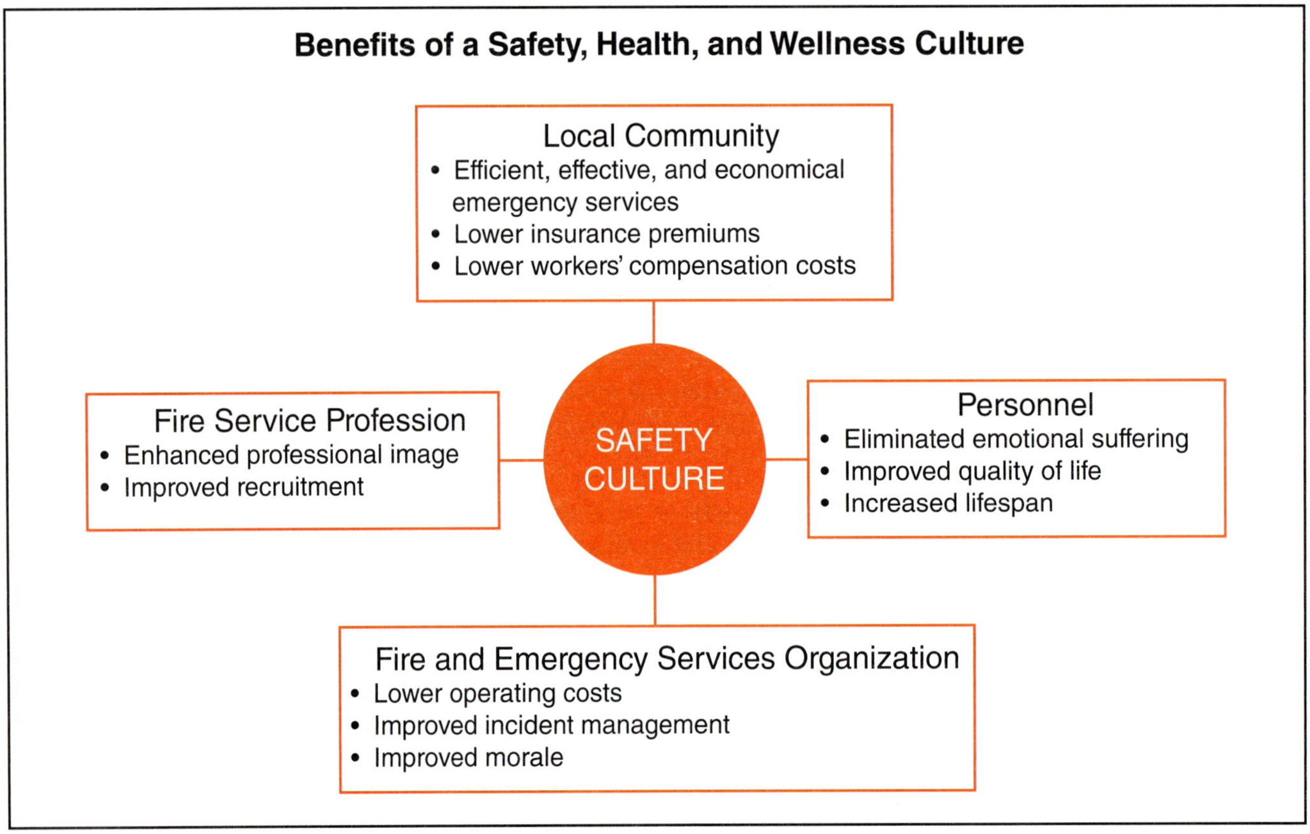

Figure 2.20 A safety, health, and wellness culture offers wide-ranging effects.

These benefits are not strictly the result of changes at the top of the organization or in the community. They require that everyone fully support the new culture and make internal behavioral changes. Everyone must change unsafe behaviors, have a proactive attitude toward safety, and relate ethically to subordinates, peers, supervisors, and others in authority. As firefighters identify disparities between safety values and behaviors, they must take responsibility for their actions and base those actions on the safest and healthiest choices possible.

It can be an emotional experience to visit the National Fallen Firefighters Memorial at the National Fire Academy in Emmitsburg, Maryland **(Figure 2.21)**. The names listed there often commemorate acts of heroism that resulted in the ultimate sacrifice. Too often, analysis reveals that reasonable safety precautions could have prevented these deaths. These include maintaining good physical conditioning, having regular physical examinations, staying off the roof or not staying under a roof when a building has experienced extensive fire, and driving defensively.

Figure 2.21 The National Fallen Firefighters Memorial is located on the campus of the National Fire Academy. *Courtesy of NFFF/USFA.*

Most firefighters would rather see their names on a service pension check than on a memorial. The primary benefit to active participation in a safety, health, and wellness culture is the ability to live a long and active life free of disabilities. An added benefit is the elimination of emotional suffering for the individual and their family.

Organizational Benefits

The fire and emergency services organization will also benefit from the shift to a safety-based culture in the following ways:

- **Financial***:* The organization will save money through reduced insurance costs, apparatus and equipment repair and replacement costs, and liability payments to victims. The organization will also save money by avoiding citations and fines from the Occupational Safety and Health Administration (OSHA) for infractions of federal safety requirements.
- **Operational:** Benefits will include improved emergency incident site management, reduced or eliminated fatalities and injuries, and increased membership participation and morale.
- **Morale:** The organization will adopt safety, health, and wellness as a core values that will determine strategic goals, establish objectives, and influence decision-making both at emergency incidents and in nonemergency activities. Professional development at all levels of the organization should include safety as a focus.
- **Other:**
  - Recruitment, both for career and volunteer personnel **(Figure 2.22, p. 76)**, can be made easier as the organization projects a safety-conscious and employee-centered image.

Figure 2.22 Community recruitment and training can include both career and volunteer fire fighters.

— A partnership for safety will improve labor/management relations.

— Training will be more consistent with a focus on the crew resource management approach to safe operations (See Chapter 3, Behavioral-Based Safety).

Community Benefits

Communities will benefit in many ways. They will receive more efficient, effective, and economical service. Apparatus and equipment maintenance costs will decrease when organizations eliminate or reduce accidents. Workers' compensation premiums can be reduced as the *experience modifier* is reduced (see information box). Overall, public trust in the fire service's reliability will grow, and increase the number of services agencies can offer. Public spending usually reflects public trust.

Workers' Compensation Experience Modifier

Public and private organizations purchase workers' compensation insurance policies for their employees from commercial companies. Premium rates are based on an experience modification rate (EMR). The experience modifier is a formula that compares the organization's accident and injury record over a three-year period with average data for all organizations in a similar industry, such as fire departments. The median receives a rating of 1.0. Companies that have an EMR greater than 1.0 pay a higher than average premium, and those with less than 1.0 pay a lower than average premium. A low EMR drives a private industry's overhead costs and gives those industries with a lower EMR a competitive advantage.

Professional Benefits

The fire service will always have the image of heroism and sacrifice. The public will continue to look at members with admiration and respect for the job. Shifting the culture to one based on personal safety will enhance the fire service's image with the public. An enhanced professional and public image makes the career fire service more attractive to potential applicants and builds support for the fiscal needs of the agency. The increased emphasis on safety helps to attract volunteers who associate safety with professionalism.

Safety Initiatives

Initiatives are specific projects that reach specific objectives, usually within a specific time frame. Initiatives have been developed to assist both organizations and individuals in changing their cultures. These initiatives are described in the following section.

National Safety Culture Change Initiative

In April 2015, the International Association of Fire Chiefs (IAFC), through a partnership with the USFA, published the *National Safety Culture Change Initiative*, a study of behavioral motivation and its role in the reduction of risk-taking behaviors in the fire and emergency services. The partnership aspires to identify both positive and negative firefighter behaviors and eventually to recommend culture and climate changes for the fire service. A copy of the NSCCI report is available on the USFA website.

National Fallen Firefighter Foundation-Everyone Goes Home®

Since the 16 Firefighter Life Safety Initiatives were jointly developed by representatives of the major fire service constituencies in 2004, the NFFF has been tasked with making them known throughout the fire service. The Initiatives continue to inform the emerging safety culture in the U.S. fire service, providing a foundation for the work of fire department and EMS organizations who want to ensure the safety of their firefighters and medics. See **Appendix B** for a list of the 16 Firefighter Life Safety Initiatives. Additional resources, research, and news involving the Initiative are available on the Everyone Goes Home website.

Strategy for a National EMS Culture of Safety

In 2009 the National Emergency Medical Services Advisory Council (NEMSAC) recommended that the Department of Transportation's National Highway Traffic Safety Administration (NHTSA) develop a strategy to create a safety culture in EMS. The three-year strategy-development process was finalized in 2013 with six core elements to advance a safety culture in EMS. More information on these health and safety initiatives is available on the EMS website.

IAFF, IAFC Wellness-Fitness Initiative

The Fire Service Joint Labor Management Wellness-Fitness Task Force has developed three programs to educate firefighters and promote wellness and fitness in the fire service. The three programs are: The Fire Service Joint Labor Management Wellness/Fitness Initiative (WFI); the Candidate Physical Ability

Figure 2.23 Candidates wear a weight vest to simulate the weight of protective clothing and an SCBA during a test. *Courtesy of Mesa (AZ) Fire Department.*

Test (CPAT) **(Figure 2.23)**, and the Fire Service Peer Fitness Trainer certification program. The WFI Resource provides additional information on successful program models. These programs represent a commitment by labor and management to positive, individualized wellness and fitness. More information is available on the IAFF website.

Other Initiatives

Other groups have developed initiatives to assist organizations and individuals in changing their cultures. Among these initiatives, concepts, and resources are the following:

- **NFPA Electric Vehicle Safety Training** — This training helps prepare firefighters for the increasing number of electric vehicles on the road in the U.S. Their web page offers up-to-date information about recent car models and much more. NFPA also offers fire departments a free field guide.
- **Safety Stand Down** — This joint venture between the IAFC; the IAFF; the Volunteer and Combination Officers Section of the IAFC; and the Safety, Health, and Survival Section of the IAFC focuses attention on safety-related issues.
- **Seatbelt Pledge** —The National Fire Service Seatbelt Pledge was created in 2006. After the death of a firefighter who fell from the apparatus on the way to a call, Dr. Burton Clark of the National Fire Academy spearheaded a challenge to firefighters to improve their safety by wearing their seatbelts. More than 850 departments and 150,000 firefighters have signed the National Fire Service Seatbelt Pledge and committed to wearing their seatbelts. The reach of the Seatbelt Pledge has also spread beyond the United States. Firefighters in Australia, Canada, the United Kingdom, and U.S. military bases around the world have recognized the importance of wearing seatbelts. Since 2011 the NFFF, through Everyone Goes Home®, has led the program and included the use of seatbelts in its 16 Firefighter Life Safety Initiatives. More information on this program, relaunched as the International First Responder Seatbelt Pledge program, is available on the Everyone Goes Home® website.

NOTE: In addition to initiatives, the U.S. Department of Homeland Security and OSHA have developed regulations in this area. They address respiratory protection training programs, confined-space requirements, hazardous materials training, national credentialing, and implementation of the National Incident Management System-Incident Command System.

Chapter Review

1. What three characteristics can be used to demonstrate and measure culture?
2. What three things will continue to happen if an agency adheres to negative cultural influences?
3. What are the differences between psychological and organizational barriers?
4. What similarities exist among common change models?
5. How can a safety-based culture can save an organization money?

Discussion Questions

1. How can the portrayal of firefighters in films and mass media influence people already in or thinking about joining the fire service?
2. How does your ethnicity, gender, or age influence personal and department safety culture?
3. How can an agency or an individual firefighter overcome psychological barriers to change?

Behavioral-Based Safety

Chapter Contents

chapter 3

Key Terms

Communication Model 92
Freelance 97
Health and Safety Officer (HSO) 86
Line-of-Duty Death (LODD) 90
Situational Awareness 95
Strategic Safety Plan................................. 88
Workload Management 102

FESHE Learning Outcomes

Occupational Safety and Health for Emergency Services

4. List and describe the components of risk identification, risk evolution, and incident management.
6. Apply the knowledge of an effective safety plan to preincident planning, response, and training activities.
10. Describe the responsibilities of individual responders, supervisors, safety officers, and incident commanders, safety program managers, safety committees and fire department managers as they relate to health and safety programs.

Behavioral-Based Safety

Learning Objectives

After reading this chapter, students will be able to:

1. Describe the evolution of behavioral-based risk management practices.
2. Explain the purpose and development of strategic safety plans.
3. State the primary goal of the Crew Resource Management (CRM) model.
4. Identify factors that affect CRM.
5. Apply the fundamentals of CRM to individuals, crews, and organizations.
6. Identify the characteristics of a High Reliability Organization (HRO).

Chapter References

K.A. Wilson, C.S. Burke, H.A. Priest, E. Salas. 2005. "Promoting health care safety through training high reliability teams," Quality and Safety in Health Care, 14: 303-309. Accessed online at qshc.com.

Chapter 3 Behavioral-Based Safety

Establishing and maintaining a safe environment depends on everyone in the organization. Administrators, supervisors, and individuals are all responsible for safety.

Behavior-based safety finds its origins in applied behavior analysis, which is the scientific study of behavior. It involves identifying practices that could reduce the risk of injury through the study of behavioral psychology. Over the past century, social scientists and safety professionals have taken various approaches to creating safe work environments. The fire and emergency services use two practices covered in the chapter:

1. Crew Resource Management (CRM)
2. High Reliability Organization (HRO)

This chapter will also provide an overview of the evolution of risk management.

History of Risk Management

In the early 1900s, the goal of risk management focused on equipment operators. An operator's physical, mental and moral shortcomings were considered to be the main reasons for poor safety performance. In 1911, Frederick Winslow Taylor wrote *The Principles of Scientific Management* which stated that the route to safer performance was determining the best practice through scientific methodology. If workers were compliant and followed procedures, safer operations would result. Early discoveries in the United Kingdom and Germany identified the enemy of safety as human error, whether from mental or moral defect or from being procedurally noncompliant.

When he published *Industrial Accident Prevention: A Scientific Approach* in 1931, Herbert William Heinrich laid the foundation for behavior-based safety. His research concluded that unsafe acts caused 90 percent of workplace accidents. After reviewing accident reports, he concluded that supervisors typically blamed employees for injuries, failed to conduct detailed investigations, and rarely determined the cause of the accidents.

The methods Heinrich advocated usually resulted in barriers placed between the employee and the risk. For example, a metal guard could be installed on a machine to prevent human contact with a moving part that could cause an injury. In fire and emergency services, adding layers of heat-resistant fabric to protective clothing places a barrier between the firefighter and the heat source.

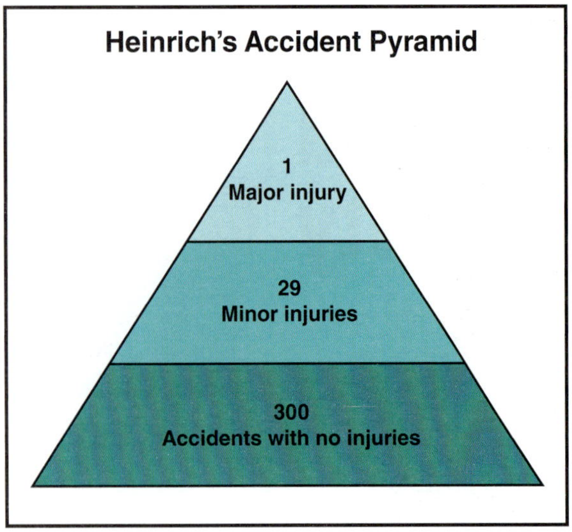

Figure 3.1 Herbert Heinrich's Accident Pyramid was developed in the 1920s.

Heinrich's research generated other behavior-based safety conclusions. Essentially, employees were a problem to control and safety was best defined as the absence of negative events or near misses. The conclusion of this initial stage of behavior-based safety was that measuring negative events and looking for causes could control behavior.

One of the most important safety perspectives Heinrich developed was the importance of the human factor in accident prevention. Heinrich further believed that people's errors often result in a near miss and/or sometimes in an injury. His theory led to the *Accident Pyramid,* which suggested that hundreds of near misses occurred prior to the disabling injury **(Figure 3.1)**. He suggested that the near misses be investigated to the same detail as injury-producing accidents.

In response to Heinrich's research, fire and emergency services have focused on fatalities, as the fallen firefighter reporting system illustrates. In fact, many training agencies display the number of fallen firefighters for the current year prominently on the classroom bulletin board. However, in the last decade, the near-miss reporting system has become a means to anonymously share incidents online. The IAFC's Near Miss Reporting System considers a near miss to be "an unintentional unsafe occurrence that could have resulted in an injury, fatality, or property damage if not for a fortunate break in the chain of events." According to Heinrich's Accident Pyramid theory, a focus on near misses should result in fewer injuries and a reduction in fatalities.

NFPA statistics (Firefighter Injuries in the United States) show a decline in firefighter injuries since 1988. This data supports the practices of departments that have similar injury reductions. However, caution is necessary where production, financial, and public perception pressures continue to push firefighters to do more with less. Employees often think that working fast is more important than working safely. It is difficult for some to grasp that personal safety does not cost valuable seconds. Instead, it maximizes the ability to work efficiently and longer in dangerous situations. When supervisors begin to understand that safety is human-driven, the obvious need to proactively manage work through a risk management system becomes a reality.

In the early years of the fire and emergency services, safety promotion was primarily the job of the training division or training officer. The training officer balanced teaching people to be aggressive in attacking a fire and how to work safely. The very idea of teaching people to work safely in a high-risk environment seemed like an impossible task. Other than limited personal protective equipment (PPE), experience, and intuition, training officers had little to work with.

Even today, the federal government, through agencies such as OSHA, recognizes that regulations do not cover all work activities at a fast-moving, ever-changing emergency scene. Because of this, training safely is a challenging and critical part of the fire service **(Figure 3.2)**. Some safety organizations have existed for over a century. The development timeline of safety organizations is shown in the following information box.

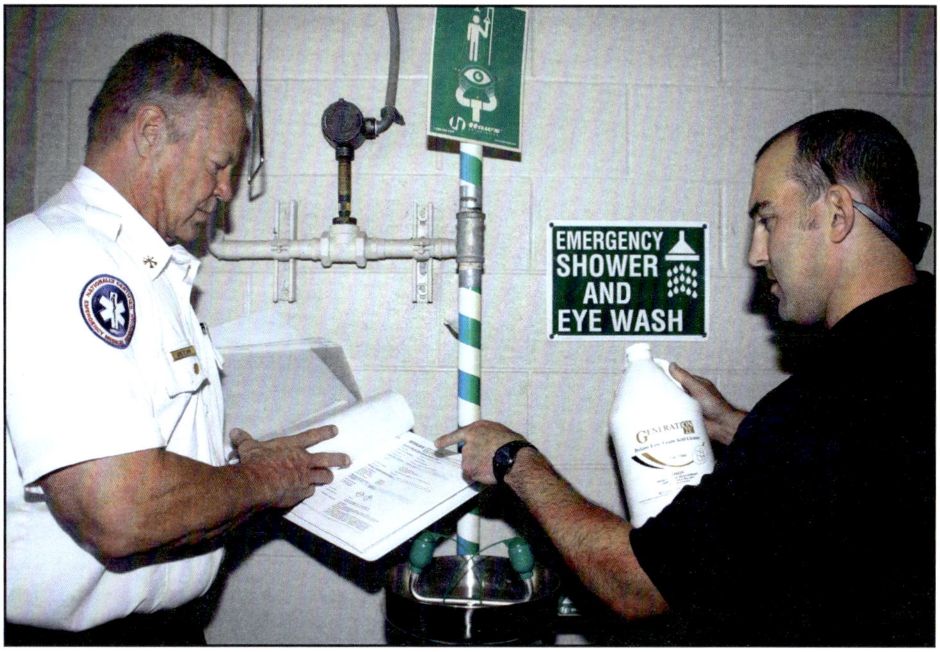

Figure 3.2 A Health Safety Officer reviews proper procedures for using an emergency eye wash inside a fire station.

Development of Safety Organizations

- 1896 — National Fire Protection Association
- 1913 — National Safety Council
- 1911 — American Society of Safety Engineers
- 1938 — American Conference of Governmental Industrial Hygienists
- 1939 — American Industrial Hygiene Association
- 1970 — National Institute for Occupational Safety and Health
- 1971 — Occupational Safety & Health Administration

Today's training officers have many tools to work with, from significantly improved PPE to thermal imagers and state-of-the-art simulation centers **(Figure 3.3)**. A training officer's work has evolved from teaching tasks and tactics to include teaching situational awareness. However, training officers cannot accompany each recruit into the field and ensure that risk-taking culture does not influence new firefighters' thinking. Cultural change, and therefore behavior change, must accompany safety-oriented training in order for it to take hold in a department. Additionally, many of the traditions and artifacts that support unsafe practices will need to be challenged. This evolution in our traditions will require an unwavering focus throughout the entire fire and emergency services.

Figure 3.3 Thermal imaging cameras are increasingly used in live-fire training.

A significant step in accident reduction was fostered by OSHA with the requirement for an increase in employee involvement through health and safety committees. The need for employee involvement was based on the following three factors:

1. Employees are exposed to hazards daily.
2. Group input can create more effective solutions to problems.
3. Employees are more likely to support programs where input has been provided.

As of 2015, twenty-five states, Puerto Rico, and the U.S. Virgin Islands have OSHA-approved state plans. Twenty-two state plans (twenty-one states and one U.S. territory) cover private and state and local government workplaces. The remaining five state plans (four states and one U.S. territory) cover state and local government workers only. Most states adopt updated OSHA federal regulations within six months of any rule change. In line with this concept, many states have adopted specific safety committee requirements that are more stringent than OSHA regulations.

If accident, fatality, and injury statistics plateau, the fire service Health and Safety Officer should not consider adding more levels or functions of bureaucracy in their departments to make their departments safer. Bureaucratic organizations tend to add rules, look to people unfamiliar with the actual work in the department, and make accountability to management more important than to the firefighters who encounter risks. Before adding more bureaucratic steps to any program, the **Health and Safety Officer (HSO)** should consider asking the question: "Will the accidents, injuries, or close calls we have be reduced through more bureaucracy?"

Health and Safety Officer (HSO) — Officer designated by the chief of department as defined by NFPA 1521. Officer in the fire department primarily responsible for the management of the health and safety program.

The anonymous suggestion box served as an early effort to increase employee involvement. With the development of labor representation in the fire and emergency services, they have been replaced with labor-management committees, commonly referred to as safety committees. However, the suggestion box remains an important tool.

The NFPA recognized the importance of safety and the creation of safety committees with the development of NFPA 1500, *Standard on Fire Department Occupational Safety, Health, and Wellness Program*. It requires the establishment of a safety committee and the position of HSO in fire and emergency service organizations.

While safety committees are one of the most effective tools to promote a health and safety program, many committees are used inappropriately or underutilized. A primary function of traditional safety committees has been to review accident and injury reports **(Figure 3.4)**. However, accident review is an after-the-fact approach. In some cases, it can promote a *fix-blame* approach to accident prevention, rather than a *fix-the-problem* approach.

In well-run and supported health and safety committees, knowledge obtained from incident reviews can help departments develop and adopt better safety awareness and training practices. Well-run safety committees enlist the help of peers and cross-professional support to review or audit work. A good safety committee will empower members to stop any operation they feel is unsafe.

PITT MEADOWS FIRE & RESCUE SERVICE INCIDENT
ACCIDENT/INCIDENT/NEAR MISS INVESTIGATION REPORT

EMPLOYER NAME PITT MEADOWS FIRE RESCUE SERVICE	Firefighters Attending Investigation # Firefighters Absent #
ADDRESS WHERE INCIDENT OCCURRED:	

INCIDENT OCCURRED:

PLACE	DATE YY/MM/DD	TIME
		☐ AM ☐ PM

INJURED PERSON(S) or NON-INJURY ☐

LAST NAME	FIRST NAME	JOB TITLE
1)		
2)		

NATURE OF INJURY

1)		
2)		

EQUIPMENT FAILURE/DAMAGED

YES/NO DESCRIBE:

MEMBERS INVOLVED AND/OR WITNESSES

NAME	RANK	ADDRESS IF NOT FD MEMBER	TELEPHONE
1)			
2)			
3)			
4)			

ACCIDENT/INCIDENT DESCRIPTION

Briefly describe what happened, including the sequence of events preceding the incident

CONTRIBUTING FACTORS AND CAUSES

List of any unsafe conditions, acts, procedures or factors that in any manner contributed to the incident

RECOMMENDATIONS

Identify any corrective actions that have been taken and any recommended actions to prevent similar incidents

RECOMMENDED CORRECTIVE ACTION	ACTION BY WHOM	ACTION BY DATE
1)		
2)		
3)		
4)		
5)		
6)		

JOSH INVESTIGATION COMMITTEE

NAME	SIGNATURE	TYPE OF REPRESENTATIVE			DATE
		☐ EMPLOYER	☐ WORKER	☐ OTHER	
		☐ EMPLOYER	☐ WORKER	☐ OTHER	
		☐ EMPLOYER	☐ WORKER	☐ OTHER	
		☐ EMPLOYER	☐ WORKER	☐ OTHER	
		☐ EMPLOYER	☐ WORKER	☐ OTHER	

Figure 3.4 Having templates for various reports, such as this accident reporting form, are important for standardizing reporting procedures. *Courtesy of City of Pitt Meadows Fire and Rescue Service.*

Safety committees can serve within any division of the fire service and combine with other regional fire departments to strengthen safety awareness and cooperation. The behavior-based safety concept needs personnel committed to the process. Committee members have input into the development of safety policies and procedures that they must adhere to.

An Effective Safety Committee

- Receives active administrative support for the committee
- Receives active labor organization support for the committee
- Meets on a regular schedule
- Shares results with the organization's membership
- Ensures continuity by rotation of 50 percent of members
- Conducts or monitors post-incident analyses
- Generates three to five goals for each year
- Routinely evaluates its overall effectiveness
- Encourages reporting of near-miss incidents
- Reviews all of its near misses as well as accidents and injuries.
- Monitors the progress of safety inspection results
- Helps develop a climate of positive safety peer coaching
- Helps develop an overall health and wellness program
- Helps evaluate the effectiveness of safety-in-training programs
- Helps evaluate the effectiveness of health and safety programs

Strategic Safety Plans

NFPA 1500, section 4.2, Risk Management Plan, recommends fire departments adopt comprehensive risk management plans that address all facets of operations and those having to do with facilities, training, and administration. Section 4.3, Safety and Health Policy, addresses the need for a department occupational and health policy that states goals and objectives for the prevention of event outcomes that negatively affect the health and well-being of firefighters. A **strategic safety plan** outlines broad safety goals for a fire department and should be developed with diverse representation of the entire fire department. **(Figure 3.5)**

Strategic Safety Plan — Plan that identifies specific goals that support and enhance the safety and risk management activities of the department.

Developing Strategic Plans

A natural progression for the safety committee should be to assist the HSO in developing the strategic safety plan. Strategic plans help focus the safety committee's efforts by identifying specific goals. This proactive approach permits the safety committee to become an integral component of the department's risk management program.

Safety committee members should evolve from being safety cops to being part of a collaborative team focused on obtaining positive results and moving away from blaming individuals. Generally, the movement is away from enforcement toward advocacy. This focus is especially useful when other factors such as organizational drift, normalization of deviance, and poor management accountability are identified as contributing causes to negative events. Safety committees that operate within this type of environment rarely have difficulty obtaining members.

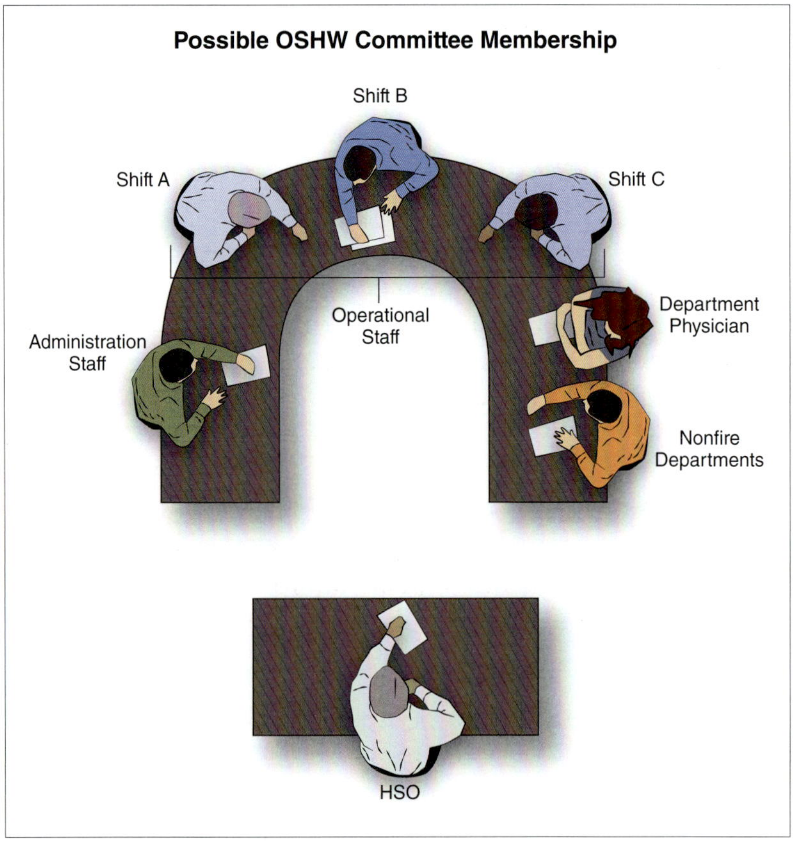

Figure 3.5 In order to include all stakeholders in an organization's safety culture, the occupational safety, health, and wellness committee membership should reflect individuals at all levels of the organization and any partners outside the organization.

Elements of Strategic Plans

Strategic plans should have the following elements:

- Introductory statement
- Vision statement
- Mission and values
- Objectives of the plan (examples)
  - — Establishment of policies
  - — Goal of no accidents or no lost time injuries
  - — Cultural issues to be addressed
  - — Responsibility of staff members
  - — Ongoing national standard compliance
- Initiatives or principles
- Leadership commitment
- Record keeping and reporting
- Responsibility and authority
- Hazard and risk management
- Planning
- Facilities and equipment
- Training and developing competency
- Communication
- Incident/accident management
- Member wellness
- Legislative compliance
- Reviews and audits

Strategic plans map an organization's course into definite time components. The plan attempts to account for external factors such as changes in the economy, demographics, service requirements, hazards, and technologies. Strategic plans are based on trends and expectations, but external factors can be unpredictable. Therefore, strategic planning, while relying on the best judgment of the planning committee based on available information, must be responsive to change.

Strategic safety planning can be effectively applied to the organization's safety needs. The safety committee uses the same process to analyze the safety trends of the organization and develop responses.

SWOT Analysis

Strategic plans are often created through a process called SWOT analysis **(Figure 3.6)**. It may be best to have an experienced facilitator lead the SWOT analysis. The acronym stands for the following:

- **Strengths** — Identify the strengths of the organization's safety program.
- **Weaknesses** — Identify safety program weaknesses.
- **Opportunities** — Recognize the opportunities to improve safety levels by reducing the number of near-misses, accidents, injuries, or fatalities.
- **Threats** — Identify unfavorable internal and external threats posed by lack of training, lack of funding, and unsafe behaviors.

The administration and the employees must support the plan, implement it evenly and fairly, and review the plan periodically. If the review indicates that the plan must be altered to meet new threats, the development process must restart.

SWOT Analysis	
Strengths What does the organization do well? What resources does the organization have to draw upon? What do citizens see as your strengths?	**Weaknesses** What needs to be improved? What resources does the organization lack? What do citizens see as weaknesses?
Opportunities What opportunities are open to the organization? What trends could the organization take advantage of? How can the organization use its strengths to take advantage of the opportunities?	**Threats** What threats are confronting the organization? What trends are a threat to the organization? What threats do the organization's weaknesses expose it to?

Figure 3.6 Example of a SWOT analysis.

Crew Resource Management

In order to achieve the goals of the 16 Firefighter Life Safety Initiatives and to meet ongoing initiatives of the National Fallen Firefighters Foundation, the International Association of Fire Chiefs, and International Association of Fire Fighters initiatives, fire departments need to focus on improving individual situational awareness and organizational culture and trust. One approach adopted by the IAFC is the Crew Resource Management (CRM) Model.

While the primary goal of CRM in the fire and emergency service is to reduce **line-of-duty deaths (LODD)**, the model applies to reducing all types of injuries and accidents in emergency and nonemergency situations. This section describes the background of CRM, factors that CRM addresses, and the application of CRM concepts.

Line-of-Duty Death (LODD) — Firefighter or emergency responder death resulting from the performance of fire department duties.

CRM Background

The roots of CRM can be found in the airliner crashes of the 1960s and 1970s. Crash investigations into those disasters indicated that human error was a significant contributing factor. These findings spurred the National Aeronautics and Space Administration (NASA) to develop an air crew workshop in 1979. It was tasked with addressing the specific elements of human behavior that were found to contribute to these accidents: decision-making, lack of interpersonal communications, and leadership.

The basic CRM Model used for training in the commercial airline industry as well as military aviation has continued to evolve. According to the National Transportation Safety Board, disasters have dropped from approximately 20 per year to as few as 1 or 2 per year. Other organizations that have adopted CRM have also noticed an improvement in their operational safety records.

In the IAFC *Crew Resource Management* publication, the authors describe the similarities between flight crews and fire and emergency services crews, including the following:

- **Crew structure** — Both crews have a leader/follower organization or hierarchy.
- **Teamwork** — Both crews work together toward a common goal.
- **Crew composition** — Both crews may be composed of personnel who are permanently assigned to work together. They may also be assembled on short notice from a variety of sources.
- **Workload** — Both crews may experience long hours of mundane activities that are punctuated by periods of intense, high-stress activities.

Modern aircraft cockpits are equipped with multiple safety warning devices, redundant controls, and computers that can fly the aircraft without a pilot's immediate attention. A flight deck and apparatus cab have similarities, specifically with the broad array of information being directed at the Captain/Lieutenant and the crew's ability to assist with situational awareness and interpreting the information.

While the commercial airline industry and the military have seen improved technology reduce accidents, injuries, and fatalities, the fire and emergency services continues to experience unacceptable levels of injuries and fatalities.

CRM Factors

The five factors that the commercial airline industry has worked to improve with CRM are as follows:

1. Communications among crew members
2. Situational awareness
3. Decision-making
4. Teamwork
5. Implementation barriers

Fire and emergency service CRM training should also include techniques for workload and error management. Training emergency personnel to effectively and correctly overcome the negative aspects of each of these factors will result in a change in the work environment.

Communication Model — A representation of the process that people use when interacting with others on a daily basis.

CRM Communications

The CRM Model attempts to correct one of the major problems uncovered during analysis of airliner accidents: lack of communication among flight crew personnel. To correct this problem, crew members must use an effective **communication model** which consists of five components:

1. **Sender** — Person who initiates the message. The sender develops the idea; translates it into oral, written, or nonverbal symbols; and sends it to the receiver.
2. **Message** — Content or idea as it is contained in the medium.
3. **Medium** — Vehicle that carries the message to the receiver. It may be oral (sounds), written (words on paper or in electronic format), or nonverbal (physical gestures or facial expressions among others).
4. **Receiver** (also known as ***Listener***) — Person or persons who is/are the recipient(s) of the message and responsible for interpreting it based on personal knowledge and bias.
5. **Feedback** — Confirmation from the receiver to the sender that the message was received and understood. Effective communication is not complete without feedback. Feedback may use any medium suitable to assure the sender that the receiver has full understanding of the message. When feedback occurs, the participants exchange the roles of sender and receiver, and the feedback contains a new message.

A sixth component may be present in the communication model: interference. Interference may be external or internal to the participants in a conversation; descriptions are as follows:

- **External interference** — Includes all audible and visual distractions in the environment. *Examples:*
  - — At an emergency incident, sirens, apparatus or power-tool engine noises, radio static, and noise created by a fire **(Figure 3.7)**
  - — Visual distractions such as the sight of injured victims, massive devastation, flashing lights, and media presence

Figure 3.7 Two Chief Officers communicating at a noisy fire scene. *Courtesy of Ron Jeffers, Union City, NJ.*

- **Internal distractions** — Includes items unique to the receiver. Generally speaking, internal distractions may be more difficult to overcome than external distractions. *Examples:*
  — Personal prejudice toward the sender or message
  — Hearing deficiency
  — Lack of knowledge of subject under discussion
  — Inflated ego on the part of the receiver
  — Lack of self-esteem on the part of the receiver

Successful communications depend on the sender and receiver overcoming external and internal interference. The external interference may be the easiest to control: Move the conversation to a quiet, less-distracting area; use headphones; turn off audible warning devices.

Internal distractions may be more difficult to overcome because they require some behavior changes, training, or use of hearing aids in case of hearing loss. One way of overcoming internal interferences is to apply common communications traits and tasks suggested in the CRM Model.

Common Communication Traits

Internal distractions result in the receiver misunderstanding or not receiving a message. The result can be tragic and fatal. To prevent such a result, both the sender and the receiver must improve their speaking and listening skills. Through research, speech and interpersonal communications professionals have identified a series of common communications tasks that can aid the sender and receiver. **(Table 3.1, p. 94)**

Communications Tasks

CRM is based on the application of the basic communication model. The CRM Model suggests the message sent between sender and receiver must be respectful, clear, concise, and complete. To further expand and improve on the communication model, CRM gives each member of a crew additional responsibilities in the communications process:

1. **Inquiry** — In situations where safety is an issue, it is the ability and opportunity to ask *why*. The purpose is not to question routine tactical issues but reconcile any difference between what is occurring and what should be occurring. A crew member is *not* questioning authority, however. Firefighters should take care when using this technique because it can be easily overused in emergency situations.
2. **Advocacy** — Ability to state one's position, based on facts as the crew member perceives them. It may be helpful to communicate those perceived facts during advocacy. Advocacy may help to increase situational awareness, improve understanding, and avoid a disaster.
3. **Listening** — Use of effective listening skills as described previously.
4. **Conflict resolution** — Ability of each member of a crew to stay focused on a situation and how to best resolve conflict. Conflict is normal because humans have individual points of view. Conflicts are not about *who* is right or wrong, but about *what* is right. The focus should be on the resolution of safety issues and not tactical decisions.

Table 3.1
Tasks for Improving Communications

Sender	Receiver
To ensure that a message is understood and heard, the sender must perform the following tasks: • *Establish a frame of reference for the receiver* — Sender and receiver must both understand the subject of the message. • *Provide a clear and complete message* — Message must contain all the information required to comprehend the situation or accomplish the mission. • *Omit any subjective or biased information* — Sender must adhere to the facts and eliminate any opinions that might alter the understanding of the message. (This "task" may not be invariably applicable to the fire service as "opinions" are at times very useful until more complete "facts" are available.) • *Understand the importance and effect of nonverbal communication* — Nonverbal of communication carries up to 93 percent of the message. As little as 7 percent of the message is carried by words that are spoken. • *Repeat the message to ensure that it is understood* — Repeating and paraphrasing the message helps the receiver to hear and understand the message, provide correct feedback, and act on the information. (This task may not be invariably applicable to the fire and emergency service as the situation does not always allow time for repeating.) • *Be respectful* — Never speak in a disrespectful manner. Insulting, humiliating, or defaming the receiver is the best way to prevent effective communication.	The receiver is equally responsible for ensuring that the communication process is complete and successful. The receiver must perform the following tasks: • *Use effective listening skills* — Essentially, this task means that the receiver must be mentally ready to listen, listen to the complete message before providing feedback, and maintain eye contact with the sender (when possible). • *Put aside preconceived notions* — The receiver must listen to the actual message and not allow personal bias affect the message. • *Properly interpret nonverbal communications* — If the verbal and nonverbal messages do not agree, the receiver asks the sender to clarify the message. • *Ask for clarification of the message* — Through feedback, the receiver tells the sender that the message was not understood. Ask for more information or to have the message simplified or rephrased. • *Be respectful* — Show respect for both the message and the sender.

5. **Feedback** — Confirmation that the message is fully understood and the communication process is complete. CRM depends on good feedback by the receiver. If the message was not understood, the sender restates the message, and the process continues until confirmation is received.

With these responsibilities, crew members feel empowered to have an active role in the communications process. They actively contribute. Department members must be aware that the supervisor's authority is not being questioned. Crew members may question decisions when they do not fit a situation and may result in life-threatening conditions.

When considering communication in the context of the Incident Command System (ICS), it is important to remember that it is a two-way street. Command gives direction based on external observations, resource availability and the incident timeline. Functional or geographic teams provide up-to-date information on their situation. In many incidents, the team or group leaders will have different views of the situation than those managing larger groups or geography, such as Division or Group Supervisors. In effect, the small-group leaders will be the eyes and ears for the Division/Group Supervisors.

Situational Awareness

Negative outcomes can occur when we lose **situational awareness**. If there is too much focus on getting to the seat of the fire, we may not notice the loss of structural integrity or the change in fire behavior that indicates that flashover is approaching.

Situational Awareness — The ability to comprehend the implications of surrounding conditions and apply that understanding to help predict how and when the situation may change in order to take appropriate action.

Situational awareness is sometimes referred to as a process that exists on three levels:

Level 1: **Perception** — Perceive the situation around us.

Level 2: **Comprehension** — Apply our knowledge and past experiences to our perception and develop an understanding of the meaning of the situation.

Level 3: **Application** — Take our understanding of the situation and apply it to the future, thereby predicting how and when a situation will change and what action is appropriate.

CRM requires that all members of a crew maintain situational awareness and communicate any changes in the environment or conditions or behaviors of other crew members to the crew leader and each other. The loss of situational awareness creates an opportunity for making improper decisions. Eight factors that may lead to the loss of situational awareness include the following:

- **Ambiguity** — Information received is confusing or unclear. For instance, the building is two stories at the front, three floors at the rear.
- **Distraction** — Loss of focus on the original mission without appropriate rationale. An example would be attempting to provide exposure protection before rescuing a confirmed victim.
- **Fixation** — Too focused on a single element of the situation. An example would be raising an aerial into electrical wires while focused on a rescue situation on upper floors.
- **Overload** — Tasks or information overwhelm crew members or they attempt to perform all the tasks themselves. For example, a single company attempts to carry out a fire attack, search and rescue, and ventilation at the scene of a high-rise fire.
- **Complacency** — False sense of comfort based on seemingly similar experiences. "We come here twice a week on false alarms, so this must also be a false alarm."
- **Improper procedure** — Policies or procedures are violated or ignored without justification; for example, a first-arriving company starting initial operations without giving a size-up report.

- **Unresolved discrepancy** — Two or more pieces of information that do not agree; for example, the visual indicators of smoke coming from a structure do not match the crews' knowledge of a building's contents.
- **Lack of comprehensive hazard surveillance** — Crew members become so fixated on one detail that they ignore everything else. An example would be focusing on a vehicle accident and ignoring the hazard of the highway traffic flow nearby.

In the CRM training provided by the US Coast Guard (USCG), proper situational awareness depends on performing the following actions:

- Maintaining effective communications
- Recognizing and making others aware of any deviation from standard operating procedures (SOPs) or policies
- Monitoring crew member performance
- Providing information in advance of an operation or mission
- Identifying any potential problems or existing hazards
- Communicating the desired course of action
- Communicating the mission's status continuously
- Evaluating the situation for any changes continually
- Clarifying expectations of crew members continually

Decision-Making

The ability to make correct, well-informed decisions is important in nonemergency situations and critical in emergencies. The IAFC provides an example of another decision-making model developed by Ludwig Benner, Jr. of the National Transportation Safety Board (NTSB). The model uses the acronym *DECIDE* to symbolize each of the steps in the process:

- **D**etermine the problem.
- **E**valuate the scope of the problem.
- **C**onsider options for mitigating the problem.
- **I**dentify the most appropriate option.
- **D**o the most appropriate option.
- **E**valuate the effectiveness of the actions.

Benner's model separates some of the elements of the individual steps and adds an evaluation of the effectiveness of the decision. When applied to the CRM process, decision-making moves from an individual task performed by the crew leader to a group effort involving all crew members. CRM is based on the effective use of all resources. Therefore, the knowledge and observations of the crew, which are part of these resources, can be accessed by the crew leader to help make decisions.

Teamwork

The fire and emergency services have long depended on teamwork to accomplish their assigned goals and objectives. Crews, like teams, have specific tasks they must perform together. At an emergency incident, crews may be involved in suppression, rescue, overhaul, ventilation, or a myriad of other tasks **(Figure 3.8)**. Teamwork occurs when the crew effectively and efficiently performs its designated tasks.

Figure 3.8 Teamwork is crucial to firefighter safety at an emergency incident. *Courtesy of Ron Jeffers, Union City, NJ.*

The lack of teamwork can result in loss of crew cohesion, failure to accomplish the objective or goal, and potential injury or death. Symbolic of a lack of teamwork is the individual or crew member who **freelances** and performs the first task that comes to mind without coordinating with crew members or the Incident Commander (IC).

Freelance — To operate independently of the Incident Commander's command and control.

Teamwork requires certain skills and actions on the part of crew members and the leader. Some social and communications skills are required, including the following:

- **Listening** — Effective listening by the leader and crew members is critical to the success of a team. The leader must be open to the ideas and concerns of the crew, and the crew must be ready to hear and act upon the leader's decisions.
- **Questioning** — Crew members must be able to ask questions in order to better understand team objectives. The question must be concise and focused on the objective and personnel safety.
- **Persuading** — Crew members must be able to freely provide and defend ideas and suggestions without fear of criticism. Ideas and observations must be concise and focused on the objective and personnel safety.

- **Respecting** — Interacting respectfully is basic to all interpersonal communication.
- **Helping** — Crew members must be encouraged to help coworkers. They also must be encouraged to ask for help.
- **Sharing** — Crew members must be willing and able to share with the rest of the group. Sharing includes resources, workload, information, and personal concerns.
- **Participating** — Participation in a team's activities means being a part of the team. Crew members participate in the decision-making process and share in the success or failure of an operation.

Most teams consist of two basic elements: leaders and followers. In the traditional approach, the leader had complete authority and sometimes, especially on the fireground, used an autocratic style. The followers carried out the leader's orders without question or comment. Autocratic leadership has led to errors because the leader erred.

The CRM Model redefines the relationship between leader and follower to create a new approach to teamwork. A description of each position is as follows:

- **Leader** — Leads others through formal authority or informal characteristics; therefore, the position may be assigned or assumed, depending on the circumstances.
- **Follower** — Accepts the formal authority of another person or subordinates themself to the skills of a formal or informal leader; may become the informal leader of a group or unit.

Effective CRM teamwork depends on both the leader and follower possessing certain skills. These skills are briefly addressed in the sections that follow.

Leadership Skills

Books and publications on leadership in business, industry, and the fire and emergency services are many and varied. They can provide a leader with models and examples to develop strong leadership skills. The CRM Model provides the following skills important to a leader:

1. **Authority** — A leader must have the legitimate authority to command and control the operation and resources to perform a mission. A leader has the responsibility for all decisions and crew safety and must be able to define and communicate mission goals to the crew. Formal authority does not prevent the leader from soliciting input from crew members and altering command decisions if warranted **(Figure 3.9)**.
2. **Mentoring** — A mentor is defined as a trusted guide or adviser. A leader teaches skills and transfers knowledge to crew members and advises them on professional topics from career advancement to operational procedures. A leader also uses mentoring skills to motivate the crew and monitor its progress to ensure success. In some situations, a crew member may mentor a leader.

Figure 3.9 A Chief Officer and an Incident Commander discuss strategy at an incident. *Courtesy of Ron Jeffers, Union City, NJ.*

3. **Conflict resolution** — Through the use of situational awareness and communication skills, a leader recognizes conflict within the group and resolves it to the best effect. At the same time, a leader must recognize the concerns that led to the conflict. These concerns may be varied, valid, and real, based on the experiences and observations of the crew. A leader must be ready to reassess the situation based on crew concerns.
4. **Mission analysis** — This resembles the size-up process used by emergency responders when they arrive at an emergency incident. Mission analysis requires the leader to perform the following tasks:
   - Gather information on the situation.
   - Evaluate the level of risk against the potential benefit of an action.
   - Identify, define, and communicate objectives and goals. Create and implement an incident action plan (IAP). This step applies to the Incident Commander (IC).
   - Evaluate the effectiveness of the plan.
   - Develop contingency actions.
5. **Mental aspects:**
   - Consider that people can generally remember only ten items or tasks at one time.
   - The leader must provide objectives easily remembered by followers.

A follower must have an attitude that includes a can-do spirit, an optimistic and cooperative nature, a willingness to listen, and the ability to be open-minded. By developing each of these, the follower (and the leader) can create a strong team and overcome most adverse situations.

SMART Objectives

Incident commanders should develop SMART objectives: **s**pecific, **m**easurable, **a**chievable, **r**elevant, and **t**ime-oriented. In other words:

- What is the objective?
- Why does it need to be achieved?
- When does it need to be done?
- What does success look like?

Followership Skills

The CRM Model empowers followers to challenge the leader's decision or perception of the situation based on situational awareness and ability to analyze the situation. Because the leader's authority must be preserved, followers must recognize when to challenge or clarify orders if the situation is not as the leader has defined it.

To gain the right to question the leader's decisions, a follower must be:

- **Alert to the situation** — Focused on the mission and conditions that affect it
- **Physically fit** — Nourished, rested, and hydrated
- **Mentally fit** — Alert to the surroundings, his/her personal condition, the condition of the leader and other crew members, and potential stress factors

Implementation Barriers

Barriers prevent the implementation of the CRM model by the organization. Barriers may halt effective communication, situational awareness, effective decision-making, and good teamwork. Factors that may create barriers to implementation of the CRM Model include:

- **Resistance to change** — Most people like the status quo and distrust any type of change.
- **Fear of looking foolish** — People do not want to look stupid, feel foolish, or appear to be wrong.
- **Blame for misunderstanding** — Some people will blame others when there is a misunderstanding, and this can lead to conflict.
- **Intentionally withholding information** — Leaders and followers can intentionally withhold information. This situation may occur for many reasons, including the fear of being wrong or losing control of the situation.
- **Complacency** — This occurs when a person is comfortable with the current situation or condition and does not want to risk the results change may cause.
- **Physical, emotional, and mental fatigue** — When a crew member cannot provide the maximum effort or is unable to concentrate on the situation, a barrier is created.

- **Halo effect** — This occurs when people allow a personal trait or characteristic to influence their perception of a leader or crew member. By seeing a leader as infallible, others do not question that person's decisions.
- **Being outside the group** — When crew members feel that they are not part of a group, they will not listen, make comments, or work effectively with the unit.

Disruptive Attitudes

Personal attitudes that create barriers to a cohesive and safe crew relationship can harm the mission of the unit. The Federal Aviation Administration (FAA) identified the following five negative or disruptive attitudes that can increase the risk to flight crew safety:

1. **Anti-authority** — Sign of independence and lack of teamwork. A symptom is the *Don't tell me what to do* response to an order or a crew member suggestion.
2. **Impulsivity** — Stress-induced desire to *Do something!* Symptomatic of a desire to act, even though people have not gathered all the facts, analyzed the situation, gathered resources, or developed an action plan. This attitude can lead to disastrous and sometimes fatal decisions.
3. **Invulnerability** — *It won't happen to me* thinking leads to taking risks, ignoring safety policies, not using safe procedures, and not waiting for additional resources. Sometimes associated with teenagers but the syndrome certainly applies to members of the fire and emergency services.
4. **Bravado** — Belief that *I can do anything* . . . even without the training, personal protective equipment (PPE), or resources necessary to tackle the hazard. Belief became part of the culture of the fire and emergency services described in Chapter 1. An example is the traditional view that respiratory protection was not needed for fire fighting. The mandatory use of self-contained breathing apparatus (SCBA) occurred as a result of fatalities and chronic respiratory diseases among firefighters.
5. **Resignation** — Belief that all action is useless and that the individual or group has no control over a situation. Strong, motivated leadership is required to overcome this attitude.

The IAFC CRM publication includes two other negative attitudes present in some firefighters. These attitudes are based on research done by U.S. Air Force Major Anthony T. Kern in his writings on flight safety and CRM. They are as follows:

- **Pressing** — Attitude that results from external pressures causing the leader and crew to feel an urgency to accomplish the mission simply to finish quickly and return to quarters. The pressure may be as simple as the desire to finish an interrupted meal, catch the end of a game on television, or get more sleep. Generally, these external pressures are trivial. However, they can result in a leader making poor decisions and crew members taking shortcuts that compromise safety.

- **Air show syndrome** — Attitude that results from complacency with the belief that *we have always done it this way and never been hurt,* which causes a crew to push the envelope and attempt more dangerous activities. This sometimes fatal attitude can take one of the following forms:
  - Desire to be the center of attention by performing a heroic (if not stupid) act for the benefit of onlookers, media, or peers.
  - In training situations, this syndrome may be apparent in individuals or units attempting to outperform others.
  - Some company officers or unit leaders may also take the approach that the way the company or unit performs a task is superior to the way the task is taught at a training center.
  - Some company officers or unit leaders may also take the approach that performing a task in a nonstandard way will get them recognition.

The IAFC publication suggests that these negative attitudes can be prevented by using positive thoughts or statements to counteract them.

Workload Management

People have limited capacities for remembering information and performing tasks. A few people can multitask. Some people have Type A personalities that allow them to focus their energies on getting a job done. For most people, it is a challenge to remember which responsibilities must be accomplished and in what order.

Workload Management — The Crew Resource Management factor that attempts to reduce stress by reducing a job it its most basic tasks and assigning those tasks to individuals and groups.

We regularly hear about people who are overloaded, emotionally exhausted, and stressed due to heavy workloads. **Workload management** is a CRM factor that attempts to reduce stress and still meet unit objectives. It can and should be applied to emergency and nonemergency activities.

Basically, workload management involves reducing a job to its most simple tasks and assigning the tasks to individuals and groups. If a crew has trained together, the tasks will be easier to assign and accomplish during an emergency. For instance, if a crew responds to a motor vehicle accident requiring extrication, each crew member will have an assigned task such as patient stabilization, fire or vapor suppression, and power extrication equipment operation **(Figure 3.10)**. In career and volunteer agencies, crews are often composed of different personnel due to vacations, holidays, travel distances for volunteers, and a decrease in volunteers. Crews may be composed of personnel from other units, shifts, or simply the first people to arrive at the scene.

Figure 3.10 One crew member uses a power saw to create an opening in a school bus roof, while another sprays soap and water to keep the blade cool during cutting.

If all personnel are trained to perform tasks in the same manner, the officer can make specific task assignments with confidence that personnel can perform them effectively and safely. An example of this could be the need to raise a long ground ladder (40 to 50 feet [12.2 to 15.2 m]) when no single company is staffed with sufficient personnel to do so safely. If all companies are trained in the same steps, it is possible to safely call on personnel from multiple companies.

Challenges occur when a situation changes and a crew is not prepared for it. For instance, in the scenario given earlier, if one member of the crew is absent, tasks may have to be reassigned and further prioritized. Another situation might involve the addition of a new or temporary crew member who has not practiced the tasks with other crew members and has to be told what to do and when. Or the new crew member might have a disruptive attitude that hurts the crew's teamwork.

Workload management benefits from prioritizing tasks, assigning tasks to the appropriate crew members, and training for all contingencies. When it becomes apparent that a crew may be overwhelmed by the workload, resources must be requested. When considering command and control functions, the implementation of the National Incident Management System-Incident Command System (NIMS-ICS) can significantly improve workload management.

Error Management

No matter how hard people try, mistakes occur, errors are made, and accidents happen. While management theories such as Total Quality Management (TQM) may propose that *zero error* should be part of the manufacturing goal, it is not attainable as long as humans are involved. Therefore, it is necessary to manage or control errors to reduce the potential for their occurrence. Further, when considering emergency situations such as fire control, it is unlikely that commanders will initially have adequate information to approach zero-error levels of decision-making.

The CRM Model includes a process for error management intended to prevent, reduce, or repair errors. The process is based on the following three layers or defenses:

1. **Error avoidance** — Requires adherence to safety standard operating procedures (SOPs), training to a high level of proficiency, preincident planning, situational awareness, effective use of resources, and minimizing personal and crew distractions. This layer has the greatest potential to reduce accidents, injuries, and fatalities due to errors.
2. **Error trapping** — Requires that crew members recognize that an error has occurred and take steps to stop it and prevent it from becoming a disaster. Crew members must be able to analyze the error, mentally project it into the worst consequences, and take action to reverse the process.
3. **Error mitigation** — Involves repairing the damage caused by an error, performing a post-incident analysis and critique, and providing training, equipment, or SOPs to lessen the potential for the error to occur again. This layer is last and takes place once errors have occurred.

CRM Application

The CRM Model presented in this chapter can be applied by an individual, a crew, or an organization. It can be applied to any situation. While the material presented can convey an idea of how CRM works and its advantages, it is best to participate in a formal training program. Training classes usually include role-play, decision-making scenarios, and practical skills evolutions that will require participants to apply CRM skills. The sections that follow are examples of how individuals, crews, and organizations can apply CRM.

CRM Training

To implement the CRM Model, the fire and emergency services organization must provide training in the CRM factors. Training should follow an accepted teaching format such as the one used by the commercial airline industry and includes the following components:

- **Awareness** — Learn about CRM. The awareness component includes the following activities:
  - — Describe the safety culture.
  - — State why the safety culture must be changed.
  - — Describe the benefits of a new safety culture.
  - — Describe crew resource management (CRM).
  - — Explain how CRM applies to tasks and duties of fire and emergency responders.
  - — Explain how CRM will help create and sustain a new safety culture.
  - — Explain the limitations of CRM when applied to emergency situations.
- **Reinforcement** — Apply CRM factors in controlled scenarios. Firefighters are required to solve problems, make decisions, and work as a team to accomplish a set of goals. These practical training evolutions may be based on any type of emergency or nonemergency situation. Positive reinforcement is important to the learning process because it fixes the problem but does not assign blame.
- **Refreshment** — Provide refresher training periodically for members. This training, which may include a review of the CRM factors, role-play, or practical training evolutions, emphasizes the importance of CRM and the organization's commitment to it.

Training for personnel must apply to individuals who may act alone, crews led by a company officer, and entire organizations led by fire chiefs. Six CRM areas should specifically address each of the three categories of personnel. They include communications, situational awareness, decision-making, teamwork, workload management, and error management.

Communications

Communications are internal and external for the individual. Internally, firefighters must be aware of their ability to concentrate on tasks and maintain an awareness of their physical capabilities to complete tasks. Externally, they must listen and speak on radios and talk to other firefighters under a variety of circumstances.

Company officers must be open to crew comments **(Figure 3.11)**. Crew members must have a sense of trust if they feel the need to question the company officer's decisions. Care must be taken to exercise good judgment in emergency environments.

Organizations must have free flow of information from the top and to the bottom. At emergency events, IC's must give clear objective assignments, make sure they are understood, and gain critical feedback from officers in frontline areas.

Figure 3.11 This chief officer is using good listening skills to help him receive and understand feedback from his personnel.

Situational Awareness

Whether firefighters are working alone or with a crew, they must always maintain situational awareness and be alert for changes. Leaders must be aware of the external and internal conditions of the organization. Customer feedback, media comments, and political agendas represent external forces that may affect funding or political support for the department. Internally, labor leaders, officer groups, divisions such as training or fire prevention, and informal leaders have needs and concerns that can affect the organization's day-to-day operations.

At an emergency incident, the IC as well as all NIMS-ICS officers at the Company, Task Force, Division, Group, Branch, and Section levels must constantly monitor a situation, including the weather, fire behavior, and the physical conditions of the operational crews. Contingency plans must be available for any changes. The IC must be prepared to change plans at a moment's notice **(Figure 3.12)**.

Figure 3.12 The assigned ISO and the IC work together to monitor safety at emergency incidents.

Decision-Making

Before responding to an emergency call, individuals and crews begin making decisions about what PPE to wear and responsibilities on the scene. Routing is done collectively with the company officer looking at maps and the driver steering the apparatus.

If a crew leader makes an error in judgment, the crew has the opportunity to catch the error quickly. It is important that the crew leader not feel threatened by the crew's input in the decision-making process. At emergency incidents, decisions must be based on the most timely, accurate information available, preplanning that follows department SOPs, and the opinions of other personnel when time permits.

Teamwork

Individuals performing tasks are part of a system of response and should view their responsibility to complete an assignment as essential. The unit applies teamwork by working together toward the same goal, according to the IAP and the crew leader's tactical directions. Crew members perform assigned tasks and assist others with theirs. The group works as a team and stays aware of each other's locations and situations **(Figure 3.13)**.

Each division (such as fire suppression, training, or fire prevention) works as a team, helping the entire department to work toward the common goal stated in the organization's mission statement.

Figure 3.13 Crew members work together at an incident scene. *Courtesy of Chris Mickal, District Chief, New Orleans (LA) FD Photo Unit.*

Workload Management

A crew leader assigns tasks based on a crew member's ability and training. They should also assist other crew members and be ready to ask for help. The proper application of teamwork should ensure that each division within an organization has the resources (personnel, funds, time, and equipment) to perform tasks. When one division is overwhelmed by a task, other divisions can provide the resources to accomplish it.

The same approach can be taken at emergency incidents. Training officers and fire inspectors may be used for nonsuppression tasks such as scene control and staging to free suppression personnel for fire fighting or other emergency activities.

Error Management

When an accident occurs, firefighters must know the procedures for reporting the accident and placing the unit back in service as soon as possible. A crew must be able to recognize potential errors or mistakes and avoid making them, such as ensuring that an apparatus is parked downwind and downhill from a hazardous liquids spill. When an error does occur, crew members must be prepared to either implement actions that will lessen the effects of the error or withdraw from the hazardous area.

Fire departments should strive to eliminate errors and mistakes. Creating, implementing, monitoring, and altering the strategic plan as needed is one approach. Another is developing and enforcing SOPs for all phases of a department's operation. Analyzing injury, fatality, loss, and fire statistics provides information for taking a proactive approach to safety and fire prevention **(Figure 3.14, p. 108)**.

CRM Current Applications

CRM has proven itself in the aviation industry for over twenty-five years. Improved safety resulted from improved communications, situational awareness, decision-making, and teamwork of flight crews. Today, the CRM Model is being applied in the fire and emergency services to improve the safety and efficiency of fire crews. Individuals, crews, and organizations can apply the CRM factors, as well as the concepts of workload and error management, to help ensure that everyone goes home safely. CRM is everyone's responsibility and has the potential to change the culture of the fire and emergency services by reducing accidents, injuries, and fatalities. There is a growing need to collect data and use evidence-based research findings to direct organizations toward safety plans and initiatives that work.

High Reliability Organization

A High Reliability Organization (HRO) is an organization that performs high-risk work without catastrophic events. HRO is a total system engrained in every organizational level and every task performed. The system promotes the free flow of information at all levels and encourages people to participate in safety-related issues. Additionally, an HRO has proactive functions (plans and readiness) and reactive functions (after-action reviews and investigations). The HRO process embraces CRM as a complementary part of the safety system.

The HRO concept was developed by Karlene Roberts of the University of California, Berkeley, in the 1990s. It has been implemented by a wide variety of organizations including the U. S. Navy's carrier-based aviation program, commercial aviation, the nuclear power industry, NASA, healthcare facilities, educational institutions, and a variety of fire and emergency services organizations.

According to Roberts' research, HROs share five characteristics:

- **Process Auditing** — System of checks intended to recognize problems that can lead to accidents. This includes safety drills and equipment testing as well as a reevaluation of solutions to previously recognized problems.
- **Reward System** — Rewards motivate individuals to perform safely. A reward system can be applied to the organization as a whole.
- **Quality Degradation Avoidance** — Avoiding a degradation or loss of quality is essential to maintaining a safety culture. A universally accepted standard is used to compare the individual or organization's safe behavior.

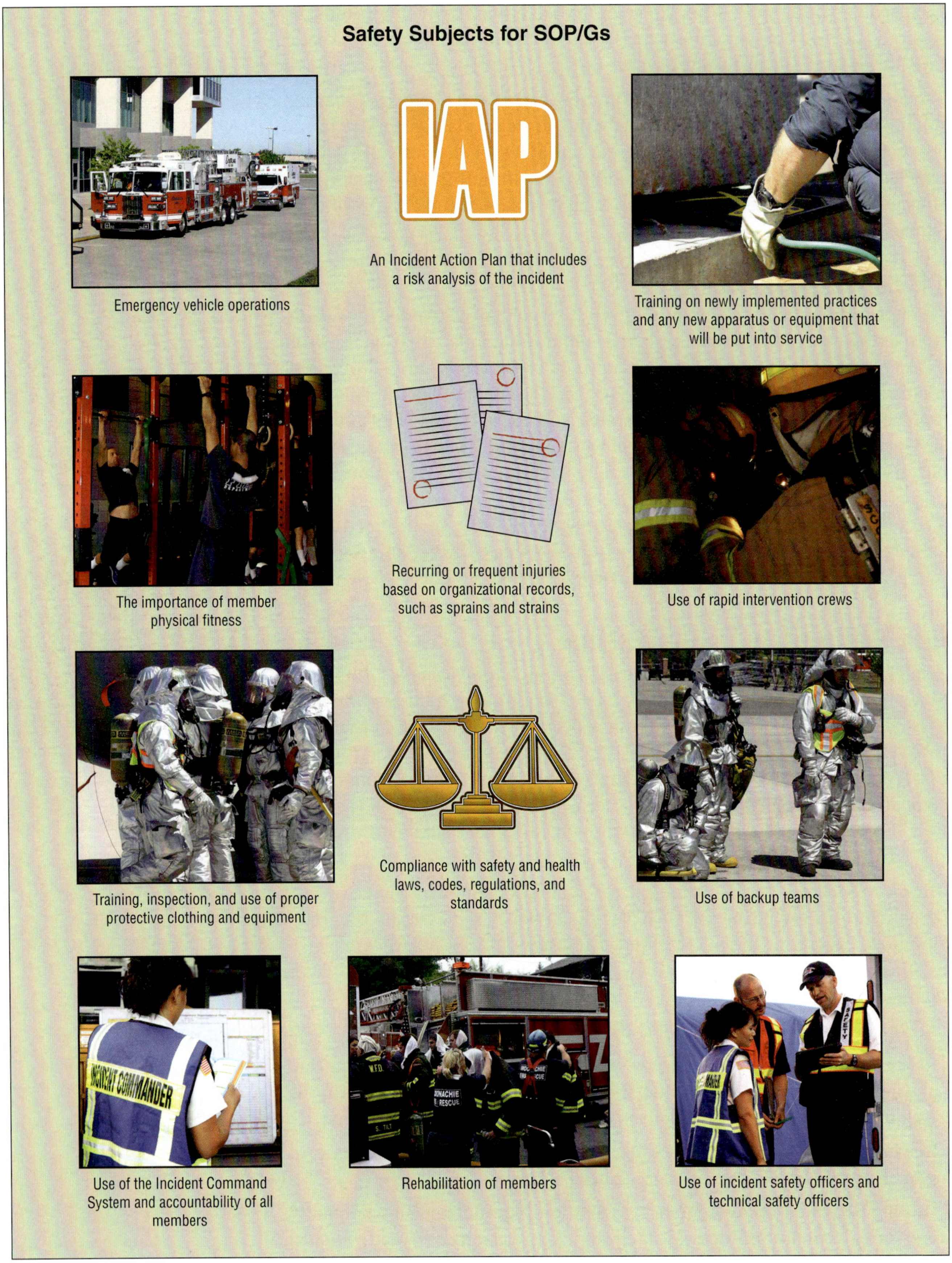

Figure 3.14 SOP/Gs should comprehensively cover safety subjects that are relevant to the department or organization and comply with all appropriate codes and regulations in the jurisdiction. *Photo courtesy of Ron Jeffers, Union City, NJ.*

- **Perception of Risk** — This consists of the following two elements:
  - — The knowledge that a risk exists.
  - — The extent to which the risk is recognized, minimized, or appropriately mitigated.
- **Command and Control Elements** — This characteristic consists of the following five elements:
  - — Migrating (delegating) decisions to the person who has the expertise to make the decision. For example, at an emergency scene, a technical expert in the ICS-Planning Section can give immediate advice to the Incident Commander or other command officers. Similarly, one of the joint ICs can initiate an ICS-Unified Command organization.
  - — Creating redundancy by providing a backup system, made up of personnel or equipment **(Figure 3.15)**.

Figure 3.15 Firefighters ideally operate in groups of two. *Courtesy of Chris Mickal, District Chief, New Orleans (LA) FD Photo Unit.*

For instance, ICS approaches this with the use of the Rapid Intervention Team or Crew (RIT/RIC) concept.

  - — Senior managers who exercise situational awareness of the *big picture* and strategic goals. This type of manager provides subordinates with the support necessary for them to make decisions and do their jobs. In ICS, the IC does not interfere in the tactical decisions of subordinate officers as long as they function within the IAP and have not created an unacceptable personnel safety issue.
  - — Formal policies and procedures that ensure predictability in the operation but also allow for flexibility in a fluid environment. The fire service uses SOPs/SOGs to provide the element of predictability at incidents, but the ICS gives all levels of Command the flexibility to deal with unexpected situations.
  - — Training that develops strong teams and creates redundancy through cross-training. For instance, fire service training is primarily accomplished in team format (typically called "companies"). Progressive fire agencies train all members to utilize a wide array of tools, not just those assigned to their company.

Research indicates safety outcomes are more positive as a result of training high-reliability teams (HRT) within HROs. HRT members - crews in the fire service - can learn to demonstrate certain behaviors consistently in normal, dynamic, and stressful environments (Wilson, et al, 2005). Guidelines for HRTs found in the high-stress environments of health care that are applicable in the fire service are:

- **Sensitivity to operations** — HRTs ensure all team members know the big picture. Various forms of communication increase awareness of situational factors.

- **Commitment to resilience** — Team members must give feedback to colleagues, monitor their performance in real time and continually practice reactions to worst-case scenarios to share knowledge.
- **Deference to expertise** — Expertise can come from any level and must be encouraged. The orientation to decision-making on the front line should be collective and incorporate minority opinions.
- **Reluctance to simplify** — Teams seek to understand complexities of work and environment, and look for how interwoven operations, personnel, and equipment.
- **Preoccupation with failure** — Encouragement of error reporting and utilizing training behaviors that decrease errors. Teams seek feedback before, during, and after events. Teams also focus on success as a liability because it leads to overconfidence.

HROs foster mindfulness in employees. Mindfulness includes the previous HRT guidelines that constitute a worker mindset. HROs and CRM are similar. Both can be found at all levels of the organization, they are compatible and useful at emergency events. Safety leadership should support and recognize these programs as mutually supportive.

Chapter Review

1. How did risk management techniques evolve throughout history?
2. What does the SWOT acronym stand for?
3. What are the five CRM factors?
4. What are the external and internal conditions of the fire service organization that could impair situational awareness?

Discussion Questions

1. How can a safety committee made up of multi-divisional members bring about cultural change?
2. How can crew members improve their communication skills using CRM?
3. How is the CRM model applied differently at the individual, crew, and organization levels?
4. How do the processes of a HRO compare to those of the CRM?

Health and Fitness Program Requirements

Chapter Contents

Key Terms

Cardiovascular disease 115
Health Program.. 115
Narcotic 128
Risk Management Plan 136

FESHE Learning Outcomes

Occupational Safety and Health for Emergency Services

2. Identify occupational health and safety programs for industry and emergency services today.
4. List and describe the components of risk identification, risk evaluation, and incident management.
9. Describe the components and value of critical incident management programs.
11. Describe the components of a wellness/fitness plan.
12. Identify and analyze the major causes involved in line-of-duty firefighter deaths related to health, wellness, fitness and vehicle operations.

Health and Fitness Program Requirements

Learning Objectives

After reading this chapter, students will be able to:

1. Explain how the basic components of a health and fitness program can be used to benefit individual firefighters and fire departments.
2. Identify causes of health and wellness related casualties.
3. Explain the legal requirements of health and wellness programs.
4. Identify occupational health hazards and risks in the fire service.
5. Identify health hazards and risks related to a firefighter's personal lifestyle choices.
6. Describe the importance of standardized and impartial hiring requirements.
7. Identify the necessary components of a successful health program.
8. Describe the wellness components of a health program.
9. Describe the value of on-scene and on-site stress management, vitals assessments, and fitness training.

Chapter References

"Fire-Related Firefighter Injuries Reported to the National Fire Incident Reporting System (2012-2014)". 2016. *Topical Fire Report Series*, Vol. 17, Issue 6/August 2016. Accessed online at Usfa.fema.gov

Hylton J.G. Haynes and Joseph L. Molis. 2015. "United States Firefighter Injuries – 2014". National Fire Protection Association. Accessed online at nfpa.org.

Heart Disease in the Fire Service. 2013. International Association of Fire Fighters. Accessed online at iaff.org

"Tackling Fire Service Suicide: A Reality Check". 2015. International Association of Fire Chiefs. Accessed online at iafc.org

"Tackling Fire Service Suicide: A Reality Check". 2015. International Association of Fire Chiefs. Accessed online at iafc.org

Chapter 4
Health and Fitness Program Requirements

A 2013 mortality study by National Institute for Occupational Safety and Health (NIOSH) of 30,000 firefighters who served from 1950-2009 on the San Francisco, Chicago, and Philadelphia fire departments showed fire fighting increased risks for cancer.

Figure 4.1 Not all firefighter fatalities are reported as line-of-duty deaths.

Linking fire fighting and cancer can be difficult for many reasons. Without better documentation of firefighter exposures and improved research linking occupational chemical exposures to firefighting tasks, it may become more difficult to make presumptive claims for workers' compensation during or after a firefighting career. Future studies will need to concentrate on establishing stronger causality.

In 2015, forty firefighters died from stress or overexertion, according to the National Fire Protection Association. Heart attacks caused thirty-five of those deaths. Chronic sleep disturbance, physical exertion, mental and emotional stress, poor diets, and lack of exercise all contribute to **cardiovascular disease.**

Cardiovascular disease (CVD) — A family of diseases that includes hypertension (high blood pressure), atherosclerosis, cardiovascular heart disease, and stroke.

Not all firefighter fatalities occur during emergency operations. Many that follow occupational illnesses are not reported as Line-of-Duty Deaths (LODDs) **(Figure 4.1)**. The illnesses may result from exposure to occupational chemicals and/or stress. They can occur years after an injury, exposure to hazardous materials, or onset of a long-term illness. Cardiac arrest and strokes may occur due to stress or hypertension on duty, during other types of duty, or during off-duty hours.

Health Program — Any program designed to assist and encourage employees to adopt and attain a healthier lifestyle.

The most effective means of preventing or reducing job-related injuries, illnesses, and fatalities is proactive and based on a comprehensive **health program**. It should address a firefighter's physical, emotional, mental, financial, family, and spiritual wellness. Common program components include annual health and fitness screenings, fitness instructors **(Figure 4.2)**, cooking classes and menus, employee assistance programs, drug and smoking cessation programs, religious services, modified work duty for recovering firefighters, PPE training, and annual PPE testing.

Figure 4.2 A fitness instructor offers sample workouts for firefighters.

The trend of health programs has been to move from maintaining general health and well-being to identifying, managing, or reducing specific and potentially costly health conditions. New disease management programs target high-risk conditions such as diabetes, heart disease, cancer, stress, tobacco use, and excess weight. Financial incentives and consequences can motivate high-risk employees. Data tracks the success and effectiveness of these programs. Organizations have realized the cost-saving advantages of offering health programs as benefits to their employees. Many labor unions support the creation of these programs and frequently include them in contract negotiations.

In 2014, the Assistance to Firefighters Grant (AFG) program awarded over $2.8 million in wellness-related equipment in response to requests from nearly 40 fire departments nationwide. The Sacramento Metropolitan Fire District used an AFG grant in the following way:

On October 7, 2010, a District veteran firefighter died from a cardiovascular event that occurred while he was fighting a vegetation fire. A subsequent LODD evaluation conducted by NIOSH recommended that the District implement annual medical evaluations for all firefighters that incorporate standardized exercise stress tests. Further, the report recommended the District should perform an annual physical performance (physical ability) evaluation on all members.

The District's ability to fund annual medical evaluations was significantly limited by the national economic crisis. In 2012, the District applied for, and was subsequently awarded, an AFG for the implementation of a mandatory wellness, fitness and injury prevention program for all uniformed employees. The program is consistent with the International Association of Fire Fighters/International Association of Fire Chiefs (IAFF/IAFC) Joint Labor Management Wellness-Fitness Initiative and NFPA Standards 1582 and 1583.

The District's grant-funded wellness, fitness, and injury prevention program kicked off in 2013. When the grant period ended in June 2014, the District had provided 239 immunizations; 65 new-employee physical assessments; 544 hours of fitness assessments and counseling; 882 hours of health and fitness training for injury prevention; 504 annual medical physicals; and a formal employee assistance program for behavioral health. Annual workers' compensation claims for line personnel decreased by 12 percent in the first two years of the program, reducing workers' compensation claims by 26 percent ($533,000) after just one year.

The goal of most health programs is twofold. First, the goal is to ensure that employees and volunteer personnel are fit to perform in physically intense fire and emergency services. Many experts believe that firefighters in good physical and mental condition use less sick leave, require less medical care, and have longer life expectancies. This helps to meet the second goal: reduced health-related costs for the employer. Less time is lost due to illnesses or injuries, which lowers workers' compensation and health insurance premiums and reducing overtime and substitute worker costs.

Modified Components of Medical Evaluation and Fitness Programs

In 2012, the fire department in Rogers, Arkansas, updated its Medical Evaluation and Fitness Program SOP to ensure that members over the age of 40 received a more-thorough cardiac evaluation. With firefighters experiencing heart attacks throughout the country, Rogers Fire Department Fire Chief Tom Jenkins approved the added evaluations. The SOP states, "1. Medical Component — This is the foundation of the program. It is the beginning of data collection, developing standards, and setting goals for the individual. This component of the program also exists to gather the information necessary for the department's physician to certify a member for use of respiratory protection and to be part of our hazardous materials response."

The health and wellness program is a cooperative endeavor between labor and management. The administration must develop, support, and implement a plan with realistic objectives, and supervisors must ensure that personnel assigned to them understand the program and participate in it. Finally, each individual must embrace the program, get involved in it, and provide accurate feedback to ensure the program is meeting its goals.

Many departments develop SOP/Gs that support the health program through mandatory participation requirements and the frequency of various tests. For instance, the firefighter wellness physical of the Rogers, Arkansas, Fire Department's *Medical Evaluation and Fitness Program*, #302 includes:

- Comprehensive health background questionnaire
- Metabolic profile analysis (blood panel)
- Lipid panel (including cholesterol, triglycerides, HDL, and LDL)
- PSA blood test (if over 40 years of age)
- Cardiovascular stress test (exercise EKG) with physician interpretation
- Cardiac CT scan for calcium ossification (one every two years if age 40 or over)

Firefighters must have a complete baseline medical physical and fitness assessment. If requested, department staff may schedule exams, in coordination with the contract vendor, during an employee's shift.

Casualties Related to Health and Wellness

More than 60,000 firefighter injuries occur each year, but the trend, as reported by NFPA and USFA, is decreasing. A 2014 USFA report on fireground-related injuries indicates the following:

- From 2012 to 2014, roughly 34,000 firefighter injuries occurred annually. Of these injuries, more than 85 percent occurred on the fireground; less than 15 percent occurred while responding to or returning from an incident.

- The majority of fire-related firefighter injuries (87 percent) occurred in structure fires. Injuries resulted in lost work time for 44 percent of firefighters with reported fire-related injuries.
- Fires resulting in firefighter injuries were more prevalent in July (11 percent) and peaked between 1 and 4 p.m.
- Of the reported fire-related firefighter injuries, 27 percent were caused by overexertion and/or strain.

Firefighter injuries have decreased during the last 30 years, according to the NFPA. The number of overall injuries in 1,000 calls compared with the number of calls has fallen from 1.21 to .48. The number of injuries at fire scenes has decreased less dramatically from 26.8 to 20.8 for every 1,000 calls. The leading causes of fireground injuries were falls, slips, and jumps (28.7 percent) with overexertion and strains a close second. While these injuries may appear less serious, many affect employment and professional longevity.

Some problems may not show up immediately. Cancer, cardiovascular and pulmonary diseases, exposure-related conditions, and skeletal disorders can appear long after a firefighter's career concludes. The NFPA tries to track injuries sustained during service, but attempts to track illnesses into retirement are rare. It can also prove difficult to directly relate latent occupational injuries to specific incidents. Exposure to carcinogens may have occurred over time, which makes correlation to a specific cancer especially difficult. More than thirty states have presumptive cancer laws that make it easier for firefighters to file workers' compensation claims after retirement. Each state has its own legislation.

To prevent or reduce injuries and illnesses from the outset of a firefighter's career, the department needs to offer members a comprehensive health and wellness program that includes education, raises awareness, and increases physical fitness. One goal of a health program is that firefighters stay in excellent health to retirement and beyond. A way to achieve this goal is by monitoring the health of retired firefighters on an annual basis.

Health fitness programs must closely monitor fitness regimens at fire stations. A recent trend in fitness uses "extreme" workouts in the stations. These can take individuals with lower fitness levels and put them in workout regimens beyond their capabilities. The risk of participating in these workouts outweighs the benefit of overall fitness for the fire service. Some overexert due to peer pressure; others get impatient and do not take time to safely develop their abilities. Either way, the result can end a career and add considerable expense to an occupational care budget.

An additional risk to fire departments from on-duty physical fitness is when crews or individuals exercise to fatigue. This elevates the risk of not being able to perform physical duties shortly after working out. First, health and safety officers (HSOs) should consider all risks related to physical fitness efforts. Then, they should offer education and set expectations for captains and battalion chiefs to support firefighter fitness programs.

Programs that educate Peer Fitness Trainers (PFTs) can ensure that station and individual workouts aid firefighting and help firefighters learn to do exercises correctly **(Figure 4.3)**. PFTs can evaluate individuals, identify weaknesses, create appropriate workouts, and evaluate progress. As individuals increase job-related fitness, they are less likely to injure themselves or others. The ACE/IAFF Peer Fitness Trainer Program is designed specifically for firefighters.

Figure 4.3 Many departments have peer fitness programs that provide exercise coaching to firefighters.

Legal Requirements

Aspects of a health program may be found in federal, state, and local government regulations. Some requirements can be found in OSHA regulation and standards.

State OSHA programs must meet the minimum requirements of the federal OSHA programs. State occupational safety and health programs will usually include more stringent requirements. Some requirements are based on NFPA standards, including 1500, 1581, 1582, 1583, and 1584, when they are adopted by the authority having jurisdiction (AHJ).

Additional laws that relate to health programs are the medical records acts. These require the retention of medical records for the duration of employment plus 30 years after separation from the fire and emergency service **(Figure 4.4)**. These acts also require that an employee's health records are kept private. Only the employee, his/her physician, and a specifically designated person have authorized access to this information. Some fire departments use the requirements of this law to report and keep files on all smoke exposures in a firefighter's career.

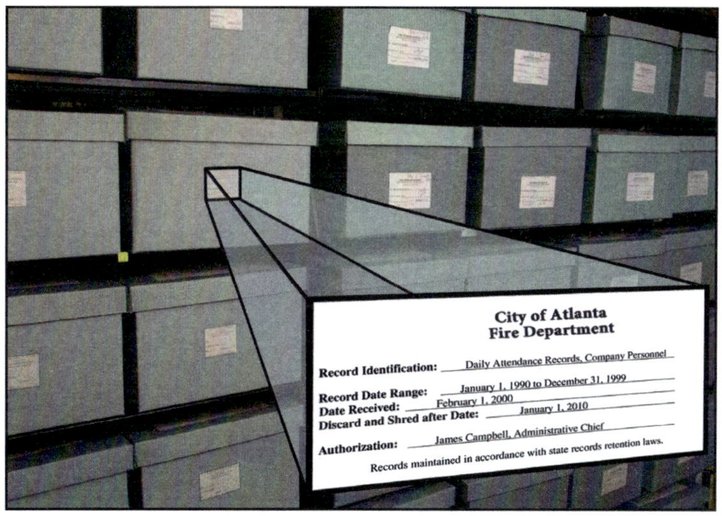

Figure 4.4 Medical records must be retained for 30 years after personnel leave the department.

Laws that regulate wrongful termination have been established to protect employees. Some of these laws apply to termination for medical reasons. These laws specify the steps that the organization must take to terminate an employee on medical grounds. The laws protect the rights of the employee while establishing the protocol that must be followed. The employee is given certain responsibilities in these cases, as well as the right to appeal a medical termination decision. Specific federal laws include the Americans with Disabilities Act (ADA), Title VII of the Civil Rights Act, the Family and Medical Leave Act, and labor, health, and safety laws. It is good practice to add these specific laws to labor/management agreements and department rules and regulations.

Health-Related Risks

Fire and emergency responders face many health risks throughout their careers. Accidents involving motor vehicles, structural collapses, or other physical injuries are covered in this manual, along with-risks that can be reduced or eliminated by a comprehensive health program. These risks can be divided into the following categories:

- Fire suppression and other emergency activity
- Exposures to chemical and biological hazards
- Lifestyle choices

Fire Suppression and Other Emergency Activity Risks

Dangers in the profession are not always obvious, and they may take a while to make an impact. The may not be unique to firefighting and emergency services, but the intense nature of the professions raises these risks.

Figure 4.5 Heat stress and smoke exposure contribute to cardiovascular disease.

Cardiovascular Disease

The physical hazards found in emergency environments can place a combination of physical, emotional, and mental demands on firefighter and result in different types of stress. The primary cause of firefighter fatalities during emergency operations is stress-induced cardiac arrest **(Figure 4.5)**.

Stress can also cause strokes that can be fatal or result in long-term disabilities. Cardiac arrests and strokes may not occur until after the emergency incident is over. Repeated exposures to stress can also weaken the heart muscles and cause repeated heart attacks. A 2012 IAFF report found that, for every fatal on-duty heart attack, there are 17 non-fatal cardiovascular disease events that happen in the U.S. Fire Service.

Besides contributing to cardiovascular disease and strokes, stress also contributes to depression, emotional fatigue, physical fatigue and professional burnout. Health programs address stressors proactively to avoid outcomes that result in CVD and heart attacks or strokes.

In the past half century, stress has become a common part of life. A 2015 American Psychological Association survey shows that 42 percent of Americans said they felt anxious or nervous in the previous month before the inquiry, and 37 percent said they were irritable or angry.

In the fire and emergency services, stress is a fact of life. Firefighters experience low levels of chronic stress, with moments of acute, physical and/or mental, and emotional stress. Earlier stress on emergency scenes and stations has been discussed; however, individuals are often unable to relax and take this stress home with them. For that reason, a stress management plan is an essential part of the wellness portion of the employee physical fitness and wellness program.

Stress – Physical, physiological, and psychological stress can result from actions at the emergency scene, as well as factors while at the station. It actually begins prior to the shock caused by the alarm upon dispatch. Firefighters are considered to be in a constant state of readiness when they are on duty. This is attributed to being on guard, never knowing what's next or when it will occur. This constant low-level arousal can eventually lead to longer term stress burnout.

Upon dispatch, the startle effect from the alarm (often in the middle of the night), initiates a level of acute stress that builds throughout the incident until firefighters return to quarters. As a result of studies done on the relationship between the use of alarm bells and stress, there has been a move in some departments to replace the bells with a gradual tone alarm.

When the body is stressed, the hypothalamus triggers the autonomic nervous system and the pituitary gland to produce epinephrine and cortisol hormones. This enables the body to engage in activities like sprinting, rapid breathing, and an increase in heart rate because all muscle tissue is elevated to higher levels of performance by the hormones. The liver responds to the hormones in the blood by producing sugar, which adds a boost of energy. If unused sugar, adrenaline, or cortisol remain in the bloodstream, it can aggravate or set the stage for developing diseases like diabetes or obesity.

Stress in the fire and emergency services can take two forms: physiological stress (work environment experiences) and psychological stress (total life experiences). Physiological stress is caused by the emergency responder's work environment and can be created by the following situations:

- Sound of the alert tone in quarters
- Shift work involving irregular schedules
- Lack of sleep or interrupted sleep cycles
- Loud noises (sirens, air horns, station radios, explosions)
- Station overcrowding
- Hazardous or toxic environments
- Exposure to extremes of heat and cold
- Long hours of physical exertion
- Carrying heavy PPE
- Fatigue

If unrecognized or left untreated, stress can lead to tragic results. The Firefighter Behavioral Health Alliance tracks firefighter suicides and reported that 88 firefighters committed suicide in United States in 2014. The Alliance lists post-traumatic stress disorder, depression, and sleep-cycle disturbances as some of the compounding variables linked to suicides. Peer-awareness training can help firefighters understand the despair that can lead to suicide and identify other firefighters at risk.

Cardiac Deaths

According to a study published in the *New England Journal of Medicine* in 2007, the death rate for firefighters between the ages of 50-59 due to cardiac arrest is twice the rate of the group from 40 to 49 years of age. The rate for firefighters who are 60 or over was four times the rate of the younger group. **Source: *New England Journal of Medicine* study published March 27, 2007**.

Since 1984 cardiac deaths have generally been declining. In 2014, fire fighting experienced one of its first increases in the number of cardiac-related deaths per total number of deaths on duty. In 2014, 56 percent of firefighter deaths were cardiac-related. NFPA statistics show that sudden cardiac death remains the leading cause of death of on-duty firefighters in the United States.

Trauma

Even with the proper PPE, firefighters can get hurt. Due to trips, slips, falls and other forceful impact, they are vulnerable to internal trauma. Wearing PPE and having complete situational awareness is necessary to avoid traumatic injuries.

Noise

An often-overlooked risk of emergency operations is noise-induced hearing loss **(Figure 4.6)**. Impaired hearing can cause serious or fatal mistakes in training or emergency operation situations. Additionally, constant or repeated exposure to apparatus and equipment engine noise, audible warning devices, and on-scene noises can cause tinnitus, which is the destruction of small nerve endings in the ear canal that normally carry sound into the hearing chambers. A symptom of tinnitus is a high-pitched ringing in the ears. While some hearing loss is the result of the aging process, the onset of noise-induced hearing loss may occur at any age.

Cancer Risk Management

Cancer continues to be one of the top threats to firefighter health. Approximately 45 percent of firefighters will be diagnosed with cancer within seven years of retirement. Meta studies in the United States and Australia show firefighters have an elevated risk for leukemia and testicular cancer.

Fire research shows carcinogens are present in almost all types of fires, due to the amount of manmade materials found in residential and commercial structures, automobiles, and manufacturing sites. Today's firefighters encounter smoke containing toxic gases, vapors and particulate matter that can permeate protective gear, increasing the risk of dermal contamination and the severity of inhalation injury. These manmade products also tend to burn at higher temperatures. As temperatures increase, so does the skin's permeability. Increased rates of absorption, combined with increased amounts of toxic chemicals, leave firefighters increasingly vulnerable to contamination in almost every type of fire.

Hearing Protection

Figure 4.6 There are a variety of styles of hearing protection. Wearing hearing protection should be promoted when appropriate. (a) *Courtesy of Alan Braun.*

When worn properly, structural gear does offer a degree of protection against dermal contamination **(Figure 4.7, p. 124)**. However, there are significant limitations in the structural PPE ensemble. For example, research has identified the structural gear hood to be the most permeable piece of PPE. The neck, with its anatomical properties, is especially susceptible to absorbing chemicals that land on it. As manufacturers develop new products to increase protection, fire departments and firefighters need to stay abreast of current research outcomes and adopt behaviors to manage the carcinogen risks. HSO's must lead in this area: sharing information to educate others, and making recommendations to improve operational safety.

Some practices for limiting exposure to contamination by carcinogens include the following:

1. Add curriculum to training academies that support best practices for reducing or eliminating exposure to contamination by carcinogens. It is equally important for Training Officers to demonstrate carcinogen avoidance in training activities and exemplify safe personal hygiene by cleaning themselves and their gear after training activities.

NOTE: More emphasis should be placed on keeping structural hoods tucked securely into coats during operations.

2. Wear full turnouts and SCBAs from the beginning of fire activities to the completion of overhaul and investigations. Full structural gear provides limited protection from dermal contamination. Although not considered a level A or B type suit, structural gear offers more protection than work uniforms.
    a. Wear NFPA 1971/1981 compliant protective ensembles for fire fighting.
    b. Consider identifying a hot zone with tape or other methods where full turnout gear and SCBAs must be worn **(Figure 4.8)**.
    c. Remain upwind from fire scenes and diesel exhaust when wearing PPE but not breathing positive pressure air from SCBAs.
    d. Provide as much natural ventilation as possible to burned structures prior to starting investigations.
    e. CO monitoring is insufficient for declaring an area to be safe after a fire. Firefighters in the hot zone must always wear full PPE and SCBA.

3. Upon exiting the structure, conduct gross decontamination of all PPE and equipment to remove soot and particulate matter. Incident Commanders should support decontamination procedures by setting up a decon-type corridor for firefighters heading to Rehab.
    a. The initial decontamination should take place when crews leave the hot zone and are assigned to rehab. Initial decon should include using a soft bristle brush and damp towel to remove large debris from PPE. Firefighters should continue to wear their PPE and remain on positive pressure air to avoid further contamination during this effort. Prior to entering rehab, firefighters should remove all of their turnout gear (remaining on positive pressure as long as possible) and hang gear on portable hangers located downwind of rehab.
    b. Use wet wipes/baby wipes to remove soot from your head, face, jaw, neck, underarms, hands, and lower legs. These wipes should be carried aboard every vehicle in the fire department fleet to ensure this practice can be conducted wherever smoke is present **(Figure 4.9)**.

Figure 4.7 Structural fire fighting protective clothing will provide limited protection against many hazardous materials.

Figure 4.8 Different colors of cordon tape or rope can be used to identify the boundaries of different color zones.

c. The final step of initial decontamination is to use a red line to hose off all PPE and equipment prior to leaving the scene. Bag the contaminated structural gear and place it on the apparatus away from the crew compartment. (NFPA recommends fire departments provide second sets of turnouts for the safe and healthy management of contaminated ones.)

4. Upon arrival back at the station, firefighters should take off all work uniform clothing, shower thoroughly, put on clean undergarments and work uniform, and wash their contaminated work clothing in station laundry facilities.

    a. After washing contaminated clothing, washing machines should run empty through a short cycle with hot water and bleach.

    b. There are no known environmental concerns that need to be addressed for washed contaminated work uniforms worn under structural PPE.

5. Clean all PPE and equipment immediately after showering at the station.

    a. Wear latex gloves and break down PPE into separate components.

NOTE: Helmet brims should be removed from helmets and thoroughly cleaned. Hoods should be placed in station washing machines and washed in warm water and dried on medium heat. Replacement brims and hoods can be ordered separately from manufacturers to serve as backup sets.

    b. Use plenty of cool to lukewarm water to rinse contaminates off all PPE and equipment. It may be necessary to use a mild detergent and water solution to scrub heavily contaminated areas **(Figure 4.10)**. Structural gear and SCBA manufacturers have recommended products and practices for cleaning their products. These procedures should be followed to maintain warranties and guarantee reliability.

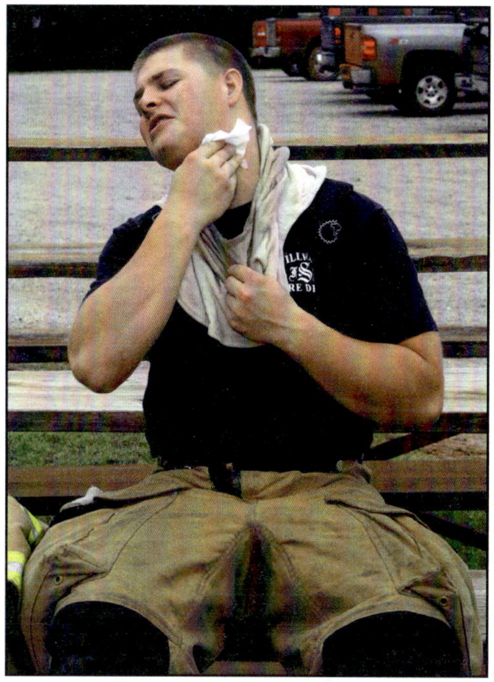

Figure 4.9 Using moist towelettes to remove soot from skin immediately following each incident is one preventive measure against cancer.

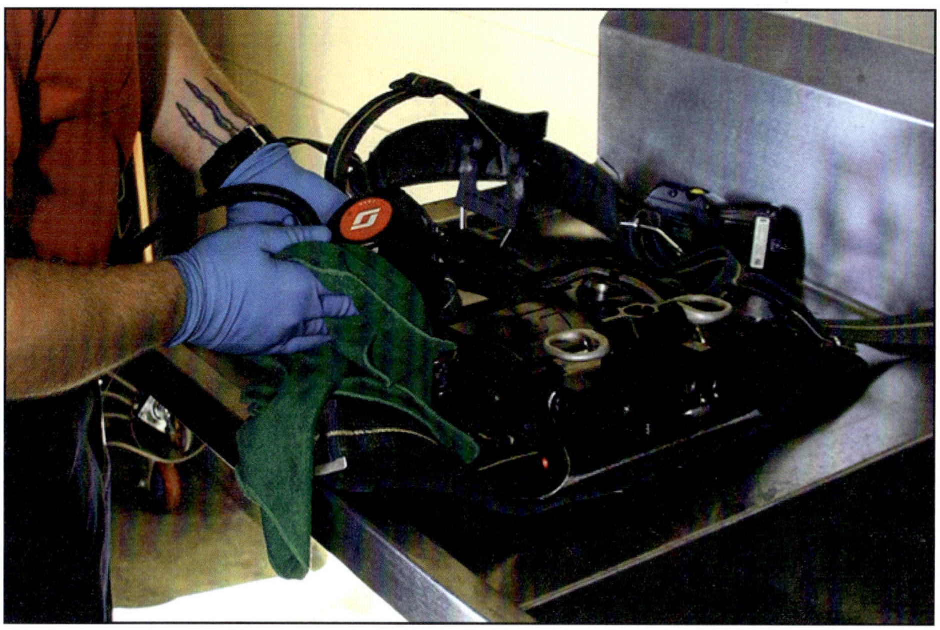

Figure 4.10 Cleaning PPE and equipment may require the use of a mild detergent and water, depending on the level of contamination.

c. Hang wet structural gear to dry in shaded, well-ventilated areas. One fire department built portable angle iron racks with metal hangers to dry their gear. To save space, the racks can be removed when not in use.

d. Do not launder structural coats or pants in residential washing machines.

e. Follow all manufacturer recommendations and NFPA standards when developing annual rotations for washing structural gear.

6. Place contaminated gear in airtight bags prior to leaving the scene in order to limit exposure to contaminated structural gear fumes and particulate matter. These airtight bags should be carried in compartments away from occupant compartments. Contaminated gear should never be carried in crew compartments of apparatus, or in passenger or trunk compartments of a personal vehicle.

7. If the apparatus is contaminated with smoke, the crew compartment must be cleaned with soap and water.

a. Wear latex gloves and N95 masks while decontaminating the inside of apparatus.

b. Keep apparatus windows rolled up around fire scenes.

c. Consider purchasing apparatus with vinyl seating to facilitate cleaning.

NOTE: Do not use high-pressure air to clean gear or compartments on apparatus, as it increases the risk of contaminates becoming airborne, inhaled, or making contact with skin.

8. Keep all structural gear out of station living quarters. Store structural gear in separate turnout rooms designed specifically for this purpose.

9. At fires and EMS scenes, limit firefighter exposure to diesel fumes. While placement of apparatus must be efficient for fire operations and safe for EMS operations, it should also consider the health of responders and patients and eliminate the hazard of diesel exhaust.

NOTE: When conducting maintenance in station bays, avoid running apparatus engines unless exhaust hoses are provided.

a. Turn off apparatus in public, unless lighting is required for safety.

b. Do not start apparatus until all firefighters are on the truck and all doors are closed.

c. Do not place any gear or clothing near diesel exhaust pipes.

10. Keep records of all exposures to smoke. More than 30 states have cancer presumption laws that help firefighters file workers' compensation claims if, later in life, they are diagnosed with certain forms of cancer. HSOs should inform firefighters in their departments of state laws pertaining to cancer claims.

a. Departments may implement automatic exposure documentation reminders that appear as units become available on mobile data terminals.

b. Forms can be completed electronically if departments include exposure documentation in EMS software.

An ongoing cooperative study between the University of Illinois Fire Service Institute (IFSI), the UL Firefighter Safety Research Institute (FSRI) and the National Institute for Occupational Safety and Health (NIOSH) is researching the cardiovascular and chemical exposure risks in modern firefighting. The interim report (Horn, Kerber, Fent, Fernhall, Smith, 2016) identified several immediate recommendations for preventing or limiting chemical exposures at fires:

- Always wear SCBA when working in or around fire damaged structures.
- Remain upwind of fires and smoke plumes. Wear SCBA if in the smoke plume or around diesel exhaust.
- Doff contaminated gear before entering the rehab area.
- Do not wear or store contaminated turnouts inside the apparatus cab, in personal vehicle compartments, or in station or home living areas **(Figure 4.11)**.
- After each fire response, decontaminate and/or launder turnouts, fire hood, and other equipment that readily contact the skin.
- Wash hands and neck immediately after the fire, and shower as soon as possible.

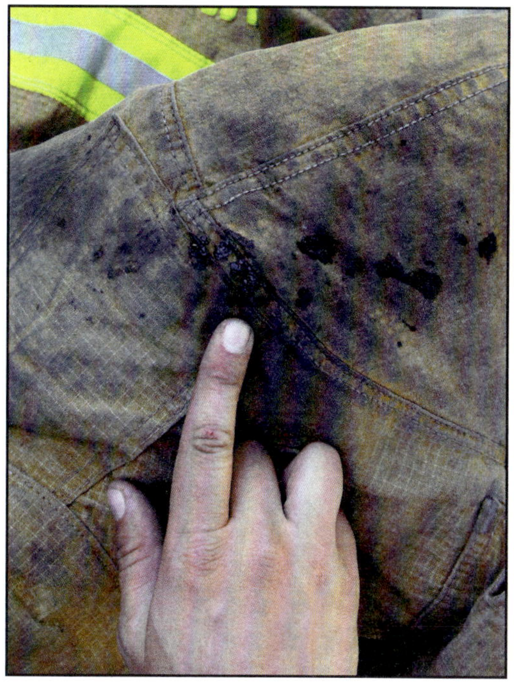

Figure 4.11 SOP/Gs should state that contaminated equipment should not be taken home. The equipment should be cleaned at the station or by an authorized third party.

Fire and emergency responders are increasingly exposed to biological hazards during medical responses. Both air and blood-borne pathogens can be present when dealing with victims and medical waste. Tuberculosis, HIV, hepatitis A, B, and C, MRSA and other easily transmitted diseases must be avoided with proper PPE and vaccinations.

Personnel can be exposed to viruses, colds, flu, lice, and contaminated food in the confines of the fire station. Improper food preparation, lack of personal hygiene, unsanitary conditions, and sharing bedding or personal protective equipment (PPE) can all result in exposure to job-related illnesses.

Lifestyle Choice Risks

Metabolic syndrome raises the risk for heart disease, diabetes, stroke, and other health problems.

Five factors comprise metabolic syndrome. They include the following:

- A large waistline
- A high triglyceride level
- A low HDL cholesterol level
- High blood pressure
- High fasting blood sugar

Having even one of these risk factors increases risk of heart disease. It is possible to prevent or delay the onset of metabolic syndrome by making healthy choices and reducing stress.

The following lifestyle choices, not influenced by occupational exposure to hazardous conditions, can make firefighters increasingly susceptible to serious health problems:

Use of Tobacco Products

Tobacco use can result in decreased lung capacity and lead to emphysema, pneumonia, and lung cancer. Smokeless tobacco products cause cancer of the mouth, throat and stomach. As a result, some organizations prohibit smoking while on duty and require recruits to sign a nonuse agreement.

Alcohol

Moderate alcohol consumption (defined as two drinks a day for men age 65 and younger, and one drink a day for women of all ages and men over 65) may provide some health benefits. But heavy drinking and binge drinking (four or more drinks for women and five or more drinks for men) offer no health benefits and can increase the risk of serious health problems like cancer, cardiovascular disease, depression, seizures, gout, hypertension and pancreatitis.

Electronic Cigarettes

The American Lung Association states the following about e-cigarettes:

- They are unregulated by the FDA.
- E-cigarettes allow the user to inhale nicotine and other substances in a vapor.
- They are generally battery-operated and use a heating element to heat liquid from refillable cartridge, which releases a chemical-filled aerosol.
- E-liquid is nicotine extracted from tobacco, usually mixed with propylene glycol.
- Nicotine is an addictive substance.
- FDA lab tests in 2009 found detectable levels of cancer-causing chemicals.

Narcotic — Any class of substances or drugs that numbs, paralyzes, or causes loss of feeling in the body. Narcotics may be addictive, alter moods or behavior, and induce sleep.

Drug Abuse

Many prescription and nonprescription drugs have harmful side effects, including physical dependence and addiction. In addition, as they affect the central nervous system, **narcotics** can induce sleep, impair judgment and lower reaction times. For that reason, both prescription and over-the-counter medications carry warnings to users about the dangers of driving or operating heavy machinery.

Anabolic-androgenic steroids can be misused or abused. Anabolic steroids have gained popularity and recognition among athletes as a means of boosting performance or improving physical appearance. Short-term effects include mental problems and extreme mood swings that can lead to violent behavior. Long-term effects of abuse include kidney problems or failure, liver damage, and increased risk of heart attack or stroke.

Marijuana Use

Although a growing number of states and the District of Columbia have legalized marijuana, it remains illegal under federal law. For firefighters, marijuana use is normally considered unlawful and against local fire department policy, as is showing up for work under its influence. However, since marijuana can be detected in the bloodstream for days or weeks after use, testing policies may require periodic review and revision.

Marijuana contains the mind-altering chemical delta-9-tetrahydrocannabinol (THC). When a person smokes marijuana, THC passes from the lungs to the bloodstream and is carried by the blood to the brain. The high users feel is produced when THC overactivates parts of the brain with the highest number of cell receptors. Other short-term effects of marijuana use include altered senses and sense of time, changes in mood, impaired body movement and memory, and difficulty with thinking and problem-solving. Long-term health effects may include reduced learning functions, mental illness, and a substance use disorder.

Poor Nutrition

Improper eating habits can result in obesity, diabetes, heart disease, and strokes. Both obesity and diabetes are considered epidemics in the United States. Uncontrolled diabetes can result in blindness, kidney disease, cardiac arrest, and nerve damage in the extremities, necessitating amputation. Some agencies, recognizing the connection between these diseases and nutrition, are providing education in proper eating habits.

Poor Hydration

Approximately 65 percent of the body is water. A reduction in water harms the body's ability to maintain its thermoregulatory mechanisms **(Table 4.1)**. Dehydration increases the firefighter's risk of heat-related injuries, such as heat cramps, heat stress, and heatstroke.

Severe dehydration may also contribute to an imbalance of electrolytes that are essential to the proper operation of the heart. This imbalance may produce dangerous dysrhythmias that can result in sudden cardiac death.

Table 4.1
Symptoms of Dehydration

Percentage of Weight Loss	Symptomology
1 percent	Thirsty
3 percent	Elevated heart rate
4-5 percent	Loss of up to 30 percent work capacity

NFPA 1584: *Standard on the Rehabilitation Process for Members During Emergency Operations and Training Exercises*, emphasises in Annex B.1.(a)that thirst is a delayed symptom of dehydration. Firefighters should rehydrate once every 20 minutes with water.
NFPA recommends firefighters involved in intense physical activity for more than an hour consume a fluid carbohydrate solution of 4-8 percent at a rate of 6-8 ounces every 15 minutes.

Lack of Physical Exercise

Inactivity and the lack of a structured physical fitness program increases the potential for obesity, physical stress, cardiac arrest, physical injury, and pulmonary exertion. A growing number of organizations now provide physical fitness programs, clothing, and equipment to their members. Results have been mixed, however. Some departments report increased rates of injury when firefighters do very strenuous workouts while on duty.

Overexertion and Overtraining

Ironically, firefighters that appear the most physically fit are often most susceptible to injury. Overtraining to the point of muscle fatigue can result in severe musculoskeletal injuries. Firefighters should strive for injury prevention by following a balanced training regimen that includes both strengthening and flexibility exercises.

Hiring Requirements and Procedures

Ensuring a healthy and physically fit fire and emergency services organization begins with hiring candidates that meet established medical and fitness criteria. The hiring criteria must be applied consistently to all applicants. Therefore, validated physical ability tests should be mandatory for everyone hired by career or combination departments. Some jurisdictions require the use of a task-oriented fitness selection test. In some states, volunteer firefighters are considered employees and must meet hiring criteria. All volunteer organizations should seriously consider using the same criteria for entry-level personnel.

Any screening device employed by an agency during the hiring process must result from a valid hiring practice. For that reason, there must be a job analysis to determine the types of tasks that fire and emergency responders perform and the level of fitness and health that is required for those tasks. To ensure valid fitness requirements, employees should be able to perform the entry-level requirements. Based on that analysis, experts develop valid screening processes. An applicant is not required to demonstrate the physical ability to perform at a level that exceeds the requirements of the essential duties of the job. This requirement applies to all departments. See **Appendix C** for additional information from the IAFF/IAFC on entry-level testing.

Hiring Requirements

NFPA 1500 states that, prior to hire, applicants shall receive a medical evaluation by the fire department physician and be certified as medically fit to perform the duties of a firefighter. Specific medical requirements are found in NFPA 1582, *Standard on Comprehensive Occupational Medical Program for Fire Departments*. Departments that operate fixed or rotary wing aircraft must also require that pilots who work for the department obtain a medical certificate following an exam by an FAA-designated Aviation Medical Examiner.

Applicants may be required to take third-party certified physical ability tests as part of the hiring process. These standardized tests are based on work-related activities. Local agencies may develop tests and have them certified or acquire a standardized test such as the Candidate Physical Abilities Test (CPAT) or National Wildfire Coordinating Group (NWCG) Pack Test.

Local political realities can influence hiring decisions. Pressure to hire an individual or give preference to a particular group can result in perceived bias or unfairness. To counteract the political pressures, it is essential to rely on a set of job-related standards. Those standards, which include medical, physical, mental, and emotional elements, must be applied to all applicants. By adhering to standards, the organization's hiring practices will be viewed by the public and politicians as highly professional and ethical. At the same time, those standards will help ensure the safety and health of fire and emergency responders.

Hiring Procedures

Once the hiring criteria have been established, a consistent set of hiring procedures must be developed. Local policy will determine the order in which these steps are taken. General hiring procedures include:

- Announcing or advertising the vacancies to be filled
- Accepting applications during a specific time period
- Administering a written personality and/or mental aptitude (psychological) test
- Administering a physical ability test
- Selecting the applicants that arc most qualified for the position
- Performing a medical evaluation
- Performing a background check

The following sections discuss two types of physical ability tests and provide an overview of the medical evaluation.

Candidate Physical Abilities Test

Physical abilities testing developed out of concern for the health and fitness of firefighters. Minimum qualifications for entrance into the fire service are designed to ensure that new hires may adequately perform duties as assigned. Most career fire departments have established some form of physical abilities testing that candidates must pass to be considered for employment.

The International Association of Firefighters (IAFF) and the International Association of Fire Chiefs (IAFC) developed a Fire Service Joint Labor-Management Wellness-Fitness Initiative roughly two decades ago. Their goal was to develop a comprehensive and positive approach to health and fitness. The process revealed that fire departments often hired people who were physically incapable of a successful fire service career. In response to this problem, the task force developed and validated a new evaluation tool for-labor management: the Candidate Physical Ability Test (CPAT).

The CPAT simulates fireground tasks a firefighter is likely to perform. While wearing a hardhat, gloves, and a 50-pound vest, the weight of SCBA and protective equipment, the candidate must perform the following eight tasks:

- Stair Climb (carrying a 25-pound simulated hose pack)
- Hose Drag (stretching uncharged hoselines and advancing lines)
- Equipment Carry (removing and carrying equipment from fire apparatus to fireground)
- Ladder Raise and Extension (placing a ground ladder at the fire scene and extending the ladder to the roof or a window)
- Forcible Entry (penetrating a locked door, breaching a wall) **(Figure 4.12, p. 132)**
- Search (crawling through dark, unpredictable areas in search of victims)
- Rescue Drag (using a dummy to simulate removing a 175-pound victim or partner from a fire building)
- Ceiling Pull (locating fire and checking for fire extension)

The CPAT consists of physical test events and use of techniques and equipment unique to the fire service. In accordance with recommended CPAT policy, and to ensure fairness, provisions must be made to offer pretest training and counseling to all applicants wishing to take the test.

Figure 4.12 Many jurisdictions use the standardized CPAT to evaluate potential recruit firefighters. Forcible entry is one task included in the test.

Some agencies have found that possessing a CPAT certification is not a reliable indicator of an applicant's ability to perform the functions of a firefighter. It is not clear whether this is due to improper CPAT testing procedures, a long interval between CPAT and agency testing, or some other unidentified factor(s).

NWCG Work Capacity Test

The Work Capacity Test (WCT), commonly referred to as the Pack Test for Red Card Certification, consists of three tests designed to measure an employee's aerobic capacity, muscular strength, and muscular endurance. The Red Card certifies that a person is qualified to complete the duties to which they are assigned. The test was originally developed by the National Wildfire Coordinating Group (NWCG) to set minimum training, experience and physical fitness standards for wildland fire positions. Like the CPAT, the WCT is designed to ensure that persons employed on the fireline are physically capable of performing the assigned tasks. The tests also ensure that even employees assigned to office duties have attained a minimum fitness level.

The Work Capacity Test **(Table 4.2)** is divided into three levels of duty: light, moderate, and arduous. The tests are described below:

Arduous Work involves above-average endurance (aerobic fitness), lifting more than 50 pounds (muscular fitness), and occasional demands for extraordinarily strenuous activities. All wildland firefighters perform arduous duty.

Moderate Work involves lifting 25 to 50 pounds and moderately strenuous activity on occasion. Safety officers and fire behavior officers have positions of moderate duty.

Light Work involves mainly office-type work with occasional field activity.

The NWCG Pack Test matches levels of fitness with the work employees perform. Fire service managers can reference this test to establish required levels of fitness for all employees, regardless of duties assigned.

Table 4.2
Work Capacity Test

Fitness Requirements	Test	Description
Arduous	Pack Test	3-mile hike with a 45-pound pack in 45 minutes
Moderate	Field Test	2-mile hike with a 25-pound pack in 30 minutes
Light	Walk Test	1-mile hike in 16 minutes with no pack

Medical Evaluation/Examination

NFPA 1582 provides an outline of medical evaluations that should be performed in examining the health and fitness of firefighter candidates. The recommended medical evaluations consist of a comprehensive examination of these functions and systems (and others):

- Head and neck
- Eyes and vision
- Ears and hearing
- Dental
- Nose, oropharynx, trachea, esophagus, and larynx
- Lungs and chest wall
- Heart and vascular system
- Abdominal organs and gastrointestinal system
- Reproductive system
- Urinary system
- Spine and axial skeleton
- Extremities
- Neurological disorders
- Skin
- Blood and blood-forming organs
- Systemic diseases and miscellaneous conditions
- Tumors and malignant diseases
- Psychiatric conditions
- Use of chemicals, drugs, and medications

NFPA 1582 also includes a comprehensive list of 13 essential job tasks and descriptions. Prior to any medical evaluation, the physician performing the examination should understand the work firefighters perform and demands imposed on a firefighter **(Figure 4.13, p. 134)**. The evaluation aims to determine health and fitness as they relate to the work. When the department has an assigned physician who conducts all medical evaluations, that physician should be experienced and familiar with NFPA 1582 and 1583, as well as the essential tasks assigned to a firefighter.

Figure 4.13 Physicians assigned to conduct medical evaluations should be familiar with the essential tasks performed by firefighters.

If the department does not have the resources to have an assigned physician, one that is selected to conduct the medical evaluations should be briefed as to the essential tasks. In addition, the physician should be provided an opportunity to observe and perhaps perform many of the tasks associated with firefighters work. The briefing should include, but not be limited to, documentation regarding the essential job and tasks as well as the work's psychological and physiological demands.

The examining physician must understand the environmental conditions a firefighter will be exposed to when performing essential tasks. Firefighters have many PPE requirements that the physician should include as part of the medical evaluation. The full ensemble of boots, pants, coat, hood, and helmet, as well as the required SCBA, tools carried, and work performed all place specific requirements on the firefighter. Firefighters assigned to enter and perform work in hazardous atmospheres may require full Level A suit protection, and the demands, especially from heat in those suits, must be considered in performing the medical evaluation.

Firefighters who perform wildland fire suppression will face additional levels of physical stress. Physicians may not be fully aware of the extraordinary demands that layers of protective clothing and safety devices place on the body. Firefighters are well aware of the heat buildup inside PPE ensembles; the examining physician may not be and should be briefed prior to performing medical examinations.

Based on NFPA 1582, any candidate determined to have a Category A preexisting condition shall not be certified for employment. The Category A medical condition would preclude a person from working as a member in a training or emergency environment because they are a safety risk. Category A conditions include deformity to the skull that would prevent use of a helmet, color blindness, hearing loss in the unaided better ear greater than 40 decibels, emphysema, pulmonary hypertension, hypertrophic cardiomyopathy, uncontrolled hypertension, and history of spinal surgery with rods still in place.

Applicants with Category B medical conditions that, based on the severity or degree of the condition, would prevent them from performing the duties of a firefighter in training or at an emergency incident shall not be certified as fit for duty. However, the AHJ can make reasonable accommodations to allow such an individual to perform other assigned functions. Category B medical conditions include (but are not limited to) the following:

- Surgical procedures to correct or improve hearing or other conditions of the ear
- Tracheal stenosis
- Cystic fibrosis
- History of surgical repair of aneurysm of the heart or major vessel
- Diseases of the kidney
- Anemia and diabetes controlled by diet, exercise, and oral hypoglycemic agents

Many hiring policies include the use of all tobacco products and/or addictive drugs as cause to deny employment. Applicants may be required to sign a statement that indicates they do not use tobacco and will not use it throughout their employment. The justification for this prohibition is the established link between tobacco and cancer. In more than 35 states, cancer is considered a job-related illness and covered by workers' compensation.

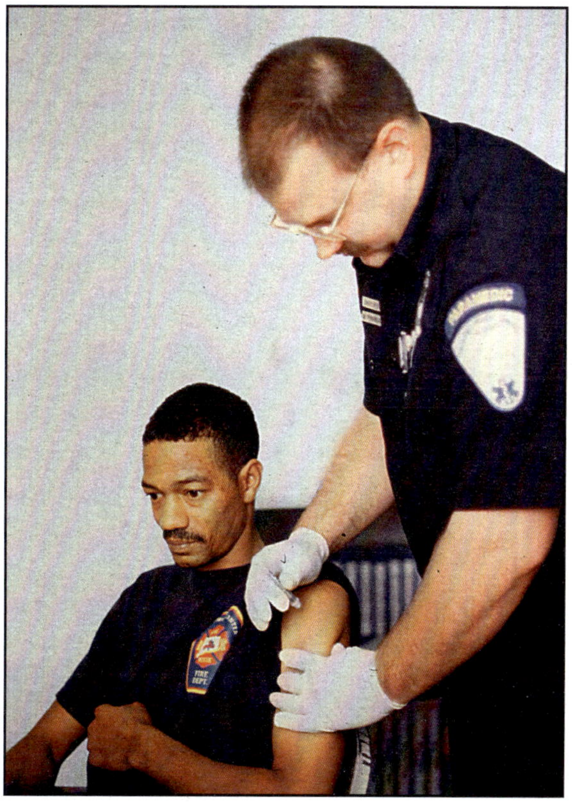

Figure 4.14 Entry-level personnel are given vaccinations to protect them from infectious diseases.

Applicants may also be required to have certain vaccinations or immunizations prior to employment. Additional vaccinations, such as hepatitis A or B, may be provided to a new hire **(Figure 4.14)**. The hepatitis B vaccine must be offered, but it can be declined with documentation.

In addition to the medical evaluation, many fire departments require applicants to complete a psychological or mental aptitude test. This test consists of two parts: a written test and oral interview. The written portion requires that an applicant answer many questions about himself or herself. Many of the questions repeat the same information in a variety of ways to determine consistency. The second portion of the test involves an oral interview with a psychologist. The psychological test aims to determine whether an applicant can handle the emotional stress that fire and emergency responders are subjected to in the line of duty.

Health Program

The employee health program seeks to maintain and improve the health of the individual firefighter or emergency responder. Individual health improvement happens through wellness education and physical fitness training. A holistic approach attempts to address potentially hazardous situations before they occur. This section discusses the overall elements of the health program.

NFPA standards require the establishment of committees and programs to ensure firefighter health. Specifically, fire departments must have an established **risk management plan**. Elements of that plan guide the department in the development of a comprehensive approach to risk management.

Risk Management Plan — Written plan that identifies and analyzes the exposure to hazards, selects appropriate risk management techniques to handle exposures, implements those techniques, and monitors the results.

An essential element of a risk management plan is the designation of a safety and health committee. The committee should include department management, individual members, and representatives of member (labor) organizations. Others may be included, but the key partnership in this committee is between the fire chief and the employees. Comprehensive management and success of a wellness program depends on the willing participation of employees. When administrative staff and members of the department work together to establish a workable and well-supported health and safety plan, success should follow.

To lay the foundation for the development of a health program, a well-defined statement of health should be established. Managers of and participants in a departmental risk management program must understand the program's mission.

A complete program includes defining health, clarifying the program mission, and communicating the process to ensure active participation and self-responsibility of department members **(Figure 4.15)**. After all, a health program must be accepted by every member of a department, and each member must take responsibility for their own health in order for a program to succeed.

Figure 4.15 Health and wellness information can be distributed to members through workplace notices.

Selecting a health program model is the starting point for changing the organization's culture. Ongoing open communication, discussion of problem issues, and agreement of a course of action all open the way to changes in policy, procedure, equipment, and apparatus.

A careful review of data relative to health and fitness by the safety and health committee is essential to proceeding with the final development of a department's health program. Such research, with recommendations, should be included in a final report to the fire chief and department.

A history of department members' injuries, illnesses, and medical issues, as well as job-related fatalities, can be researched to reveal problem areas. Such research should be expanded, particularly in small or newer agencies that may not have sufficient experience to be meaningful.

The NFPA creates annual reports and multiyear summaries of firefighter injury and fatality data and analysis, which should help any agency initiating a health and safety program. As indicated earlier, sudden cardiac arrest is the leading cause of line-of-duty death.

Committee review should also address best practices and recommendations regarding medical requirements for a firefighter. NFPA 1582 provides a basis against which the department's current practice can be compared. Any changes and recommendations for improvement should be carefully documented and reported to the fire chief. Any review of physical or medical requirements related to firefighter duties should be well-researched and include recommendations to the fire chief. Modern fire departments normally require specific physical or medical requirements that are related to essential firefighter duties, as any requirements proven not to be job-related are likely to face court challenges. The physical demands of the job must be recognized in the design of any comprehensive wellness and fitness program. Generally, an evaluation by the department's physician is requested based on a set standard. The doctor's recommendation is given to the fire chief, who makes the final decision on hiring a candidate.

Based on the work of the safety and health committee, the department should establish departmental policy and procedures regarding:

- Medical requirements
- Physical requirements
- Infection control
- Annual medical evaluation
- Fitness for duty
- Fire department physician
- Health and Safety Officer as liaison
- Incumbent physical evaluation

Firefighters, as employees, are naturally concerned that their personal medical evaluation remains confidential. While a committee may develop the evaluation criteria, they do not monitor each individual's performance. The department's administration must also be concerned about legal violations of the employee's privacy.

Firefighters also need to ensure that an individual's medical condition does not expose that person or other firefighters to additional risks. Firefighters need to be forthcoming regarding pertinent medical information. Properly executed, a health and wellness program allows employees to talk to a physician about their health and to develop a treatment or wellness program to improve their physical fitness for the job. Benefits of a successful program include reduced healthcare costs, improved employee longevity, reduced turnover, and reduced absenteeism.

Medical Requirements

Policies must be in place to define the medical criteria for hiring and continued employment. Usually associated with the labor/management agreement, these policies protect the employee and establish management rights. Medical requirements should include:

- Annual medical evaluations
- Return-to-duty criteria
- Medical leave criteria
- Disability separation criteria
- Limited-duty guidelines

NOTE: When firefighters are injured, on or off the job, health programs should direct HSOs or the department physician to place these firefighters in limited duty or modified worker positions.

Physical Requirements

Physical requirements are based on the task analysis used for hiring personnel. In some areas, incumbent members of the organization must be able to pass the same physical test required of hires. Annual fitness tests, sometimes adjusted for age or medical conditions, are generally included in labor/management agreements. The test must be based on job-related tasks and resemble the test used for candidate selection. Physical fitness requirements should be the same for career, combination, and volunteer departments.

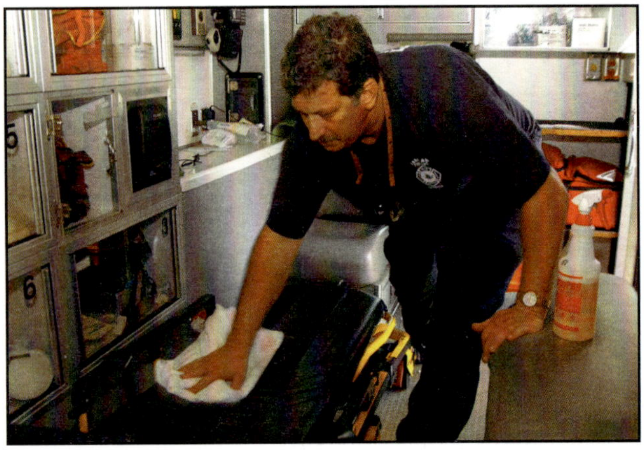

Figure 4.16 Infection control includes the cleaning and decontaminating of ambulance interiors.

Infection Control

Infection control begins with a written plan that clearly explains its intent, benefits, and purposes. The plan must cover standards of exposure control such as the following:

- Education and training requirements
- Vaccination plans and requirements for potential threats such as anthrax or hepatitis B virus
- Documentation and record-keeping requirements
- Cleaning, decontamination, and disinfection of personnel and equipment **(Figure 4.16)**
- Exposure control and reporting protocols

The Infection Control Officer ensures that an adequate control plan is developed and that all members are trained and supervised in the plan. The Health and Safety Officer (HSO) may fill this position, or it may go to another officer who has the knowledge, training, and skills to fulfill the office's duties. Local hospitals should be included in the development of any infection control protocol, as well as the department's medical control physician.

The Safety Committee should review exposure reports and maintain them in a record-keeping system. As some symptoms may not become apparent for many years following exposure, employee medical and exposure records must be preserved and maintained for at least the duration of employment plus 30 years and may only be released to the employee or his/her designated representative. Records should also be maintained for training on the proper use of PPE, exposure protection, post-exposure protocols, and modes of transmitting infectious diseases.

The HSO and safety and health committee members must annually review the infectious disease plan, updates, protocols, and equipment. Policies should also be developed that specify the types of protection that must be used to prevent an exposure.

Annual Medical Evaluation

Confidentiality is required of any established physical evaluation of an employee and the corresponding medical counsel to the individual by a provider. The physical exam, the results of that exam, and the recommended follow-up are between the firefighter and physician. The fire chief need only be concerned with the medical opinion that the firefighter is fit for duty or not.

Fitness for Duty

A firefighter's performance should be constantly under review by immediate and secondary supervisory officers with the expectation that the performance will meet standards. Any observed failure to perform can then be evaluated to determine the cause. Should the initial evaluation reveal that there is a possible physical, emotional, or psychological problem related to performance, a fitness for duty evaluation may be required. The procedures for requiring a fitness for duty evaluation must clearly indicate that the decision is based on observed objective factors of the firefighter's inability to perform the essential job duties.

The medical authority trusted with the evaluation should be thoroughly familiar with the essential duties of a firefighter and the associated physical and physiological demands of the work. The referral for the evaluation must be handled with strict confidentiality. In addition, in case of injury or illness, a determination must be made as to whether the issue is job-related or not.

Injuries or illnesses that are not job-related must be managed by the employee and a personal physician. The organization's primary concern is the firefighter's fitness for duty when he/she is medically released to return to work. Policies and procedures relevant to managing the return to full duty should be in place. The attending physician must fully understand the demands of the work.

A return to duty policy could also incorporate some form of a restricted or limited (light) duty assignment that will return the firefighters to some capacity of work while they continue to rehabilitate from an injury or illness **(Figure 4.17)**. The policy must be well defined, coordinated, and equitably applied with relevant sick leave, injury leave, and family leave policies. Work assignments need to be carefully evaluated and compared to the restrictions established by the fire department physician to ensure that the activities will not aggravate existing injury or illness, or contribute to new injuries or illnesses. Fire departments have a substantial investment in the training and experience of firefighters. Developing a light duty policy allows an employee to transition back into the workforce and stay engaged with the department, while still performing valuable work.

Figure 4.17 Limited or light duty assignments can be used for personnel on injury leave who can still perform some duties for the department.

If a firefighter is unable to perform essential duties due to a line of duty injury or illness, the department should also have a well-defined policy to manage workers' compensation issues. In these cases, the medical treatment and rehabilitation may be the same. However, the department is obligated to support the treatment and rehabilitation of that firefighter. Return to duty and light duty may be treated in much the same manner for either on-or off-duty illness or injury. The use of sick/injury leave may not be required of the employee for on-duty illness or injury. Laws differ regarding workers' compensation issues. The organization should consult an experienced risk manager or legal advisor to coordinate the agency's policy with state law.

Fire Department Physician

Proper management and direction help develop and implement an effective health and fitness program. Selecting and appointing a fire department physician or medical group is a crucial element. The requirements for the fire department physician or medical group are found in NFPA 1582. They include the following:

- The fire department shall officially designate a physician responsible for advising, steering, and counseling the members regarding their health, fitness, and suitability for various duties.
- The fire department physician shall provide medical guidance regarding the management of the occupational safety and health program.
- The fire department physician shall be a licensed medical doctor or osteopathic physician qualified to provide professional expertise in occupational safety and health relating to emergency services.
- The fire department physician shall be available for consultation.

In order to provide an NFPA-compliant medical certification program, the fire department physician must have a full understanding of the services that the department provides. Fire department work frequency combined with task analysis helps in providing this understanding. It may be necessary to train the fire department physician about the fire department tasks, operating procedures, and evolutions. Participating in the health and safety committee should help a physician become familiar with fire department operations. Smaller departments may elect to jointly hire a physician to provide services for several organizations.

Health and Safety Officer as Liaison

While the fire department physician is responsible for the medical aspects of the program, the HSO provides a liaison between the administration and the fire department physician. The HSO also ensures that firefighter exposure information is collected by department officers, and members collect firefighter exposure information and complete the appropriate forms.

Incumbent Physical Evaluation

Following employment with a fire and emergency services organization, periodic medical evaluations shall be performed on each employee involved in emergency response operations. This evaluation can be administered in conjunction with the annual physical ability or fitness testing, or scheduled separately. The employee must also receive a medical evaluation before returning to duty following a lost-time injury or illness and associated rehabilitation. The workers' compensation program has strict guidelines to follow regarding an employee's occupational injury and return to duty. An organization should also have procedures for reporting nonoccupational injuries and illnesses, and for returning to work after those incidents.

Periodic medical evaluations should document the following items:

- Interval medical history
- Interval occupational history, including significant exposures
- Height and weight
- Blood pressure
- Heart rate and rhythm

The frequency of these periodic medical examinations depend on the age of the employee. The schedule required by NFPA 1582 is as follows:

- Age 29 and under, every three years
- Age 30 to 39, every two years
- Age 40 and over, every year

Annual medical examinations, including blood tests to establish a baseline. Annual exams are required for personnel assigned to hazardous materials teams.

In addition to complete medical physicals, firefighters may be required to participate in annual hearing and pulmonary function testing. Audiometric tests or hearing tests are usually required during the hiring or admittance phase to determine a baseline. If a firefighter is repeatedly exposed to a given level of sound, an evaluation is performed annually to determine occupational hearing loss. Annual spirometry or pulmonary function testing and respiratory medical evaluations are required of all firefighters who must wear respirators, including SCBAs.

As the field of medicine progresses, newer tests will help physicians to predict and evaluate medical conditions before they become more serious problems. Departments should annually evaluate their physical assessment protocols and contracts with private providers.

Wellness and Fitness Components

Two of the essential components of a health program are wellness and fitness. Taken together they provide the behavioral and physical aspects needed for good health. Wellness provides the knowledge required to recognize the need for change in a person's behavior, such as improved dietary habits or improved coping mechanisms to manage stress. The fitness component provides the required level of structured physical activity that will help maintain the strength, stamina, and flexibility needed to safely perform assigned tasks. It will also improve cardiovascular health, lung capacity, and weight management.

Figure 4.18 Nutritious meals are a key ingredient to good health and avoiding chronic diseases.

Wellness Components

A comprehensive wellness plan is a necessary component of a holistic approach to health and safety. Wellness involves educating firefighters about the cause-and-effect of lifestyle choices on the human body. Employees have the opportunity to see the benefits of making changes in their lifestyles through:

- Nutrition **(Figure 4.18)**
- Hydration
- Spine and joint care
- Heart and lung health
- Stress recognition

Wellness training should help to avoid, identify, and understand issues that impact the firefighter on relational, financial, educational, spiritual and physical levels. HSOs must have training to recognize various symptoms, and to be able to offer resources to assist firefighters who need a helping hand **(Figure 4.19)**. Topics for employee training programs might include tobacco-cessation plans, weight control, anger management, marriage and family counseling, and financial planning.

Figure 4.19 A Health Safety Officer discusses station wellness concerns with another officer.

Nutrition

Studies have linked poor diet to heart disease, cancer, diabetes, high blood pressure, high cholesterol, and other chronic diseases. As part of the employee wellness and fitness program, the HSO can provide members of the organization with training on the importance of good nutrition. Training topics can include eating for performance, guidelines for a balanced diet, results of good nutrition, safe and effective weight-control diets, and the risks of fad diets. Some organizations provide a recommended daily meal plan for each station via e-mail. HSOs should research state regulations for these plans **(Figure 4.20)**.

An effective educational plan on nutrition can have the added bonus of altering the off-duty lifestyles of employees and their families. Information on nutritional issues can be obtained from the U.S. Centers for Disease Control and Prevention (CDC), state/territorial/provincial or local health departments, and the American Heart Association (AHA).

Figure 4.20 A Health Safety Officer trains firefighters on healthy eating choices.

Health Effects of Overweight and Obesity

Overweight and obesity are known risk factors for:

- Diabetes
- Coronary heart disease
- High blood cholesterol
- Stroke
- Hypertension
- Gallbladder disease
- Osteoarthritis (degeneration of cartilage and bone of joints)
- Sleep apnea and other breathing problems
- Some forms of cancer (breast, colorectal, endometrial, and kidney)

Obesity is also associated with:

- Complications of pregnancy
- Menstrual irregularities
- Hirsutism (presence of excess body and facial hair)
- Stress incontinence (urine leakage caused by weak pelvic floor muscles)
- Psychological disorders such as depression
- Increased surgical risk
- Increased mortality

Hydration

Although it is frequently overlooked, hydration is important before, during, and after any strenuous activities. The body can lose more than 2 quarts/liters per hour of fluid through sweat, urine, stool and exhaled air. As this amount is greater than the body can absorb through ingestion — approximately 1 to 1.5 quarts/liters per hour — this imbalance can result in serious injury or death. To ensure proper bodily functions, firefighters must continuously monitor and maintain hydration both on and off duty.

Firefighters should monitor their intake of food and water as well as the frequency, color, and smell of their urine. Infrequent urination, dark colored urine, or urine that has a strong odor may suggest an inadequate level of hydration.

The process of hydration should begin at least two hours prior to scheduled events where the firefighter will knowingly participate in activities requiring heavy exertion. Intake of fluids should be in small sips throughout the various activities and in larger quantities during scheduled times of rehabilitation. Finally, firefighters should consciously monitor their hydration for a minimum of two hours after the event **(Table 4.3, pg. 144)**.

Table 4.3
Effects of Dehydration on the Human Body

Fluid Loss	Potential Effect
1% body weight	Decreased heat transfer to skin surface where heat can be dissipated
1-2%	Increase heart rate, decreased blood pressure
2-3%	Decreased muscle endurance and energy level
> 3%	Decreased concentration and ability to focus

The 2007 Orange County Fire study on hydration, core body temperature and performance offered the following chart for evaluating how fluid loss is related to performance.

Water is generally sufficient for the purpose of maintaining hydration prior to and during events that last for less than one hour. However, research suggests that a commercial sports drink containing both electrolytes and carbohydrates should be provided for events extending beyond one hour. Cold drinks should be served cool and hot drinks be served warm for best absorption. Alcoholic drinks or those containing caffeine contribute to dehydration and should be avoided.

Spine and Joint Care

Back and joint injuries are very common for firefighters. Most injuries to the spine are a result of poor lifting technique. Injuries to the shoulders, elbows, knees and ankles vary according to the joint involved. Knee injuries are frequently due to a ligament strain or sprain, and a rupture to the rotator cuff is a common shoulder injury. Ankle injuries are usually some form of closed fracture, and tennis elbow is also a common complaint. Organizations should look carefully at joint injuries among firefighters, and provide both early assessment and continuous evolution of physical training to prevent them.

The cost of treatment for joint injuries varies from state to state, as do the type of treatments commonly offered to patients. Some states have much higher averages for surgical treatment of injuries, while others have higher rates of treatment with physical therapy. Several departments in the Phoenix area worked with local medical colleges and physical therapy providers to provide recruits with functional movement screenings and specific strengthening exercises. Periodic assessments during their academy training tracked injury rates of these firefighters to evaluate and compare their rates of injury with those of firefighters who attended previous academies.

Heart and Lung Health

As part of the cardiovascular system, the heart and lungs are susceptible to diseases that can be prevented with education and action. The organization must implement a policy for the use of SCBAs or supplied-air respirators (SARs) in all toxic atmospheres and in suspected or unknown atmospheres. All facilities must provide a smoke-free environment, or designated smoking areas with proper ventilation. Wellness training should include the following information:

- Importance of unimpaired lung capacity and unobstructed blood circulation
- Causes of heart and lung diseases
- Methods for reducing the potential for these diseases

Stress Recognition

Hazards at work go well beyond physical factors. Spotting and addressing internal issues can pay dividends for the firefighter and service as a whole.

Psychological stress is more individual in nature and is created by the total life experience of the individual, including work **(Table 4.4, p. 146)**. A short list of psychological stressors that emergency responders experience in their daily lives and take to work are as follows:

- Personality conflicts with coworkers
- Perceived lack of respect from managers, administration, or the public
- Boredom
- Lack of job satisfaction
- Concerns over promotion, layoff, or retirement
- Feelings of inadequacy
- Fear of failure
- Personal injury to self or coworkers
- Death of coworker or family member
- Personal family problems

As part of the holistic employee physical fitness and wellness program, the HSO includes activities that help members deal with stress. The National Fire Academy (NFA) has developed a stress management model program that can assist in the training of members. It defines stress, outlines the symptoms, lists the physical and psychological reactions, and provides activities for stress management. Three basic strategies for managing stress are as follows:

1. Minimize the stress-causing factors by avoiding, eliminating, or reducing them
2. Change how the individual perceives or views problems
3. Learn methods for relaxing both the mind and body

The HSO (in conjunction with the training division) develops training sessions that assist in meeting these strategies. Recommended topics for training in stress reduction include the following:

- Time management
- Role management
- Conflict resolution
- Assertiveness training
- Relaxation methods
- Diet and nutrition guidelines

Some fire departments have educational programs that help firefighters recognize when issues like health, finances, and marital difficulties will impact cognitive thought during high-risk activities. Firefighters are asked to self-report without any retribution. The Employee Assistance Program (EAP) works together with the firefighter and the department to get the individual back to work when appropriate. This type of program recognizes the dangers of the job and the barrier stress can be to conducting safe operations.

Table 4.4
Warning Signs and Symptoms of Stress

Cognitive Symptoms	**Emotional Symptoms**
• Memory problems	• Moodiness
• Inability to concentrate	• Irritability or short temper
• Poor judgment	• Agitation, inability to relax
• Dwelling on the negative	• Feeling overwhelmed
• Anxious thoughts	• Feeling lonely and isolated
• Constant worrying	• Depression or general unhappiness
Physical Symptoms	**Behavioral Symptoms**
• Diarrhea or constipation	• Eating more when you are not hungry
• Nausea, dizziness	• Sleeping too much or too little
• Loss of sex drive	• Isolating yourself from others
• Frequent colds	• Procrastinating or neglecting responsibilities
• Pain in the chest, shoulders, neck, or low back	• Using alcohol, cigarettes, or drugs to relax
• Stomach/abdominal pain	• Nervous habits (nail biting, pacing)
• Muscle tension, spasms, or nervous tics	• Feeling frustrated at having to wait for something
• Unexplained rashes or skin irritations	• Feeling restless
• Sweaty palms	• Being easily confused
• Sweating when not physically active	• Negative self-talk
• 'Butterflies' in stomach	• Feeling you can't cope
• Indigestion	• Difficulty making decisions
• Inability to sleep or excessive sleep	• Having emotional outbursts
• Shortness of breath	• Generally feeling upset
• Holding breath	• Lack of sense of humor
• Loss of energy	

Peer Support Management

Stress has always been a part of the emergency responder's life due to the high level of uncertainty, limited control over the work environment, and the psychological effect of repeated emergency calls. Critical stress levels develop when major events exceed the normal level of stress and the ability of the body to cope. For this reason, a stress management plan must be part of the organization's employee physical fitness and wellness program. This plan aims to manage the stress experienced by fire and emergency service personnel after these types of incidents:

- Mass casualties
- Fatalities involving children
- Serious injuries or fatalities involving members of the organization
- Suicides
- Serious injuries or fatalities involving close friends, relatives, or colleagues

- Violence directed toward firefighters or other emergency responders
- Deaths of civilians as a result of emergency operations
- Excessive media attention generated after incidents

Signs and Symptoms

Common symptoms and signs of stress include a rapid heartbeat, difficulty breathing, tense muscles, and headaches. Excessive stress, however, results in more severe responses. The symptoms associated with excessive stress include:

- Difficulty concentrating or staying focused
- Temporary loss of short-term memory
- Obsessive thoughts
- Loss of mental flexibility
- Tendency to withdraw or become isolated
- Abuse of alcohol and drugs

Building a Team

In order to address these symptoms, a peer support team is established and trained. The team consists of emergency services personnel, health care professionals, and clergy who have specialized training in dealing with acute and chronic stress. This team provides the necessary support for members experiencing excessive stress. It also provides annual pre-incident education for emergency personnel about the causes and results of stress, stress reduction methods, and sources of professional assistance. At major incidents, team members may provide guidance to the Incident Commander and the Command staff, as well as support for individuals at the scene **(Figure 4.21)**.

Follow-up services, consisting of additional individual sessions and referrals, can occur over several months following the incident. Throughout the process, team members maintain strict confidentiality. Many states have statutes that protect conversations with a peer supporter as privileged communication, which is only subject to disclosure for mandatory reporting.

Figure 4.21 Members of the team assist an Incident Commander at a scene.

Training

Although participation in the stress management plan is not mandatory, pre- and post- incident training can be used to prepare personnel for the possibility of a large stress incident. The first step in reducing the effects of stress is to manage it by training personnel about the psychological hazards of the job. Stress management training should include information about acute and chronic stress, including causes, symptoms, and some effective coping techniques. This type of training is different from fire fighting training and is not exclusive to entry-level firefighters or emergency responders.

All fire and emergency services personnel, including chief officers, should be included in stress management training. It is strongly advisable to have personnel who have experienced more severe levels of stress as members of the training team. This makes their personal experience and observations available and creates credibility.

Member/Employee Assistance Program

The levels of stress normally encountered by fire and emergency personnel can result in dependence on tobacco products, abuse of alcohol or drugs, domestic violence, excessive gambling, and financial difficulties. In recognition of this fact, NFPA 1500 mandates the establishment of a member assistance program within the fire and emergency services organization **(Figure 4.22)**.

Figure 4.22 A member assistance program provides private counseling services for firefighters.

A member assistance program is essential in the holistic approach to personnel health and wellness. As it is impossible for an organization to provide professional counseling services to meet all needs and ensure complete confidentiality, a member assistance program often includes outside contract counselors to provide these services. The HSO is responsible for prevention, education, and referring employees to program counselors. All officers must also be aware of the symptoms of alcoholism, drug abuse, and other types of abuse in order to provide direction and care as soon as possible.

The effectiveness of the member assistance programs cannot be denied. Studies indicate that as many as 80 percent of those individuals who receive counseling return to full productive status within the work force. Member assistance programs can also reduce the cost of prolonged medical care and lost-time benefits. In addition to fire and emergency service members, NFPA 1500 provides for assistance to immediate family members through the program. Finally, the use of member assistance programs can result in improved employee morale.

Counseling. The member assistance program outlined in the organization's policy should provide a wide range of counseling services for employees and their families. In addition to assistance with tobacco-cessation and substance-abuse assistance, domestic violence, child abuse, and family and financial counseling services should be available. Success of the program depends on the quality of the services offered, support of the administration, involvement of the members, and confidentiality.

Substance abuse. Drug and alcohol abuse can have devastating effects. As drugs and alcohol impair judgment and slow reaction times, they affect not only the individual. Those who work and live with the individual user, as well as the general public, are also affected by drug and alcohol abuse.

To offset the danger of this type of abuse, the fire and emergency services organization would be aided by a written policy on the subject in the employee physical fitness and wellness program. The HSO's duties are to:

- Develop and present educational sessions on the negative effects of alcohol and drug abuse
- Assist the administration in developing a drug and alcohol policy
- Direct members to the member assistance program when necessary
- Act as a liaison between the member assistance program and the administration.

The policy must include control mechanisms to provide rehabilitation procedures for employees, as well as reporting and privacy procedures. Some organizations have established periodic random drug testing for all employees. The medical department of the jurisdiction usually contracts health care facilities to perform this testing, and the results must be kept confidential. A policy should be established within the program whereby members who refuse to participate or who become repeat offenders following rehabilitation may be separated from employment.

Physical Fitness Component

According to the CDC, a comprehensive physical fitness plan curbs risks from the following:

- Dying prematurely from all causes
- Developing adult-onset diabetes
- Contracting high blood pressure
- Increased blood pressure
- Developing colon cancer
- Depression and anxiety

A plan also:

- Controls weight
- Builds healthy bones, muscles, and joints
- Promotes psychological well-being
- Improves sleep habits
- Reduces injuries

To develop a comprehensive, holistic physical fitness plan, the HSO must prepare an analysis of the tasks performed by organization members. The analysis serves as a basis for developing the physical fitness plan and for establishing hiring criteria for new personnel.

The physical fitness component of the employee wellness and fitness program must address flexibility, cardiovascular fitness, muscular fitness, and body composition. The HSO lists specific exercises, using appropriate equipment, to improve an individual's deficiencies in each of these areas. Criteria necessary for determining the effectiveness of the plan must be established. Effectiveness is determined through periodic monitoring of the physical fitness component and the organization's illness and injury reports (collected before and after implementation of the component).

Task Analysis

Preparing the task analysis and developing the employee physical fitness component are the responsibilities of the HSO, the occupational safety and health committee, or a physical fitness subcommittee. Members of that subcommittee would include the HSO, members of the organization's administration, representatives of the member organization, emergency response personnel, the organization's physician, and a qualified/certified exercise physiologist. The exercise physiologist provides the professional knowledge necessary to analyze the tasks in terms of physical exertion and can recommend the appropriate test criteria.

If the subcommittee is assigned to development, the first step is for committee members to perform the task analysis and develop a list of the basic services that the organization provides. These services may include, but are not limited to, structural fire fighting operations, wildland fire fighting operations, emergency medical services, light and heavy rescue operations, hazardous materials responses, training functions, and building inspections and surveys. Next, committee members determine the types of tools and equipment that each service activity requires. The tools and equipment (hoselines and nozzles, pike poles, ground ladders, and axes) indicate the types of physical tasks that each activity requires. The tasks must be performed while wearing the appropriate PPE and respiratory protection. These physical tasks may include the following:

- Lifting an inert weight such as an unconscious victim
- Pulling a hoseline, both charged and uncharged (empty)
- Operating a hoseline and nozzle
- Climbing a ladder
- Climbing a flight of stairs in full protective clothing with SCBA

- Entering and operating in a confined space with hand tools
- Pulling a ceiling with a pike pole
- Using an axe to cut through a substantial timber

Once the list is complete, the tasks can be categorized into groups of similar activities such as lifting, pulling, and climbing. A suitable physical fitness plan can be devised to meet the needs of each task **(Figure 4.23)**. The physical fitness plan is used not only to improve the physical abilities and stamina of current members, but also to provide testing criteria for candidates and applicants.

Figure 4.23 A physical fitness plan is designed to help firefighters safely perform assigned functions, such as overhaul.

NFPA 1500 requires the fire and emergency services organization to establish and provide a physical fitness plan that meets the requirements of NFPA 1583. This plan is designed to enable members to develop and maintain an appropriate level of fitness to safely perform their assigned functions. The maintenance of fitness levels will be based on fitness standards determined by the organization's physician. These levels must reflect the members' assigned functions and activities, as well as the severity of occupational injuries and illnesses associated with these activities. Results are compiled in a personnel file for each employee and maintained for analysis.

All members must be required to participate in the physical fitness plan, which may be addressed by labor/management agreements. They should be evaluated annually and certified to perform their assigned duties in emergency operations. Members who cannot meet the physical performance requirements must be required to participate in a physical rehabilitation program to assist them in meeting their designated levels. If it is determined that a firefighter does not meet organizational standards, the HSO, physician, or physical fitness officer will devise a set of exercises and schedules to assist him/her in meeting his/her goals.

A physical fitness plan for firefighters below standard should be holistic, positive, rehabilitating, and educational; not used as punishment. The goal of physical fitness is to improve the quality of life for all emergency service personnel and to help them live long, healthy lives. Good physical fitness also improves the quality of services provided to the community and the professional image of the organization. Adhering to a complete physical fitness plan helps to reduce the potential for fatigue, heart disease, stroke, shortness of breath, and stress.

It is important to recognize that purchasing workout station equipment and placing it in a station workout room does not make a wellness program complete. There must be an analysis, training, and follow-up components in any well-structured program. Numerous curriculums exist for firefighter physical fitness. HSOs should consult documents like the IAFF Peer Fitness Curriculum to understand which components are essential for a workable program.

Exercise Plans and Equipment

The exercise plan is designed to provide exercises that will improve body flexibility, cardiovascular fitness, muscular fitness, and body composition. **Table 4.5** provides a quick reference of each of these areas and lists the appropriate exercises. Warming up before and cooling down after each workout reduces the potential of injury during the exercise session. The four exercise plan areas are described as follows:

- Flexibility — Flexibility (or stretching) exercises improve the body's mobility and range of motion. Divide the flexibility training plan into attainable objectives that ultimately reach the goal established in the assessment. Attainable objectives may include a reasonable degree of improvement on a sit-and-reach test. Increase these objectives periodically (weekly) until the maintenance-level goal is attained.
- Cardiovascular fitness — Cardiovascular exercises improve the body's ability to supply oxygen through the circulation system to the muscles while performing vigorous activities over an extended period of time. Once the assessment is complete, develop an exercise program to maintain or improve the individual's ability. Playing sports is not a recommended substitute for any of the recommended exercises, although it may be used to supplement the fitness plan and make it more interesting. However, the USFA has determined that many injuries to emergency personnel occur while playing sports.
- Muscular fitness — Good muscular fitness allows personnel to perform assigned tasks more effectively and efficiently, reducing the potential for personal injury. To increase muscular fitness, the plan needs to address the following items:
  - Endurance: Ability to perform the activity repeatedly
  - Strength: Maximum amount of force a muscle can generate
  - Power: Ability to exert strength quickly
  - As an individual's strength increases through exercise, his/her power and endurance also increase. These exercises are directed at six muscle groups: midsection, legs, arms, chest, shoulders, and back.

Table 4.5
Exercise Plan Comparison Table

Exercise Plans	Desired Result	Exercise Type
Flexibility	Improved mobility and range of motion in back and legs	• Posterior Thigh • Calf Stretch • Anterior Thigh • Inner Thigh • Iliotibial Band and Lateral Thigh • Soleus
Cardiovascular fitness	Strengthened heart muscles	• Weight-bearing exercises: — running — stair climbing — rope jumping • Non-weight-bearing exercises: — bicycling — rowing — swimming
Muscular fitness	Increased strength and endurance	• Weight training: — bar bells — leg presses — bench presses — dead lifts • Endurance training: — push-ups — dips — curls — squats — sit-ups — pull-ups
Body composition	Weight loss and reduction in body fat or mass	• Jogging • Swimming • Dancing • Cycling • Brisk walking • Aerobics

- Body composition — Body composition is the relationship of lean-to-fat tissues in the body. It can be roughly estimated by using waist circumference measurements. Special equipment, such as skinfold calipers, can provide more accurate measurements. The resulting number is converted to points that are added to points achieved during the exercise portion of the assessment. Exercise helps to improve the muscle fitness of the individual. However, a healthy diet reduces fat tissue in the body.

The equipment necessary for a physical fitness plan varies according to the finances available to the organization. The HSO and safety and health committee provide the administration with a list of recommended equipment necessary to implement the plan. The organization must determine what equipment will be provided and how to fund the purchase.

Workout mats need to be available in each facility or at a central location where the exercising takes place. Free weights, exercise machines, powered stair-climbing equipment, and stationary bicycles may also be provided **(Figure 4.24)**. Initially, the organization might find it more economical to contract with a local health care facility or gymnasium for the use of its weights and machines. A physical fitness plan can be economical or costly to the organization depending on the approach taken. Either way, the health and well-being of the members and the reduced cost of lost-time injuries justify the expense.

According to a 2016 report from the National Institutes of Health, the cost of surgery to repair a rotator cuff was estimated at $5,904. This does not include diagnosis, rehabilitation costs, or time off work. A National Council on Compensation Insurance research brief (2015) evaluated the differences between states on the diagnosis and treatment of various joint injuries. The differences between states on shoulder injuries ranged from $2,397 on the low end to $6,841 on the high end. (NCCI Research Brief, 2015, State Differences in the Diagnosis and Treatment of Joint Injuries. Barry Lipton, John Robertson, Dan Corro.) It is easy to see the benefit derived from injury prevention.

Figure 4.24 A station fitness center provides cardio fitness machines and free weights.

Physical Fitness Plan Effectiveness

The physical fitness plan can be monitored through three methods:

1. Using daily or weekly exercise logs to list member participation **(Figure 4.25)**
2. Comparing lost-time injury data before and after the start of the program
3. Conducting annual physical fitness tests based on the original testing criteria to determine effectiveness

If the plan is mandatory, participation data can be used as a monitoring tool. However, the program must be flexible enough that nonparticipation for legitimate reasons is not punished. Alternative programs should also be available to members. For instance, some members may belong to a health club and prefer to perform their exercise activities there. Others may prefer bicycle riding to jogging. In any case, the alternative to participation in the organization's plan should be noted in the log.

If the program is effective, job-related injuries and occupational illnesses related to physical fitness should decrease over time. It is important to remember that USFA statistics show that injuries related to the physical fitness plan itself are leading causes of job-related lost-time injuries. There may be two reasons for which this alters the data collected on post-implementation injuries.

First, the injuries incurred in the physical fitness plan should not be included in the comparison data. Only injuries that occur at incident scenes or in non-exercise work are compared to preplan data. Secondly, the data on physical fitness-related injuries should be used to determine the need for change in the exercise plan. The lack of warm-up and cool-down exercises adds to exercise-related injuries. Poor nutrition, smoking, or personal health problems can be contributing factors.

Figure 4.25 The organization's physical fitness plan must have a means of tracking the exercises performed by the members on a daily or weekly schedule.

Any alterations in the physical fitness exercise plan should be made with the knowledge and assistance of the organization's physician and exercise physiologist. Changes may also be indicated in the wellness portion of the employee physical fitness and wellness program.

A fitness program works to increase strength, endurance, flexibility, and other physical attributes. The objective is to increase performance and to reduce muscle strain, sprains, and other related physical injuries. The definition of fitness that is most appropriate for firefighters acknowledges established fire service standards and is consistent with entrance standards.

Sleep Deprivation Management

Sleep deprivation among firefighters has been associated with cardiovascular disease, cancer, increased cortisol secretions, depression, and poor diet. There are numerous studies concerning safety issues resulting from lack of sleep.

In 2015, an academic study(1) looked at sleep disorders and their effects on members of the fire service. It showed 37.2 percent of firefighters in the study of 66 departments and nearly 7,000 participants reported sleep disorders, including sleep apnea (28.4 percent of respondents), insomnia (6 percent), problems related to shift work (9.1 percent) and restless legs syndrome (3.4 percent). Other results for those with sleep disorders included an increased risk of cardiovascular disease, diabetes, depression and anxiety. More sleep disorder sufferers were also likely to report poor health status. Sleep disorders made their biggest impact on motor vehicle crashes among firefighters. The study said motor vehicle crashes were the number two cause of death among on-duty volunteer firefighters and stood third among professionals.

Approaches to correcting sleep deprivation in the fire and emergency services are inconsistent. Some career organizations are moving toward shorter work shifts, while others are adopting 48-hours work shifts with 96 hours off in between. That approach aims to reduce the amount of time employees spend commuting. Still, other organizations sanction naps during the work shift to increase overall productivity and reduce fatigue.

Sleep research indicates the number of sleep disruptions, whether from physiological causes or from night-time call disruptions, leads to sleep deprivation and fatigue. Sleep cannot be banked, but 90-minute daytime naps can help firefighters stay more alert during nighttime call volumes. Overall, firefighters must learn that sleep cycles are as essential to good health as oxygen, food, and water. The consequences of sleep deprivation and the feeling of fatigue can only be masked by stimulants like caffeine. Research on sleep deprivation is emerging, and HSOs should watch for published studies pertaining to the topic.

On-Scene Rehabilitation

Figure 4.26 Firefighters should receive medical examinations during rehab before being allowed to return to duty at an emergency scene.

Environmental stress can be reduced by effective rehabilitation policies, education, and prevention. Policies should be established to control the length of time a fire or emergency responder works during an incident and to specify how often rehabilitation must occur. Personnel must be educated to recognize their physical limitations while using respiratory protection equipment and to not exceed them.

At emergency incident scenes or training evolutions, the Incident Safety Officer (ISO), HSO, Company Officer, or any individual firefighter can notify the Incident Commander when members have reached their limits of physical effectiveness. Incident Command must be ready for these requests with an appropriately staffed rehab location. In the best-case scenario, the Incident Commander proactively sends crews to rehab to avoid injuries.

Personnel are considered fit for duty when their body core temperatures and vital signs have returned to acceptable levels. The standards would be set by locally adopted medical protocols based on NFPA 1584 or other publications like the USFA's Emergency Incident Rehabilitation manual (2008).

- Paramedics and Emergency Medical Technicians (EMTs) make an assessment of a person's physical condition and general appearance. The assessments and rehabilitation practices include:
- An evaluation by EMTs regarding a person's physical limitations by establishing a baseline and continuing to monitor vital signs **(Figure 4.26)**.
- An establishment of mental status and level of orientation. Confusion, disorientation, or agitation without apparent cause may be symptoms of a serious life-threatening condition and require advanced medical care.

- The need for a transfer for those who are injured, severely dehydrated, severely exhausted, or those with other significant complaints to where they may receive additional medical monitoring, definitive care, and/or transportation to a receiving hospital.
- Providing vital sign data to the ISO.
- Providing nutritious food, replacement fluids, and a rest area to help personnel return to acceptable physiological levels **(Figure 4.27)**.
- When core temperatures are elevated, both passive and active cooling methods should be considered. Passive cooling would include shade and mechanical fans. Active cooling would include wet towels on the neck and arms or forearm immersion into ice water.

Figure 4.27 A rehab area allows firefighters to hydrate and receive oxygen if needed. *Courtesy of Ron Jeffers, Union City, NJ.*

Fire or emergency units/companies are not returned to service until the ISO or designee releases them as fit for duty, as described by local SOP/Gs that have been reviewed by the agencies' medical control physician. In some cases, it may be necessary for the emergency unit/company to return to quarters for a change of uniform and dry protective clothing before returning to service.

Chapter Review

1. What wellness areas should a proactive and comprehensive health program address to prevent or reduce firefighter injuries and illnesses?
2. What are some of the specific laws that govern health programs and employee medical rights?
3. How can stress cause physiological and psychological changes in the body?
4. Why is it important for all agencies to adopt standardized hiring practices?
5. What are the eight simulated fireground tasks in the CPAT?
6. What are the requirements of a return to duty program?
7. What resources can HSOs use to build wellness training programs?
8. How is task analysis used by a fitness subcommittee to create a firefighter physical fitness plan?

Discussion Questions

1. How can education and awareness about health and wellness and a commitment to physical fitness benefit firefighters beginning, during, and after their career?
2. Discuss how a firefighter's personal lifestyle can affect the goals and safety of the crew.
3. How can peer support management and member-assistance programs reinforce and improve the goals of the fire service?

Training Safety

Chapter Contents

chapter 5

Key Terms

FESHE Learning Outcomes

Occupational Safety and Health for Emergency Services

4. List and describe the components of risk identification, risk evaluation, and incident management.
6. Apply the knowledge of an effective safety plan to preincident planning, response, and training activities.

Training Safety

Learning Objectives

After reading this chapter, students will be able to:

1. Describe the need for fire service training.
2. Recognize the correlation between the types of injuries and injury prevention on the training and fireground.
3. Describe the different types of fire service training.
4. Explain how different departmental positions can ensure safe training environments.
5. Differentiate between cognitive and psychomotor training evolutions.

Chapter References

Rita F. Fahy. 2012. "U.S. Firefighter Deaths Related to Training, 2001-2010". National Fire Protection Association. Accessed online at nfpa.org.

Chapter 5
Training Safety

The goal of any training division is to prepare firefighters to prevent, respond to, and mitigate emergencies. Firefighters need training for basic medical and trauma emergencies, vehicle extrication, the many kinds of fire fighting, natural gas leaks, and more. This range of emergency response usually means entry-level firefighters gain a basic level of knowledge and skills in recruit academies for the most common types of emergencies. Less common emergencies, like airport crash fire and rescue, technical rescue, and hazardous material response, require a higher skill level. The increased medical training required to become a certified paramedic and new frontiers in community paramedicine are available to more experienced firefighters.

The task of explaining to fiscal decision makers why equal or more training is needed for fewer high-hazard call responses is difficult. This is especially true when results from NFPA Fire Analysis and Research of U.S. firefighter injuries (U.S. Firefighter Injuries, 2016) show that firefighters have the lowest number of injuries per emergency call than any time in the prior 35 years. Yet, the same research indicates that the rate of firefighter injuries of 24 per 1,000 fires has remained relatively unchanged for the last three decades (21.6 in 2016). As more fiscal resources are sought for training, training injuries have increased as a percentage of all firefighter injuries, from 6.9 percent in 1981 to 11 percent in 2015. Fire departments are challenged to gain the resources to provide realistic training that translates into safer fireground operations without injuring firefighters during the training activities **(Figure 5.1)**.

Figure 5.1 Live fire training procedures should be in accord with NFPA standards.

Today's firefighters need to spend more time training for high-risk emergencies that are less likely to occur and keeping credentials current with local, state, and federal regulatory agencies. Firefighters need training to be safe and avoid injury. Fire departments are challenged to find the fiscal resources and talented instructors to meet these training needs.

Training-Related Injuries and Fatalities

NFPA analysis and research for 2015 (NFPA Firefighter Injuries, 2016) showed 7,560 injuries were recorded during training.

A 2015 NFPA Fire Analysis and Research Study on firefighter injuries said almost 60 percent of training injuries fall in the category of strains, sprains, or muscular pain. Minor wounds (cuts, bleeding, and bruises) account for about 19 percent of training injuries.

The correlation between training injuries and fatalities is very similar to injuries and fatalities experienced on the fireground.

Reducing injuries in training should result in fewer injuries on the fireground. Training Officers must ensure firefighters are physically fit, strong levels of accountability are maintained with consideration for work/rehab cycles, and the integrity of designated teams is maintained before, during, and after training events. Training divisions must make sure the level of PPE, incident command, safety, accountability, and rehab are equal to or better than actual frontline responses. This ensures all members understand the optimum level of performance.

Types of Training Required

Training requirements fall into three categories: entry-level training, specialized training, and recertification training. The initial training in each category is provided by the organization's training division or a state, regional, provincial, or national training agency. For many skills, company- or unit-level training will be continual and will occur on a schedule established by the organization to fulfill state/provincial or national mandates.

Entry-Level Training

In states and provinces that conform to the NFPA professional qualifications standards, entry-level or recruit training will provide the knowledge and skills required for Fire Fighter I certification. Some municipalities may also include requirements for Fire Fighter II in their training curriculum, resulting in longer training time. All basic firefighters must complete the Awareness and Operations Levels of hazardous materials (hazmat) training as described in NFPA 472. Many organizations include either a basic first responder course or Emergency Medical Technician-Basic (EMT-B) training as part of the entry-level course.

Recruit school introduces fire suppression tactics, and live-fire evolutions are critical to this training. Live-fire training is considered high-hazard training and has claimed the lives of a number of recruits, as well as veteran instructors. Prior to any live-fire training, all Fire Fighter I candidates must receive respiratory protection training. This training includes all types of respiratory protection equipment, especially self-contained breathing apparatus (SCBA) **(Figure 5.2)**.

Figure 5.2 Recruit training is structured to provide the new firefighter with the skills necessary to perform a variety of tasks safely, such as the use of respiratory protection equipment.

General Knowledge Requirements (NFPA 1001, *Standard for Fire Fighter Professional Qualifications*, 5.1.1) are essential to all entry-level training. These include: "the value of fire and life safety initiatives in support of the fire department mission and to reduce firefighter line-of-duty injuries and fatalities; the role of other agencies as they relate to the fire department; aspects of the fire department's member assistance program; the importance of physical fitness and a healthy lifestyle to the performance of the duties of a firefighter."

Other types of training provided in the entry-level course include the selection, donning, use, and care of personal protective equipment (PPE); hoseline selection; moving as a team in vision-obstructed environments; and firefighter self-rescue or survival training. Practical training evolutions are used to reinforce this training and provide an opportunity to practice the skills. Personnel who may engage in wildland firefighting must be trained and tested annually in the proper deployment of the fire shelter used by the organization.

Large industrial complexes, such as refineries or chemical plants, usually provide training for personnel who are expected to extinguish incipient fires or respond as part of the permanent industrial fire brigade **(Figure 5.3)**. Similar to the skills and training provided to career and volunteer firefighters, requirements for this type of training are included in the following sources:

- IFSTA manuals: **Industrial Emergency Services Training: Incipient Level, 2nd Edition** and **Industrial Exterior and Structural Fire Brigades, 1st Edition**.
- NFPA 600, *Standard on Facility Fire Brigades*
- NFPA1081, *Standard for Facility Fire Brigade Member Professional Qualifications*

Figure 5.3 Large industrial complexes usually provide training for industrial fire brigades.

Specialized Training

Depending on state/provincial or local policies, specialized training may be provided to personnel. Most specialized training creates specialists to meet specific needs or hazards. It includes:

- **Driver/operator training** — Apparatus driver/operator training may be provided to personnel who will be assigned or promoted to the position responsible for the operation of an ambulance or fire-fighting apparatus **(Figure 5.4)**.
- **EMT-Advanced** — Higher levels of medical training provided to meet local requirements.
- **Extrication** — Entry-level course covers basic vehicle or machinery extrication. More advanced training is required if the firefighter or paramedic is assigned to a vehicle that deploys specialized power extrication tools **(Figure 5.5)**.
- **Technical Rescue** — Technical rescue training may include any or all of the following types of rescue:
  - — Rope rescue **(Figure 5.6)**
  - — confined space
  - — Trench rescue
  - — Structural collapse
  - — Water or dive rescue
  - — Ice rescue

Figure 5.4 Driver/operator training is intended to provide firefighters with the skills needed to drive the apparatus, respond safely to emergency incidents, and operate the pumps and aerial devices mounted on the vehicle.

Figure 5.5 More advanced vehicle extrication training includes using specific power tools.

Figure 5.6 Rope rescue is included in technical rescue training.

- **Marine** — Training in shipboard fire fighting is provided for personnel assigned to marine vessels, fire department fireboats, and land-based units that protect harbor facilities and respond to shipboard emergencies.
- **Aircraft Rescue and Fire Fighting (ARFF)** — ARFF training is required by the U.S. Federal Aviation Administration (FAA) for all personnel assigned to certain airports and other personnel who may respond to aircraft-related incidents on airport property **(Figure 5.7)**.
- **Hazardous Materials** — Training required for units designated to respond to hazardous materials incidents. While fire department personnel must meet the Awareness or Operations Levels training, hazmat team personnel must be trained and certified to Technician Level.

Figure 5.7 Aircraft rescue and fire fighting training for personnel who respond to or are stationed at airports in the U.S. is mandated by the Federal Aviation Administration.

Recertification Training

Fire and emergency services organizations are required by the federal and state/provincial government regulations to provide refresher and recertification training in a variety of topics. These include respiratory protection (SCBA); blood-borne pathogens and infection control; hazmat; and medical emergencies. The number of annual training hours for each topic is also mandated.

Another recent requirement is to provide training to meet national credentialing criteria. The purpose of national credentialing is to ensure that personnel trained in one area of the country can respond to and provide assistance at an emergency anywhere in the country.

Ongoing Training

Both career and volunteer departments require continual in-service training. In volunteer and career departments, there is usually an agency-set schedule, using agency-approved lesson plans and periodic testing. In volunteer organizations, this training typically takes place on a weekly basis at evening or weekend drills. Career departments train during work shifts at the fire station, at a training facility, or within the response area **(Figure 5.8)**. All training should comply with the standards set out in NFPA 1401, *Recommended Practice for Fire Service Training Reports and Records*. Following this standard helps to protect the organization from litigation and maintains a level of professionalism for training.

Figure 5.8 Both career and volunteer departments train weekly so firefighters can maintain safety, work efficiently, and learn new skills. *Courtesy of Ron Jeffers, Union City, NJ.*

Training should occur when any type of new equipment is purchased and issued or when there is a change in protocol, procedures, or policies. This type of training is provided by training officers, manufacturer's representatives, the Health and Safety Officer (HSO), or company officers or crew members who have received initial training (sometimes referred to as train-the-trainer).

Train-the-Trainer

Training programs that help key members of the organization, such as company officers, so they can train other company personnel in a specific topic, procedure, or policy are often used as a means to reduce costs. An effective **train-the-trainer** program will teach trainers such skills as how to facilitate a meeting, how to help students who are having trouble, and how to teach effective note-taking.

Train-the-Trainer — Training program in which key members of the organization train other company personnel in a specific topic, procedure, or policy.

Company Training

Often training takes place on a more informal basis between members at the company level. More experienced members will assist newer members with the finer details of performing tasks. Reinforcement of recruit training is a very common opportunity for this type of training. This is sometimes called **On-the-Job Training** or OJT.

On-the-Job Training — System of training firefighters that makes full use of personal contact between firefighters and their immediate supervisor; trains firefighters both physically and psychologically for the position they will perform.

OJT is often a formal, recordable type of ongoing training that can be delivered as a training packet for probationary firefighters or through online delivery systems that are also formalized by training divisions. Overall, Company Officers are responsible for the safe conduct of company-level training in the field and need to maintain safe practices during company training **(Figure 5.9)**.

Figure 5.9 Company Officers ensure that safe practices are used during ladder training.

Creating a Safe Training Environment

An organization must provide a safe environment while making training as realistic as possible. Administrators, supervisors/instructors, and participants must know safety takes precedence over realism in all training. The role that each plays in creating a safe training environment is described in this section.

Administration's Role

The administration must provide the resources necessary for a safe training environment. An administration must understand local, state, and federal training requirements, and it needs to provide funding and time for the training. Additionally, the administration must store and maintain training records. Other responsibilities include the following:

- Leadership and commitment
- Risk management program
- Qualified and certified instructors
- Props, equipment, and acquired structures

Maintaining records of the type of training and who is giving the training are part of record management. This includes the subject being taught and the instructors' qualifications, hours of instructions, and results of any testing.

Leadership and Commitment

It cannot be overemphasized that safety depends on the commitment of the organization's leaders. From the department chief to the training officer(s), all officers must understand their responsibility for the safety of their employees.

They must be willing to change the culture, establish another based on the 16 Firefighter Life Safety Initiatives (FSLI), and nurture that culture to allow it to spread throughout the organization. They cannot permit unsafe acts to occur during training, and they must enforce safety policies and procedures.

Risk Management Program

To reduce the potential risks to personnel, all fire and emergency services organizations regulated by legally adopted NFPA standards are required to have a risk management plan. After the plan is implemented, it must be monitored for effectiveness. This plan is designed to accomplish the following objectives:

- Identify risks **(Figure 5.10)**.
- Evaluate the potential based on frequency and severity of the risk.
- Establish appropriate controls to minimize or eliminate the risk.

The risk management plan includes all job-related activities in which fire and emergency services personnel (emergency and nonemergency) normally participate, including emergency, non-emergency, training, and support activities. Chapter 11 describes the risk management plan process in more detail.

Figure 5.10 Exterior fire attack is sometimes an example of risk avoidance; if it is unnecessary to enter the structure, then doing so would be an unacceptable risk. *Courtesy of Bob Esposito, Pennsburg, PA.*

Qualified and Certified Instructors

The administration is responsible for preparing instructors to teach the organization's training courses. Lack of preparation can result in casualties and legal liability. Instructors must not be assigned to teach courses they are not trained to teach. Instructor preparation must include:

- **Instructor certification** — Instructors must be trained, certified, and authorized to teach the courses they are assigned.
- **Instructor-student ratio** —For high-hazard training, there should be no more than five students per instructor.

- **Safety officers and technical safety officers** — Incident safety officers must be provided for all high-hazard training. Technical safety officers, who are regionally or nationally certified in the topic, must be present for technical-type training. High-angle rescue is an example of high-hazard training.
- **Instructional materials and equipment** — Instructors must be provided with the appropriate tools and equipment to teach the course.

Figure 5.11 Burn buildings and props can be very complex and include safety shutoffs and temperature sensors that must be inspected regularly and prior to use.

Figure 5.12 A firefighter does regular maintenance on power tools to ensure safe use.

Props, Equipment, and Acquired Structures

The administration must budget for and provide the props, tools, equipment, and safety materials needed for training. They must also provide live-burn buildings or acquired structures that meet the requirements of NFPA 1403. Other key elements include:

- **Props** — Props simulate some type of situation or equipment that may be encountered during an emergency incident. Props range from gas-fired, live-burn structures to wrecked vehicles **(Figure 5.11)**. Some props are designed and purchased from companies that specialize in engineering fire behavior props. Others are created locally, such as mazes used to simulate rescue situations.

NOTE: Constructing props or altering commercially purchased props places a legal liability on firefighters and the organization. Altering a commercial fire behavior prop without approval from the manufacturer will not only create liability, it will also void the warranty.

- **Maintenance** — All training props, tools, and equipment must be inspected before/after use, and periodically to ensure that they work properly **(Figure 5.12)**. Inspection records must be maintained.
- **PPE** — Some fire departments keep older turnout gear for nonfire use only. This avoids unnecessary damage to personal turnout gear during training. Members use their personal turnout gear during live burns. In accordance with the manufacturer's recommendations, personal protective equipment used by participants and instructors must be inspected and cleaned after each use **(Figure 5.13)**. The administration and supervisors must not tolerate the use of contaminated PPE during training. This provides a negative message to participants, especially recruits.
- **Safety information** —The HSO is generally responsible for providing safety training. This position must be given the resources to provide safety training on the topics of hearing and eye protection, PPE care, and infectious disease control.

Instructor's Role

Fire and emergency services personnel work in high-risk environments. Although it is impossible to remove the risks, it is possible to reduce them and increase safety. Through the efforts of fire and emergency services organizations and instructors, agencies can reduce risks through the following practices:

- Provide a safety briefing prior to activities with inherent risks **(Figure 5.14)**.
- Empower all participants with the responsibility to call a STOP when an activity feels unsafe.
- Teach the safest methods for accomplishing skills and tasks.
- Adhere to safety regulations and standards both in training and at emergency incidents.
- Establish and adhere to the National Incident Management System/ Incident Command System (NIMS/ICS).
- Use a personnel accountability system.
- Investigate, analyze, and correct unsafe activities, behaviors, and near misses.
- Establish safety policies and procedures for use in all work-related activities.
- Establish and use appropriate rehabilitation based on NFPA 1584, *Standard on the Rehabilitation Process for Members During Emergency Operations and Training Exercises.*
- Conduct a brief safety critique at the end of every training drill.

In addition to the practices above, the instructor's role is to:

- Provide students with a positive, proactive safety role model.
- Fulfill the duties of the Incident Safety Officer (ISO) when qualified and assigned.

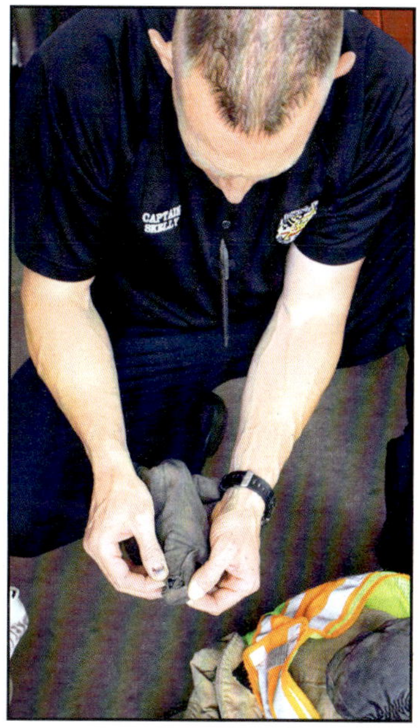

Figure 5.13 All PPE, including gloves, must be inspected and cleaned after each use.

Figure 5.14 An instructor conducts a safety briefing before training.

The most important thing to remember is that safety must be taught and practiced during training. **(Figure 5.15)** Safety should become automatic to the firefighter or emergency responder during daily activities. By requiring safety training and applying the use of safe practices, fire and emergency services personnel will be better prepared to apply those practices during emergency incidents.

Instructors play significant roles in preventing injuries and fatalities during training evolutions. As representatives of the department's leadership, they must hold the safety of training members in the highest regard. The organization must support and enforce safety, fitness, and health and wellness programs across its operations. It should commit to taking the following responsibilities:

- Ensure that all apparatus and equipment are appropriately maintained.
- Ensure that all safety equipment is properly installed and operating.
- Provide policies and procedures for the safe use of the apparatus and equipment.
- Address the fitness, health, and wellness of personnel.

Figure 5.15 SOP/Gs should require a safety officer to be present to document and evaluate any live-fire training evolution.

The NFPA report, *U.S. Firefighter Deaths Related to Training, 2001-2010,* recommends paying particularly close attention to the following types of training:

- **Physical Fitness Training** — With most training deaths being related to the health of the worker, it is important for fire departments to adopt and follow NFPA 1582, *Standard on Comprehensive Occupational Medical Program for Fire Departments.*
- **Live Fire Training** — Follow recommendations outlined in NFPA 1403, *Standard on Live Fire Training Evolutions.* Departments should develop a permitting procedure for live-fire training conducted at acquired structures that is similar to confined space entry permitting used universally in the U.S. All requirements must be met before issuing the permit and conducting the burn.

- **Water Rescue Training** — Numerous training accidents have occurred in water rescue and dive team training. NFPA 1006, *Standard for Technical Rescue Personnel Professional Qualifications* and NFPA 1670, *Standard on Operations and Training for Technical Search and Rescue Incidents*, provide guidance on conducting training and qualification levels for fire department responders.

Participant's Role

Firefighters must take responsibility for their own safety. They must consciously perform their roles during training in a safe manner. Some rules that help to ensure safety in training are as follows:

- Follow the organization's SOPs.
- Follow all directions and orders.
- Provide feedback to the instructor if something seems unsafe.
- Report unsafe acts.
- Do not participate in horseplay.
- Inspect, clean, and care for PPE according to manufacturers, recommendations.
- Report all injuries no matter how small.
- Actively participate in post-training critiques.

Practical Training Evolutions

Fire and emergency responder training is divided between cognitive and psychomotor training. Cognitive training may occur in the classroom through computer-based programs or distance-learning courses. Cognitive training provides the knowledge necessary to make sound decisions and respond quickly to changing conditions **(Figure 5.16)**.

Figure 5.16 Fire department personnel engage in a distance learning course.

Psychomotor training involves physical skills. It occurs during practical training evolutions at the training center, at the fire station, or in the field. These evolutions are intended to be as realistic as possible, and as safe as possible. When physical skills are learned, they must be reinforced by frequent practice so they will become second nature to firefighters.

To be safe and effective, practical training evolutions must include the following:

- **NIMS/ICS** — In all high-hazard practical training evolutions, the locally adopted version of the National Incident Management System/Incident Command System (NIMS/ICS) must be established. The use of the ICS not only ensures greater control of the training evolution, but also reinforces the need for it at emergency incidents. Qualified and experienced Chief Officers should serve as Incident Commanders and trainers for IC trainees during training evolutions.

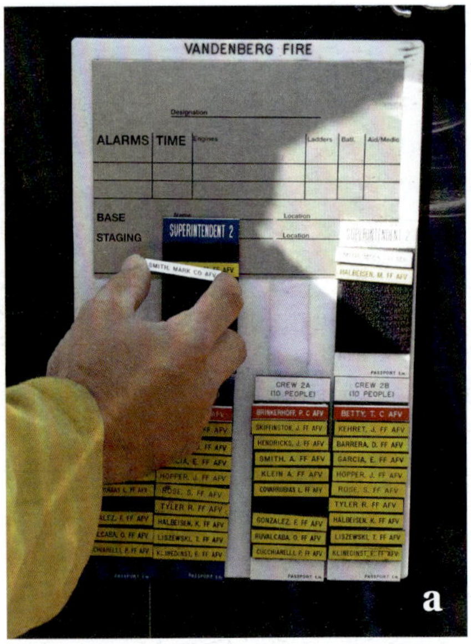

Figure 5.17a An accountability system helps the incident commander keep track of personnel at the incident. Systems must be used in training to reinforce their use at actual incidents.

Figure 5.17b Boards for attaching identification tags are common tools used for tracking firefighters at a scene. *Courtesy of Mike Wieder.*

- **Accountability system** — An essential element of the ICS is the establishment and use of the personnel accountability system. Using an accountability system ensures that the location and actions of all personnel involved in the practical training evolution is known **(Figure 5.17a and b)**.
- **Incident Safety Officer (ISO)** — An ISO must be assigned to all high-hazard training evolutions. If the evolution is large and involves multiple units spread over a wide area, assistant ISOs must be assigned.
- **Communications protocol** — The locally adopted communications protocol must be used during all practical training evolutions. A separate radio frequency is usually provided for all units involved in the evolution. Officers should designate the method by which firefighters can call for termination of the drill due to an actual emergency.
- **Personal Alert Safety System (PASS)** — Many fire agencies purchase and issue PASS devices to all personnel. PASS devices may be individual units or integrated into SCBA units **(Figure 5.18a and b)**. Unfortunately, injuries and fatalities have occurred because personnel have not activated their PASS devices prior to entering hostile atmospheres. The repeated use of the devices in training can help to make their use during emergencies more likely.

Personal Alert Safety System (PASS) — Electronic lack-of-motion sensor that sounds a loud alarm when a firefighter becomes motionless. It can also be manually activated. Also known as Personal Alert Device.

- **Proper respiratory protection** — Training in the selection and use of respiratory protection equipment reinforces the need to sample unknown atmospheres and determine the correct type of protection needed.
- **Firefighter rescue training** — Because firefighter rescue conditions are rare, training must be provided as often as possible. Self-escape or survival techniques must be developed and training must be provided so the actions will be instantaneous. Terms generally used to describe conditions that require rescue include Mayday, firefighter down, or firefighter needs assistance.

Figure 5.18a A hand held PASS unit.

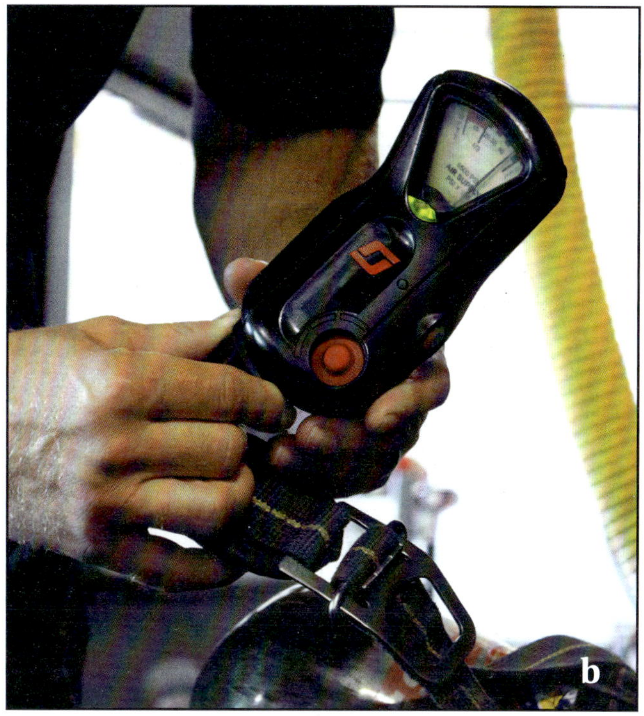

Figure 5.18b A PASS unit that is attached to PPE.

Other safety issues must also be addressed when performing high-hazard practical training evolutions. Some of these have been mentioned previously but are worth reinforcing. They include the following:

- **Live-fire certified training officer** — Many states/provinces now require that a live-fire trained and certified training officer be present at all live-fire training evolutions.
- **Training officer certified for skill** — Training officers at high-hazard training evolutions should be trained and certified or qualified in the specific skill they are teaching.

Training evolutions must also adhere to:

- **NFPA 1403 and 1500 standards** — Both NFPA 1403 and 1500 ensure fire fighter and emergency responder safety. They help only if organizations, supervisors, and participants adhere to these standards during training evolutions.
- **Occupational Safety and Health Administration (OSHA) requirements for respiratory protection** — All participants must be trained and certified to use respiratory protection equipment prior to any training evolution that involves immediately dangerous to life or health (IDLH) atmospheres, including those that may have low oxygen content. This certification includes a required medical examination and facepiece fit testing.
- **OSHA confined-space training requirements** — Confined-space training evolutions may require permits from the facility owner prior to some types of training.
- **OSHA infection-control requirements** — Medical training must meet infection-control requirements.

- **NFPA medical fitness requirements** — Prior to participating in any high-hazard training evolutions, all participants, including instructors, must meet the adopted physical fitness and medical requirements.
- **29 CFR 1910.147 lock-out/tag-out requirements** — When training in a working facility around stored energy or hazardous equipment, the facility's policies must be followed.
- **The organization's hearing conservation policies** — Hearing protection must be used during all training activities that produce excessive levels of noise (greater than 90 dB).

NOTE: When filming video for training purposes, it is important to have an additional instructor on scene. The individual doing the filming is often so focused on the subject of the video that it is impossible to see the big picture.

Chapter Review

1. Why can it be difficult to justify the need for more training to decision makers?
2. What types of training are included in the entry-level fire fighting course?
3. What rules for training participants can help ensure safety during a training evolution?
4. What requirements must training evolutions adhere to?

Discussion Questions

1. Think about the geography, industries, and general makeup of your community. What types of specialized trainings might local firefighters have to have in order to best serve the needs of the area?

Fire Apparatus Safety

Chapter Contents

chapter 6

Key Terms

FESHE Learning Outcomes

Occupational Safety and Health for Emergency Services

2. Identify occupational health and safety programs for industry and emergency services today.
6. Apply the knowledge of an effective safety plan to preincident planning, response, and training activities.
12. Identify and analyze the major causes involved in line-of-duty firefighter deaths related to health, wellness, fitness, and vehicle operations.

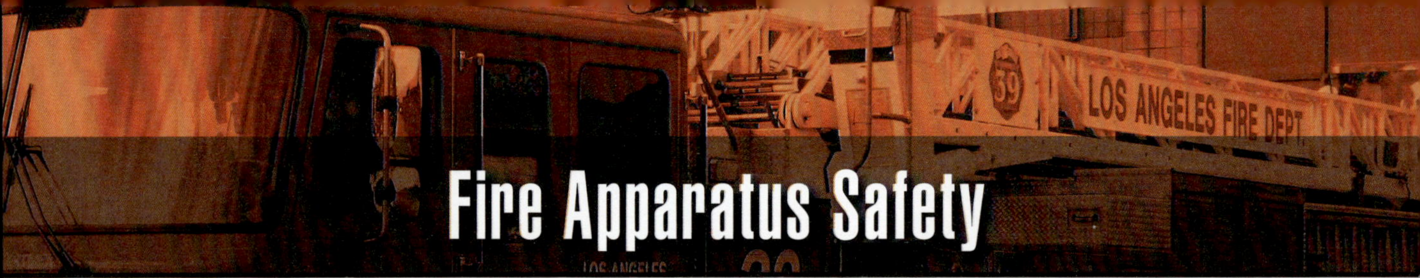

Fire Apparatus Safety

Learning Objectives

After reading this chapter, students will be able to:

1. Identify common causes and effects of apparatus accidents.
2. Recognize the legal requirements that govern apparatus design.
3. Identify apparatus safety components.
4. Review apparatus maintenance requirements.
5. Describe NFPA, state/provincial, and training requirements for fire apparatus driver/operators.
6. Identify operational concerns for in-motion apparatus and/or privately owned vehicles (POVs).
7. Describe how apparatus placement during different types of safety response can maximize safety.
8. Identify additional general apparatus safety concerns.

Chapter 6
Fire Apparatus Safety

Emergency service workers save lives and property and they protect the environment. According to the National Fire Protection Association, in 2015 firefighters responded to almost 34 million emergency and nonemergency incidents. These incidents required them to apply their knowledge and skills safely. That may seem easy, but unsafe acts by responders occur. Whether those unsafe acts are intentional or not, everyone at an incident must be responsible for their safety and the safety of those around them.

Emergency responders cannot perform their duties at an incident if they are involved in an accident while responding. The NFPA reported that fatalities from apparatus accidents exceed deaths that occur during emergency incidents. Accidents during responses eliminate the involved unit and require additional resources to assist at the accident scene. Emotionally, the responding accident stresses other emergency responders concerned with the safety of friends and coworkers.

Apparatus safety depends on some controllable factors:

- **Driver skill**. The driver/operator must be trained in the safe operation of the vehicle during emergency and nonemergency operations. The driver/operator must also be trained to operate the vehicle safely in a variety of weather and traffic conditions.
- **Personal factors**. Personal factors, including situational awareness, internal awareness, and physical fitness, also contribute to the driver/operator's ability to safely operate the vehicle. Departments should employ simulation technology for driver/operator training if possible. This technology provides the ability to present unusual and dangerous situations to drivers without risking an actual accident. However, there is no complete substitute for operating the type of apparatus that the driver/operator will be assigned to drive.
- **Condition of apparatus**. The mechanical condition of the apparatus contributes to safe operation. The organization must provide a maintenance program that ensures the vehicle is in safe operating condition at all times **(Figure 6.1)**.

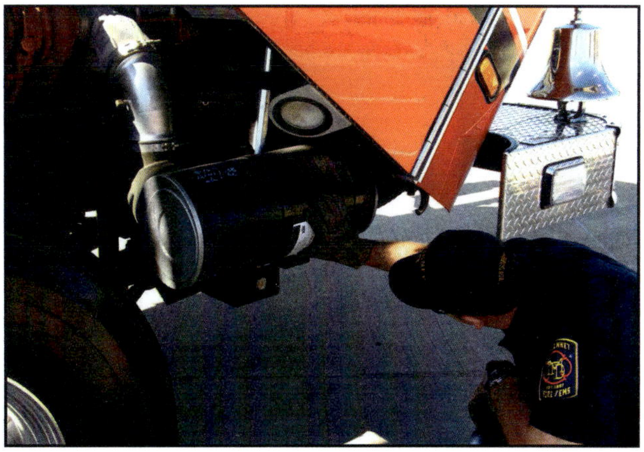

Figure 6.1 Driver/operators must be trained in the preventive maintenance and inspection of their vehicles.

- **Design criteria.** Apparatus must meet established design criteria for emergency and nonemergency vehicles. When the fire and emergency services organization creates purchase specifications for new vehicles, the specifications must meet or exceed NFPA and U.S. Department of Transportation (DOT) or Canadian requirements.
- **Standard operating procedure (SOP)**. The organization must establish, implement, and enforce a standard operating procedure for all vehicles. The policy must cover all aspects of operation for department-owned vehicles. It must take into consideration how the vehicle is driven and how passengers ride in the vehicle, including state and local laws regarding privately owned vehicles (POVs) during emergency responses.

This chapter will help organization administrators develop a preventive maintenance program and write specifications for new apparatus. It also provides supervisors with information for enforcing safe apparatus operation during all types of situations. Finally, it points out safety-conscious behaviors department members may use to operate and maintain their vehicles. Increased awareness and safety through the control of these factors will ensure the safe arrival of fire fighting units at their destination.

Apparatus Accidents

The public expects that fire apparatus driver/operators will efficiently and safely drive their vehicles. Accidents erode public confidence. Lawsuits may be filed to seek money for injuries or property damages. Some instances may result in individual criminal charges. A lawsuit can have a devastating and lasting effect on involved personnel, their families, any civilians involved, and the financial condition of the fire and emergency services organization.

Figure 6.2 Accidents involving apparatus traveling to or from an emergency incident account for the second-highest number of firefighter deaths. *Courtesy of Ron Jeffers, Union City, NJ.*

Apparatus accidents occur at a rate of about 15,000 per year, with 16,600 in 2015, according to NFPA statistics. The NFPA reports that, during the 30 years between 1977 and 2007, 406 firefighters died in vehicle-related accidents **(Figure 6.2)**. This figure constituted the second-highest category of firefighter fatalities during that period. Responding to emergency calls in personal vehicles resulted in 37 percent of the deaths.

The personnel who died fell into the following categories:

- 75 percent volunteer firefighters
- 14 percent career personnel
- 11 percent contract employees

An NFPA study determined that over 75 percent of the victims were not wearing a seat belt or other type of restraint. Responsibility for these types of fatalities must be shared by the organization's administration, supervisors, and individual firefighters. Administrators are responsible for providing seat belts or restraints in the vehicles, and for having a consistent and enforceable policy for seat belt use. Supervisors must ensure all personnel in the apparatus secure their seat belts before an apparatus responds. Firefighters are responsible for using the seat belts in the apparatus and in their vehicles **(Figure 6.3)**. Most states and all Canadian provinces have seat belt laws.

Figure 6.3 Firefighters are required to put on seat belts before riding in an apparatus.

In addition to firefighter fatalities and injuries, civilians died in collisions with fire apparatus. According to the U.S. National Highway Traffic Safety Administration (NHTSA), 73 people died as a result of collisions with emergency vehicles in 2012.

A final consideration for the effect of emergency vehicle accidents is the cost of the accident itself. Accident-associated expenses may include the following:

- Vehicle repairs
- Vehicle replacement
- Insurance claims
- Medical benefits
- Court costs
- Replacement personnel

Most apparatus accidents are preventable, yet their number remains high. So critical is the need to reduce the number of these incidents that in 2014, the U.S. Federal Emergency Management Agency (FEMA) issued its *Emergency Vehicle Safety Initiative* (FA-336). It is the culmination of a study of 12 fire and emergency services organizations and their approaches to apparatus safety. The results, ranging from the design of apparatus to the operation during responses, provide ways to reduce apparatus accidents. This document is available on the United States Fire Administration (USFA) website.

In the **Pumping Apparatus Driver/Operator Handbook**, IFSTA suggests the causes of all fire apparatus crashes can be grouped into five categories:

1. **Improper backing of the apparatus** — While they are seldom serious in terms of injury or death, backing crashes account for a significant portion of overall damage costs.
2. **Reckless driving by the public** — This category includes a variety of reckless actions, including failure to obey traffic signals, excessive speed, and failure to yield to emergency vehicles.
3. **Excessive speed by the fire apparatus driver/operator** — Excessive speed may result in the driver/operator losing control of the vehicle, being unable to stop before hitting another vehicle/object, or leaving the roadway.
4. **Lack of driving skill and experience by the fire apparatus driver/operator** — A driver/operator may have insufficient training or be unfamiliar with the vehicle.
5. **Poor apparatus design and/or maintenance** — Although poor design is seldom the case with custom-built fire apparatus, it can be a significant problem in departments that use retrofitted or home-built apparatus. A lack of adequate maintenance affects all types of apparatus.

Of these causes, only one — reckless driving by the public — is completely out of the control of the fire and emergency services, the supervisor, and the firefighter. However, good defensive driving skills on the part of the driver/operator and situational awareness by the supervisor and crew members can reduce the risk for an accident of this type **(Figure 6.4, p. 182)**.

Figure 6.4 Before driving fire apparatus under emergency conditions, all driver/operator candidates must complete a training program.

Apparatus Safety Design Requirements

Fire apparatus safety begins with the vehicle's design. Whether it is a custom-made ambulance, an aerial device or a commercially designed pickup, van, or car, the vehicle must meet federal safety design requirements. Several organizations or government agencies stipulate the minimum requirements for the design of motor vehicles. These will be discussed in the following section.

Even though vehicles are constructed to meet requirements, wear can cause them to become unsafe. The department must have a maintenance program throughout the service life of the vehicle. The maintenance program should consist of both a preventive element and a corrective element.

Fleet managers and company officers must resist the pressure to keep apparatus in service past preventive maintenance cycles in order to meet emergency response demands. Having strategically placed and reliable reserve apparatus is one way to resist this pressure. Reliable reserve apparatus fleets support apparatus safety designs by ensuring the front-line fleet is serviced properly.

Warning devices used by emergency vehicles decrease the likelihood of an accident by making their presence more obvious **(Figures 6.5 a-d)**. Updated requirements and new technology continue to improve the design and performance of audible and optical warning devices.

Legal Requirements

Motor vehicle design is regulated to some extent by the U.S. or Canadian governments. Following an alarming rise in vehicle-related fatalities, the National Traffic and Motor Vehicle Safety Act was enacted in the U.S. in 1966, giving the federal government authority to set and enforce safety standards for vehicles and roads. Headrests, seat belts, and laminated safety glass in windshields were changes that followed this mandate. As time passed and technology improved, impact-absorbing bumpers and driver and passenger airbags were required. Many of these requirements have found their way into all types of emergency vehicles.

NFPA 1901

The earliest requirements for the design of fire apparatus were created by the NFPA in 1906. Since then, NFPA standards have expanded to cover all types of fire apparatus, including pumpers, aerial devices, wildland apparatus, and aircraft rescue and fire fighting (ARFF) apparatus.

In response to firefighter fatalities and injuries in apparatus-related accidents, the NFPA continued to develop safety requirements for apparatus. Personnel were prohibited from standing on the tailboard or running board of moving apparatus, and fully enclosed crew compartments were added. The adoption of the 1991 edition of NFPA 1901, *Standard for Automotive Fire Apparatus*, saw the greatest advances in apparatus safety, requiring Vehicle Data

Figure 6.5 a-d A variety of optical and audible warning devices are required on emergency apparatus including: a) roof-mounted light bars; b) directional arrows on the rear end; c) air horns mounted in or under the front bumper; and d) sirens mounted in or on top of the front bumper.

Recorders and audible seat belt warning systems for all apparatus. Requirements for additional optical and audible warning devices, reflective trim, and improved braking systems improved safety. To reduce firefighter hearing loss, a maximum noise level was established for the cab and crew compartment. This requirement removed air horns and siren speakers from the cab roof and relocated them to the front bumper.

The 1991 edition of NFPA 1901 has been recognized as the benchmark from which improved apparatus have evolved. Only apparatus that meet that or more recent editions of NFPA 1901 or ones refurbished in accordance with NFPA 1912, *Standard for Fire Apparatus Refurbishing*, are recommended for use in emergency service.

NFPA 1451, *Standard for a Fire and Emergency Service Vehicle Operations Training Program*, continues to enhance the capabilities of drivers for all apparatus training in defensive driving techniques, backing skills, and driving in all types of weather.

Many of these standards have been adopted into law by the U.S. Department of Transportation making them mandatory for the design of all fire apparatus manufactured in the United States. The U.S. Code of Federal Regulations (CFR) contains regulations for the design of all types of motor vehicles in 49 CFR 393, *Federal Motor Carrier Safety Regulations* (FMCSR), and 49 CFR 571,

Federal Motor Vehicle Safety Standards (FMVSS). These federal government requirements are enforced by the DOT. The regulations govern such design components as the gross vehicle weight (GVW), braking power, warning devices or lights, and number of axles.

The Federal Aviation Administration (FAA) enforces regulations detailing minimum requirements for aircraft fire rescue apparatus based on NFPA 414, *Standard for Aircraft Rescue and Fire-Fighting Vehicles*. In addition, NFPA 1500, 1901, and 1912, among others, were adopted by the U.S. Department of Homeland Security in September 2007.

State departments of transportation also have laws that affect apparatus design. For instance, some states may require additional axles to carry the increasing weight of fire apparatus. Others may regulate the use of engine-retarding brake systems or the color of warning lights. Some state regulations, such as those enacted in California, are stricter than federal laws.

Federal Star of Life Ambulance Specifications

The U.S. General Services Administration (GSA) sets design standards for ambulances. In the Federal Specification for the Star-of-Life Ambulance, an ambulance that displays the "Star of Life" symbol must be compliant with the criteria in this document. Other standards and recommended practices form part of this specification, including those from the Society of Automotive Engineers (SAE) and the National Truck Equipment Association (NTEA). This document establishes five classifications of ambulances, which include the following **(Figure 6.6 a-e)**:

- **Type I Ambulance (10,001 to 14,000 Gross Vehicle Weight Rating (GVWR)** — Type I vehicle shall be a cab chassis furnished with a modular ambulance body.
- **Type I-AD (Additional Duty) Ambulance (14,001 GVWR or more)** — Type I-AD shall be a Cab-Chassis with modular ambulance body, increased GVWR, storage, and payload.
- **Type II Ambulance (10,000 and under GVWR)** — Type II ambulance shall be a van with integral cab-body.
- **Type III Ambulance (10,001 to 14,000 GVWR)** — Type III shall be a cutaway van with integrated modular ambulance body.
- **Type III-AD (Additional Duty) Ambulance (14,001 GVWR or more)** — Type III-AD shall be a cutaway van with integrated modular body and increased GVWR, storage, and payload.

NOTE: Canadian ambulances are built to the Star-of-Life standard, although no national requirement exists.

Refurbishment

NFPA 1912, *Standard for Fire Apparatus Refurbishing*, specifies the minimum requirements for refurbishing apparatus. Recognizing that small career and volunteer fire departments may not be able to afford a complete overhaul of an aging apparatus, NFPA has created two levels of refurbishment. The standard requires that safety requirements of NFPA 1901 and 1906, *Standard for Wildland Fire Apparatus*, be included on either Level I or II refurbishments.

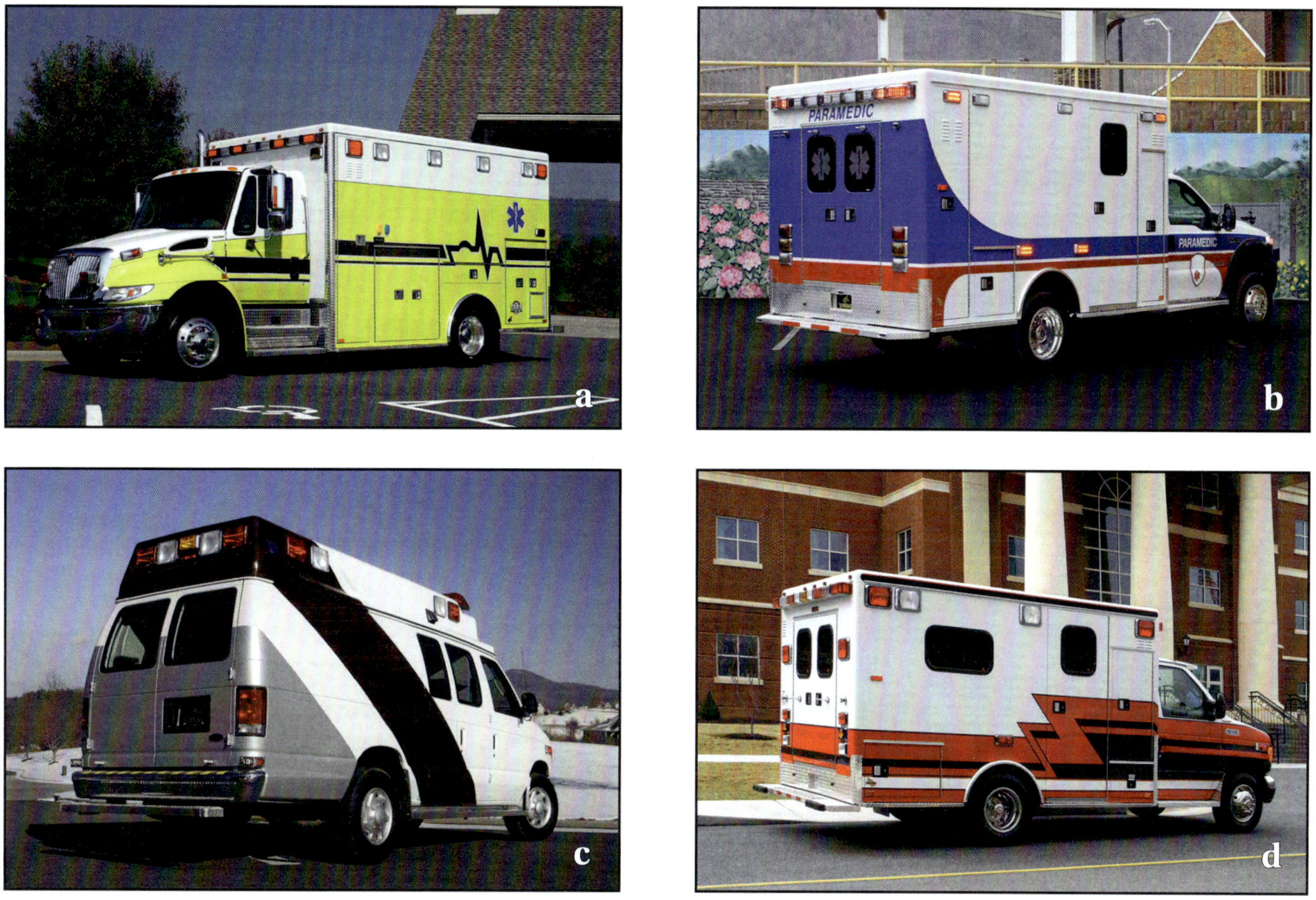

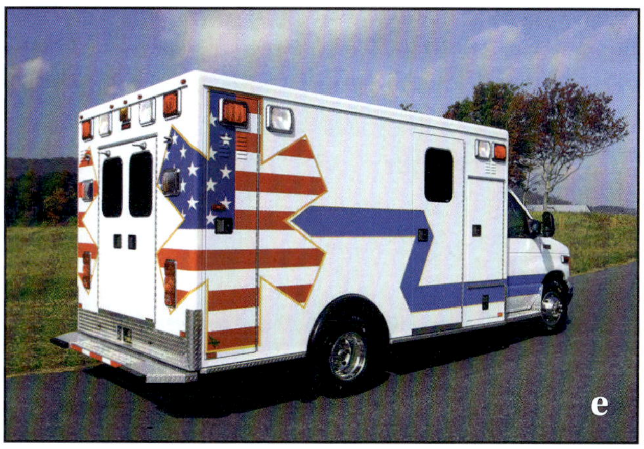

Figure 6.6 a-e Federal specifications classify five types of ambulances: a) Type I; b) Type I-AD; Type c) Type II; d) Type III; e) Type III-AD. *Courtesy of American Emergency Vehicles (AEV).*

A Level I refurbishment describes the requirements for the assembly of an apparatus using the following new components:

- Chassis frame
- Driving and crew compartments
- Front axle
- Steering and suspension components

The remainder of the apparatus may be constructed from new or existing components, such as a pump, an aerial device, a hose bed, tool compartment, and/or a rear axle. A vehicle that receives a Level I refurbishment actually receives a new make and model Certificate of Origin.

The Level II refurbishment consists of adding new components to the existing apparatus chassis. The new components, such as front axles, engines, or crew compartments, must meet or exceed the requirements for the original manufacture of the apparatus. A Level II refurbishment will retain its original title.

Fire and emergency services managers must remember that attempting to build apparatus without professional design help can lead to illegal and unsafe results. Fatalities and injuries have occurred as a result of vehicle rollovers of homemade water tenders (tankers), which were often former flammable liquid carriers. Consult an apparatus design professional before attempting to modify any vehicle for fire fighting operations.

Safety Components

When writing specifications for a new or refurbished apparatus, it is essential to consider the types of safety components required by law. At the same time, some safety equipment not mandated by law may be worth the additional funds necessary to install it. The following safety components should be included on all emergency response vehicles and apparatus:

- Audible warning devices
- Optical warning devices
- Passive visibility markings
- Rearward- and forward-facing video cameras
- Electronic data logging devices
- Position tracking devices
- Traffic preemptive control devices
- Auxiliary braking systems
- Antilock braking systems
- Bluetooth and/or Wi-Fi

Ergonomics — The applied science of workplace equipment design intended to maximize productivity and minimize potential injuries by reducing operator fatigue and discomfort.

Apparatus design should incorporate **ergonomics**. As storage compartments are located higher on the apparatus, and tools and equipment become heavier and larger, it is essential to plan for the safest location of this equipment. Skeletal and muscular injuries often result when personnel are removing tools and equipment from apparatus compartments. Apparatus manufacturers are designing apparatus that address ergonomic factors by lowering hose beds, providing steps for cab and body access, and adding pullout compartments for improved access. This issue becomes even more significant when people of smaller stature are hired by fire and emergency service organizations.

CAUTION

Firefighters must not disable emergency vehicle warning signals. The signals are there for safety reasons.

Audible Warning Devices

Two safety considerations are important when considering the installation of audible warning devices **(Figure 6.7)**. First, the device must be loud enough that the public can hear it over other ambient noises and inside vehicles with closed windows. The federal standard for ambulances states that the minimum sound level at 10 feet (3 m) must be 123 dB. The device must also project

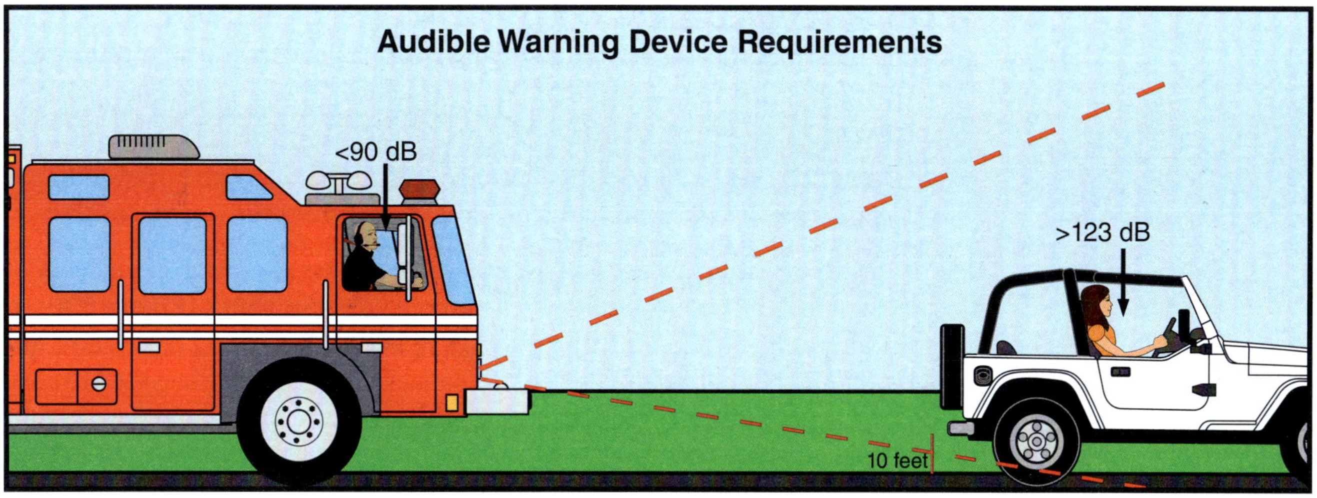

Figure 6.7 Audible warning devices must generate a minimum of 123 dB at 10 feet (3 m) while projecting the sound at an angle to be heard by vehicles approaching from the side and create a noise level not to exceed 90 dB in the cab of the apparatus.

its tone at an angle that is wide enough to be heard by drivers in cars that are not directly in front of the emergency vehicle; that is, the warning must be audible to vehicles approaching the emergency vehicle's line of travel from a side road. Finally, the tones used by the audible warning device must be easily associated with an emergency vehicle.

The second safety consideration is for the driver/operator and passengers of the emergency vehicle. The tones emitted by the audible warning device must not be so loud within the cab that they cause temporary or permanent hearing loss. OSHA has established a maximum sound level of 90 dB inside the cab of fire apparatus. The federal ambulance standard states that the maximum sound level in the cab must conform to 29 CFR 1910.95 unless the state or municipality changes the level. In the patient area, the maximum is 80 dB.

NOTE: In some jurisdictions, the use of air horns and sirens may be prohibited or restricted based on the time of day or location.

To meet these sound levels and to protect the hearing of crew members, NFPA 1901 prohibits mounting audible warning devices on the roof of the apparatus cab. Air horns and siren speakers may be located on the frame beneath the bumper, in the bumper when openings are provided, on the bumper, or on the front fenders.

Studies say civilian drivers respond better to sounds that frequently change pitch. Short bursts with the air horns and the constant up-and-down oscillation of a mechanical or electronic siren are the surest ways to catch a driver's attention. At speeds above 50 mph (80 km/h), an emergency vehicle may overtake the effective range of its audible warning device. One study indicated that a siren operating on an emergency vehicle moving at 40 mph (65 km/h) can project 300 feet (90 m) in front of the vehicle. At a speed of 60 mph (95 km/h), however, the siren is only audible 12 feet (3.6 m) or less in front of the vehicle. For this reason, some departments prohibit the use of lights and sirens while responding on freeways at speeds greater than 55 mph (86 km/h). Driver/operators must drive within the effective range of their audible warning devices.

Optical Warning Devices

Optical warning devices used on emergency vehicles have two primary purposes:

1. To alert other drivers and pedestrians of the approach of an emergency vehicle.
2. To indicate the location of the vehicle when it is stopped at the incident scene.

Optical warning devices consist of flashing lights mounted on the body of the vehicle. NFPA 1901 divides each vehicle into four zones **(Figure 6.8)**. In addition to the optical warning devices, headlights, parking lights, and tail lights can be connected to a device that causes them to flash when the primary optical warning devices are activated. Consult state/provincial laws before configuring these devices.

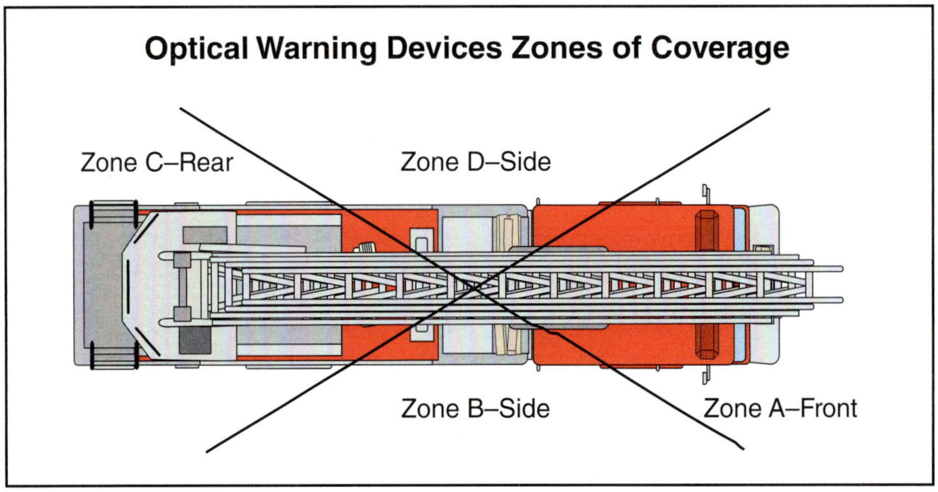

Figure 6.8 NFPA 1901 has designated four zones around the apparatus that have specific visual warning light coverage.

Figure 6.9 Studies have indicated the effectiveness of retroreflective stripes on apparatus.

Passive Visibility Markings

To improve the visibility of emergency vehicles when they are at an incident, NFPA also requires the application of retroreflective trim on the vehicle body **(Figure 6.9)**. This trim must meet the performance requirements listed in ASTM D4956-17, *Standard Specification for Retroreflective Sheeting for Traffic Control*, Section 6.1.1 for Type I sheeting. NFPA says retroreflective trim must extend across a minimum of 25 percent of the front of the vehicle and 50 percent of the sides and rear. The trim must be a minimum of 4 inches (100 mm) in width. The retroreflective trim should also be applied to the inside of cab doors so they will be visible when open at night.

The choice of vehicle or apparatus colors has been a topic of discussion for many years. The traditional *Fire Engine Red* continues in use throughout much of North America. During the 1970s, when studies on apparatus visibility at night indicated that red tended to disappear into the darkness, lime green and white replaced red. By the 1990s, red was in use again, and lime green had lost favor. Ambulances tend to be red or white, depending on local preferences. There is no national requirement regarding fire apparatus color. However, the FAA requires that Aircraft Rescue and Fire Fighting (ARFF) apparatus be "the appropriate yellow-green color" (DOT FAA No. 150/5220-10E 3.1.2.1) **(Figure 6.10)**.

Figure 6.10 The FAA requires that aircraft rescue apparatus be yellow-green.

Rearward- and Forward-Facing Video Cameras

Advances in digital video technology have resulted in the development of apparatus-mounted video cameras. These cameras provide a number of benefits, as they enable the driver/operator to see obstructions or people directly behind the apparatus. The rearward-facing camera does not eliminate the need for a radio-equipped crew member to direct rearward movement. Cameras should be installed so that the screen is not visible while the vehicle is advancing, a monitor display in the cab while moving is an unnecessary distraction.

The forward-facing camera can be used to provide a view of the emergency scene to other responding units or to a central command center. In the event of an accident, the video recording can help to determine the cause.

Vehicle Data Recorders (VDRs)

VDRs are required under NFPA 1901. They must have sufficient memory to record 100 engine hours' worth of data. They must also be-able to record the following conditions at least once per second and store them on a 48-hour loop. Required data includes the following:

- Vehicle speed
- Acceleration
- Deceleration
- Engine speed
- Engine throttle position
- Antilock braking system (ABS) event
- Seat occupied status
- Seat belt status
- Master optical warning device switch position
- Time and date

Vehicle Tracking Devices

Vehicle tracking systems can provide essential information for the telecommunications or dispatch center. By knowing the location of all apparatus, the center can dispatch the unit closest to the emergency incident. At the same time, the ability to track responding units can provide another level of safety. Responding units can be tracked and their relative positions reported to reduce the chance of a collision. First-arriving units can also be informed of the estimated arrival time and identity of other units. When their data is recorded, these devices can also be used to closely estimate a vehicle's speed prior to an accident.

Traffic Preemptive Control Devices

Some jurisdictions use traffic control devices to assist emergency vehicles during their response. Driver/operators must be aware of the devices used in their jurisdiction and how they operate. One of the simplest devices consists of placing a traffic signal in front of the fire station. This stops the flow of traffic so the apparatus can exit safely. This traffic signal may be controlled by a button in the station, by the telecommunications center, or by the alert tone or signal used to dispatch the unit. Some jurisdictions have systems that control one or more traffic lights in the normal route of travel for fire apparatus. These may be controlled by the fire station, by remote controls on the fire apparatus, or by the telecommunications center.

Another common system for controlling traffic signals for fire apparatus is the preemption device **(Figure 6.11 a and b)**. It activates sensors in the traffic lights that convert a signal to an electronic impulse. When that impulse goes to the traffic light control cabinet, it provides a green light for the direction that the apparatus is traveling.

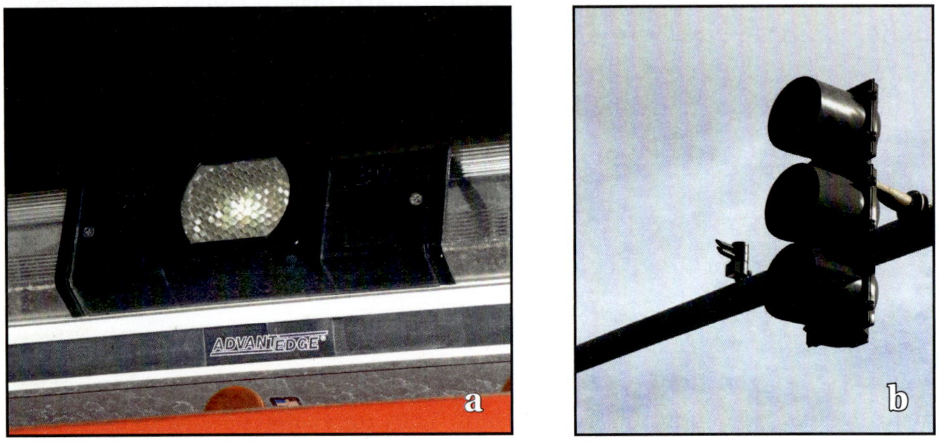

Figure 6.11 a and b Traffic preemptive control devices (a) allow approaching emergency vehicles to control intersection traffic lights (b), reducing the chance of a collision.

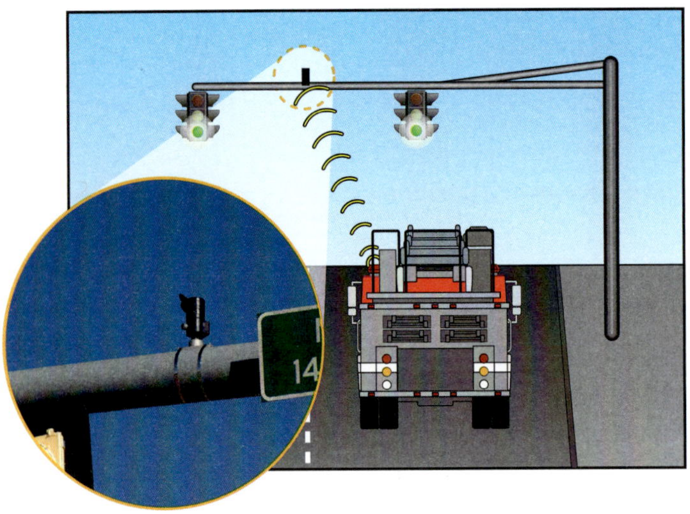

Figure 6.12 The Opticom™ emitter on the apparatus sends a signal to the detector on the traffic light support, which converts the optical signal to an electronic impulse. The system processes the signal and then manipulates the controller to provide a green light for the emergency vehicle(s) and red lights in all other directions.

In some jurisdictions, the traffic light standard may be equipped with an indicator light **(Figure 6.12)**. This light alerts the driver/operator that the signal has been received and that a green light is forthcoming. On some apparatus, the device is wired into the parking brake system. When the parking brake is set, the device will be turned off. On apparatus that do not have this feature, the driver/operator must turn off the device as soon as the apparatus is parked on the scene of an emergency. Otherwise, the device will continue to affect any traffic signals within signal reach.

Regardless of the type of traffic control devices used in a jurisdiction, there is no substitute for driving defensively. When entering an intersection with a green light, driver/operators must maintain a speed that will allow for evasive actions. If the emergency vehicle comes to a red light, the driver/operator should bring the vehicle to a complete stop and determine it is safe to proceed before doing so. If two emergency vehicles equipped with preemption devices approach the same intersection from different directions, only the apparatus whose sensor activates the system first will get a green light. The later-approaching apparatus gets a red light. The driver/operator should not assume that the system is not working if the light does not turn green. Approach the intersection with caution and come to a complete stop. When it is determined that it is safe to enter the intersection, regardless of the color of the light, proceed with caution.

Auxiliary Braking Systems

Because of their weight and increased stopping distances, fire apparatus that weigh more than 36,000 lbs (18 000 kg) are required to have auxiliary braking systems. These systems may be engine, transmission, exhaust, or driveline retarders. One version of the driveline retarder is an electromagnetic braking system that augments and works in conjunction with the vehicle's conventional service brakes. Because some of these retarders create loud noises, local ordinances may prohibit them.

Auxiliary braking systems connect to either the driveshaft or the rear axle of the vehicle. These systems can be programmed to activate when driver/operators remove their foot from the accelerator, step on the brake pedal, or use a manual selector lever mounted on the steering column. Under normal conditions, this system does not activate at speeds under 2 mph (3 km/h).

Antilock Braking System (ABS)

Most new fire apparatus come equipped with an all-wheel, antilock braking system (ABS). These systems minimize the chance of the vehicle skidding when the brakes are forcefully applied. ABS works using digital technology in an onboard computer that monitors each wheel and controls pressure to the brakes, providing straight-line braking, even under maximum application. A sensor monitors the speed of each wheel. The wheel speed is converted into a digital signal sent to the onboard computer.

When a wheel begins to lock, the sensor sends a signal to the computer that the wheel is not turning. The computer compares this signal to the signals from the other wheels to determine if this particular wheel should still be turning. If so, a signal is sent to that wheel, reducing the brake pressure and allowing the wheel to turn. Once the wheel turns, it is braked again. The computer makes these decisions more than twenty times per second, until the vehicle stops. Thus, when driving a vehicle equipped with an ABS, the driver/operator should maintain a steady pressure on the brake pedal (rather than pumping the pedal) until the apparatus is brought to a complete stop.

Some vehicles use ABS as the foundation for additional systems such as Traction Control and Electronic Stability Control (ESC). By applying the brake to the spinning wheel, traction control assists the driver of a vehicle stuck in the mud or traveling on slick roadways. Using some of the same basic principles and equipment, stability control allows the vehicle to measure wheel speed and other forces while navigating high-speed turns. The computer-operated system automatically applies specific brakes and may temporarily disable the accelerator to assist the driver/operator in maintaining control of the vehicle. Combined with the skills of the driver/operator, the ESC reduces the likelihood of a rollover.

Maintenance and Inspections

A proactive emergency vehicle maintenance program ensures that the vehicles are safe to operate. The success of the program depends on a commitment by the entire organization: the administration that develops and implements it, the supervisors who enforce it, and the members who actively participate in it.

In addition to preventive maintenance, the organization will need a corrective maintenance capacity. Corrective maintenance is reactive in nature, responding to wear, tear, and damage. A key to corrective maintenance is determining if the cost of the repair can be justified or if the vehicle should be replaced. If a vehicle is nearing the end of its service life, it may be more cost-effective to replace it than to repair it. The reliability of fire apparatus should be a key factor in this decision.

Figure 6.13 This driver/operator performs a daily maintenance inspection on an ARFF apparatus.

Preventive Maintenance

The apparatus driver/operator is primarily responsible for ensuring that assigned department vehicles perform effectively **(Figure 6.13)**. NFPA 1500 mandates a preventive maintenance program supported by accurate record keeping. The program elements are based on NFPA 1915, *Standard for Fire Apparatus Preventive Maintenance Program*.

A daily apparatus inspection checklist, similar to the Commercial Driver's License (CDL) checklist, may assist the driver/operator with documenting the information needed to make repairs. Check sheets can also help to identify problems quickly so that timely inspections and repairs can be made. For all types of maintenance, it is critically important that the driver/operator communicate clearly with the maintenance staff.

Corrective Maintenance

An effective preventive maintenance program will ensure that most repairs are made before a serious breakdown. Unforeseen damage can occur at any time, resulting from metal fatigue, collisions, or operator error. When these incidents occur, the organization must have a procedure for notifying administrative officers (including the safety officer), mechanics, and tow truck operators. Depending on the severity of the damage, it may be necessary to remove the unit from service and replace it with a reserve vehicle.

Fleet managers should also develop adequate staffing models to ensure enough qualified mechanics for the fleet size are on hand to keep the fleet operating in top condition. In a growing town or city, plans should be made to add additional staff under industry standard ratios of apparatus to mechanics, or apparatus mileage to mechanics. This is especially true if the apparatus life expectancy is higher than average fleet expectancies. NFPA 1915, *Standard for Fire Apparatus Preventive Maintenance Program*, places responsibility on the AHJ to implement a schedule of service. This schedule is based on manufacturer recommendations; it should establish criteria for when apparatus is to be taken out of service and repaired. Adequate staff must facilitate the corrective maintenance program.

Inspections and Tests

Both career and volunteer fire and emergency services organizations should have SOPs that establish the frequency and level of apparatus inspections. NFPA 1500 requires a minimum of weekly inspections. Career organizations usually require inspections at shift change or daily. A more thorough inspection may be scheduled weekly when the driver/operator performs major cleaning of the vehicle. Some departments schedule tool and equipment inspection on different days than apparatus inspection, especially for aerial ladder and rescue vehicles.

In keeping with NFPA requirements, apparatus are also inspected within 24 hours of any use, immediately after major repair, and prior to being placed into service. Semiannual or annual apparatus inspections are generally performed by maintenance personnel during scheduled preventive maintenance. In addition, some states and provinces require regular commercial vehicle inspections by certified personnel.

Acceptance Tests

Guidelines for acceptance and in-service testing are specified in NFPA 1911, Standard for the Inspection, Maintenance, Testing, and Retirement of In-Service Emergency Vehicles. Acceptance tests are performed when a new or refurbished apparatus is delivered by the manufacturer or maintenance center. These tests measure characteristics of the vehicle's mechanical, pneumatic, hydraulic, or electrical systems and compare the results to its specifications and required design standards.

Service tests are performed by a third party or by the organization annually or after any major repairs or alterations to the apparatus or component. Annual service tests are required for the following components:

- Fire pumps **(Figure 6.14)**
- Aerial devices
- Compressed air foam systems
- Foam proportioners
- Chassis components
- Breathing air compressors

In addition to the annual tests, verification that the weight of the vehicle has not changed since it was constructed is also required. The standard defines how apparatus is to be loaded; specifies that each axle be weighed separately; and then requires that the entire apparatus be weighed as a whole. If the units exceed the corresponding weight limit, changes must be made to the vehicle or the unit is taken out of service.

In addition to the required annual test, aerial devices must be evaluated following any use that may have placed them under severe strain or if it is believed that the manufacturer's operational recommendations have been surpassed. In this case, the required tests are called nondestructive tests and must be performed by a certified technician.

Figure 6.14 A driver/operator conducts a service test of an apparatus pump.

Apparatus Service Life

Some fire apparatus with 8-10 years of service are worn past the point of cost-effective maintenance and repair. Other vehicles have responded to a low number of calls, had excellent maintenance, and are serviceable after more than 20 years. The quality and timeliness of preventive and corrective maintenance can be as important as years of use in determining how well an emergency vehicle ages.

Even with an aggressive preventive maintenance program, apparatus will reach a point where continued operation is unwise from the standpoint of safety, reliability, and cost-effectiveness. NFPA 1911 does not define a specific age at which apparatus must be replaced. As safety is the primary consideration, the decision to replace apparatus is based on many factors, including the following:

- Safety
- Mileage
- Age
- Aging design or technology
- Government-mandated safety or design changes
- Cost-effectiveness of refurbishment
- Quality of the preventive maintenance program
- Quality of the original builder and components
- Availability of parts

Record Keeping

Justifying apparatus refurbishment or replacement is based on accurate and thorough record keeping. NFPA requires organizations to maintain records on all fire and emergency apparatus as part of the records management system **(Figure 6.15)**. The department should have comprehensive data on each emergency and nonemergency vehicle it owns, leases, or has been assigned. Information maintained on file for each vehicle should include, but is not limited to, the following:

- Safety-related issues
- Accident involvement
- Reliability
- Original specifications
- Acceptance test results
- Vehicle assignments
- Unit inspection records
- Preventive maintenance records
- Major overhauls, refurbishments, or repairs
- Purchase and maintenance costs
- Yearly service tests

APPARATUS MAINTENANCE CHECKLIST For use of this form, see AR 420-90; the proponent agency is ACSIM	APPARATUS	MONTH	CALENDAR YEAR

SECTION A - DAILY INSPECTION STATUS *(see legend below)*

LINE	ITEM INSPECTED	DAYS OF THE MONTH																														
		1	2	3	4	5	6	7	8	9	10	11	12	13	14	15	16	17	18	19	20	21	22	23	24	25	26	27	28	29	30	31
1	FUEL *(fill at 3/4)*																															
2	ENGINE OIL																															
3	RADIATOR																															
4	TIRE AIR PRESSURE																															
5	LIGHTS																															
6	BATTERY																															
7	FAN BELT																															
8	BRAKES																															
9	AIR PRESSURE *(drain tanks)*																															
10	CLUTCH																															
11	WINDSHIELD WIPERS																															
12	START ENGINE																															
13	WARNING DEVICES																															
14	ENGAGE PUMP																															
15	INSTRUMENT PANEL																															
16	FIREFIGHTING AGENTS																															
17	RELIEF VALVE SET *(150 psig)*																															
18	PRIMER OIL																															
19	FLUSH PUMP *(monthly)*																															
20	VACUUM TEST *(monthly)*																															
21	DRAFTING OPERATION *(annual)*																															
22	GRID MAPS																															
23	JAWS OF LIFE *(weekly operation)*																															
24	HOSE CHANGE *(pumper - monthly)*																															
25	K-12 SAW *(weekly operation)*																															
26	RADIO CHECK																															
27	APPEARANCE																															
28	OPERATOR'S INITIALS																															
29	SUPERVISOR'S INITIALS																															

LEGEND: Daily Inspection Status
(Enter status of item inspected in columns above)

1 = OK
2 = LOW
3 = DAMAGED
4 = DIRTY

5 = NEEDS ADJUSTMENT
6 = NEEDS REPLACING
7 = MALFUNCTION

8 = INCOMPLETE
9 = MISSING
10 = FIRE DEPARTMENT RESPONSIBILITY
11 = MAINTENANCE RESPONSIBILITY

DA FORM 5379-R, JAN 2017

USAPA V1.01

SECTION B - ACTION TAKEN

LINE	ITEM INSPECTED *(from Section A)* (a)	DAILY INSPECTION STATUS (b)	NAME OF INDIVIDUAL PROBLEM REPORTED TO (c)	DATE/ TIME REPORTED (d)	WORK ORDER NO. (e)	MECHANIC'S SIGNATURE (f)	REMARKS (g)

Reverse of DA Form 5379-R, JAN 2017

Figure 6.15 An apparatus maintenance form includes items for daily inspection.

Driver/Operator Requirements

Operating fire and emergency services vehicles requires the driver/operator to be in good health and physically fit. The driver must be mentally alert and able to respond to changing traffic conditions. Using defensive driving skills can reduce the likelihood of an accident.

During emergency responses in apparatus, the driver/operator must exercise even greater care. The weight and speed of the apparatus may dictate longer stopping distances and make control on rain or ice-covered road surfaces more difficult.

Responding to emergency incidents in POVs is common in volunteer and combination departments. The use of POVs presents its own special concerns, including awareness issues with the public, ability to audit vehicle condition, driver training, and liability.

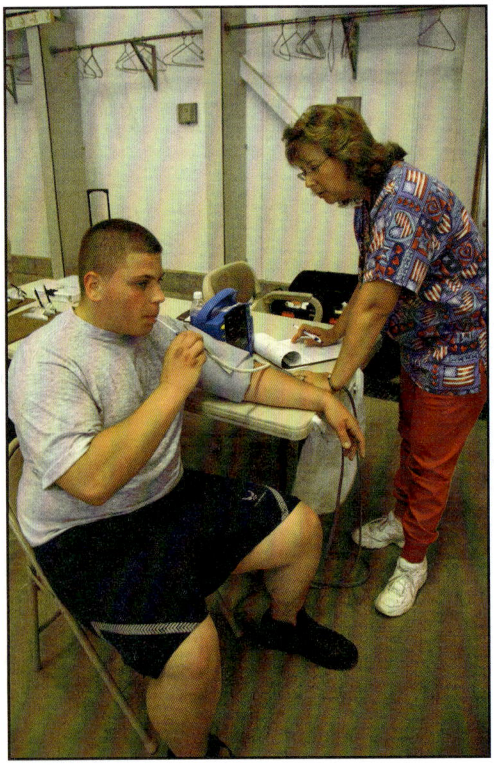

Figure 6.16 Many states require driver/operators to have medical exams as part of driver certification.

Health and Physical Fitness Testing

NFPA 1002, *Standard for Fire Apparatus Driver/Operator Professional Qualifications*, requires driver/operators to participate in periodic medical evaluations as required by NFPA 1500. Many states also require a medical exam as part of driver certification for driver/operators **(Figure 6.16)**. This testing should help to ensure the driver/operator is physically fit to perform the duties he/she may encounter during emergency operations. Emergency service organizations use many health and physical fitness programs to help personnel stay fit, including those sponsored by organizations like the International Association of Fire Fighters (IAFF) and International Association of Fire Chiefs (IAFC).

Certification

Each state/province has its own requirements for licensing emergency vehicle driver/operators. A commercial driver's license (CDL) is state/province-issued and required in order to legally drive the following:

- A single vehicle with a gross vehicle weight rating (GVWR) of more than 26,000 pounds (13,000 kg)
- A trailer with a GVWR of more than 10,000 pounds (5,000 kg) if the combined weight of the vehicle and trailer is more than 26,000 pounds (13,000 kg)
- Any size vehicle that is required to display hazardous materials placards.

NOTE: In Canada, driver's license requirements vary by vehicle weight and number of axles. Endorsements on the driver's license are required for certain types of braking systems.

At the same time, most states/provinces exempt personnel assigned to operate ambulances and fire apparatus from having to acquire a CDL, requiring only a standard driver's license. This exemption does not permit a fire department driver/operator to drive a support vehicle that requires a state/province CDL. Therefore, someone who is assigned to operate the fuel truck or supply truck may be required to have a CDL.

Many states/provinces and emergency services organizations are adopting a zero-tolerance policy toward driver/operators regarding the use of alcohol or drugs.

In response to an increasing number of alcohol-related vehicle accidents, in 2003 the IAFC called for its members to adopt a zero-tolerance alcohol policy. According to the IAFC policy, anyone who has consumed alcohol or still appears impaired eight hours after consuming alcohol, may not engage in fire fighting activities or operate an emergency vehicle.

Volunteer, combination, and career fire and emergency services organizations are encouraged to adopt a zero-tolerance policy and enforce it. When assigning an individual as an emergency vehicle driver/operator, the department should first review an applicant's driving record. Next, arrests and convictions for driving under the influence (DUI) or driving while intoxicated (DWI) should be considered. Loss of a regular driver's license is a valid reason to remove a person from the position of driver/operator.

NOTE: The state/province issues the appropriate driver's license to an individual based on the type of vehicle to be driven. It is up to the fire and emergency services organization to confirm that the license is appropriate for operation of an emergency vehicle and to ensure that the license is still in effect.

Training

The driver/operator of the unit is responsible for the safe operation and transportation of occupants riding in the unit. The company officer also shares responsibility for crew safety through the training and consistent application of rules and regulations with regard to vehicle safety, apparatus placement, vehicle inspections, and record keeping. The administration of the organization is responsible for providing training for all driver/operators, either internally or through other agencies. In some cases, company officers may be tasked with verifying the ability of personnel to operate the vehicles within their command.

Minimum qualifications for driver/operators can be found in NFPA 1002, *Standard for Fire Apparatus Driver/Operator Professional Qualifications*. Fire service organizations commonly provide a driver/operator program — a template of a local, regional, state/provincial, or federal program — designed to prepare personnel to operate emergency vehicles in both emergency and nonemergency situations. The program will provide personnel with the knowledge and skills needed to operate emergency vehicles, and detail the minimum requirements for state/federal certification.

Driver/operator training courses generally consist of classroom presentations, driving exercises, and a final driving examination. NFPA 1002 specifically defines five exercises to develop skills **(Figure 6.17 a-e, p. 198)**. They apply to all types of vehicles, not just fire apparatus. The order in which these exercises are conducted may be modified.

NOTE: Other exercises specific to tractor-trailer, tiller-operated, ARFF, and water tender apparatus should also be developed as needed by the local department.

Emergency equipment has and will continue to improve, thanks to safety features that enable the operator to better handle the vehicle in all situations.

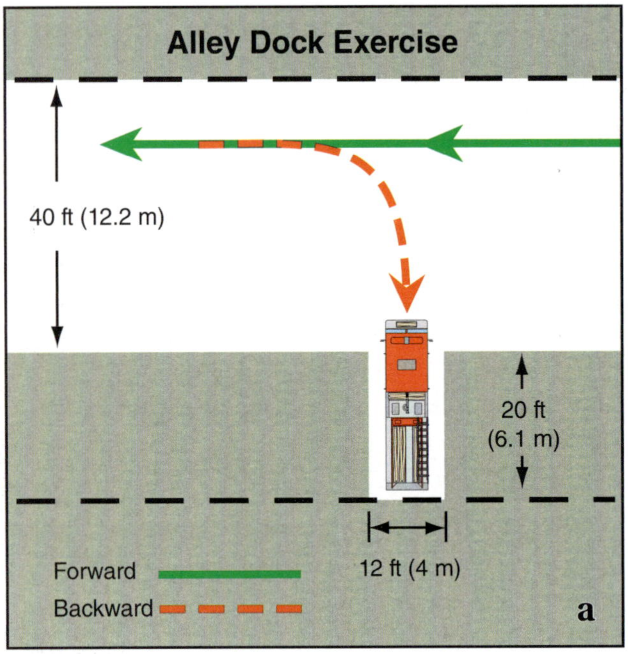

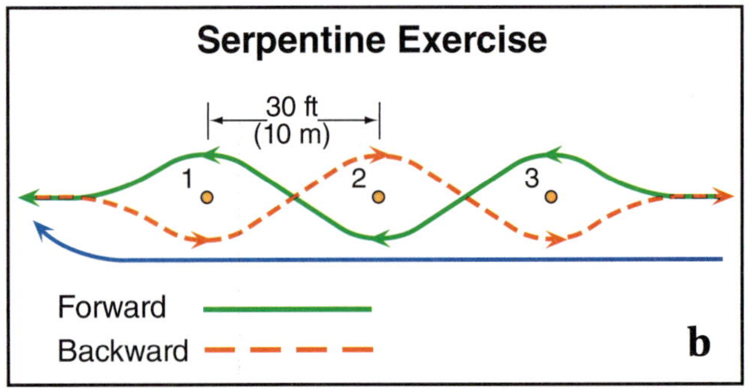

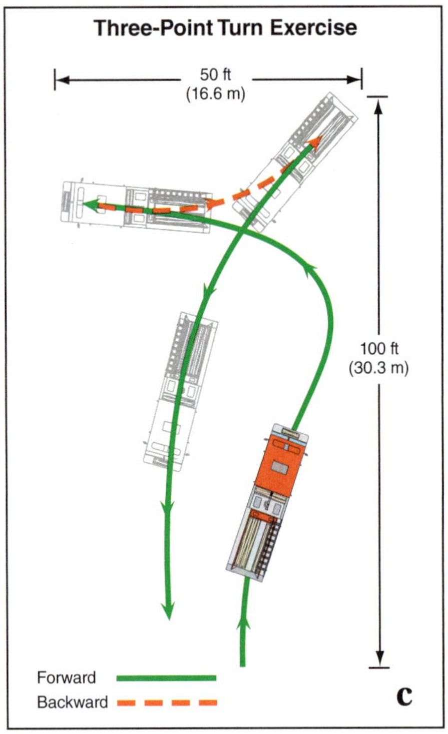

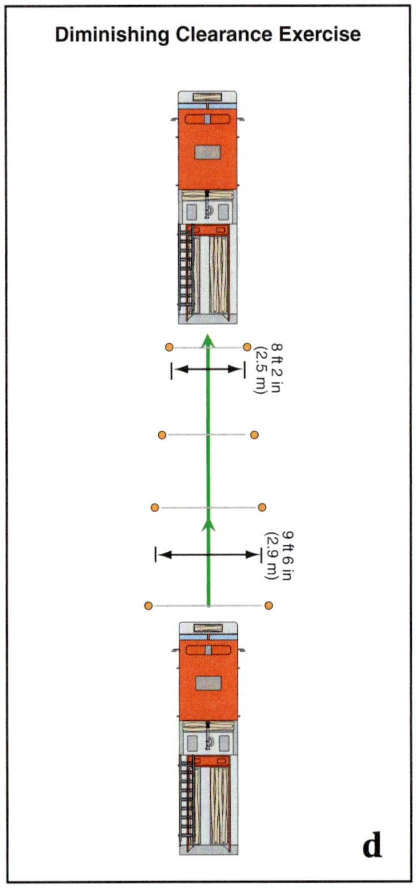

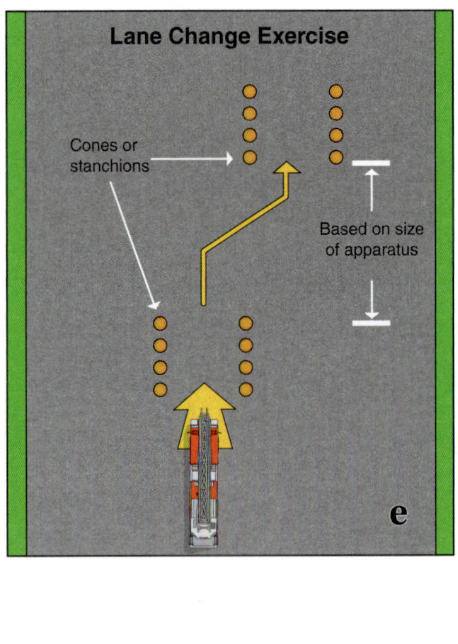

Figure 6.17 a-e NFPA specifies five driving exercises that can be used to train and test driver/operator skills: a) alley dock; b) serpentine; c) three-point turn; d) diminishing clearance; and e) lane change.

Antilock braking systems (ABS), coupled with engine braking systems, can wreak havoc on an unfamiliar operator who lacks knowledge and training. The turning radius of various vehicles may differ, as can the turning ratio (steering-wheel revolutions to angle-of-wheel turn). These differences can affect an operator's ability to gain access and egress, especially in off-road operations, steep grades, or slippery conditions. If the vehicle is not driven and operated in the correct manner, improper upshifting/downshifting can cause loss of traction, loss of power, and possible driveline failure.

Defensive-Driver Training

Driver/operators should know the rules that govern private vehicles when emergency vehicles are responding with warning lights and operating audible devices. In most jurisdictions, motor vehicle laws require that private vehicles pull to the right, stop, and remain at a standstill until the emergency vehicle has passed. This rule does not mean that driver/operators can ignore stopped vehicles — especially when two or more pieces of fire apparatus are responding in the same direction on the same street. After the first emergency vehicle passes, the citizen driver may pull back into the traffic lane without looking.

Figure 6.18 Texting while driving is a distracting and dangerous activity that can result in motor vehicle accidents.

In addition, private vehicle drivers may panic at the sound of an approaching siren and accelerate to stay ahead of the emergency vehicle, pull to the left, or stop suddenly in the traffic lane. Other drivers may not hear sirens or air horns because of loud car radios, headphones, closed windows, air-conditioning, or other noise-producing devices. Some drivers may be distracted by cell phones **(Figure 6.18)**, radio controls, or small children. Others may simply choose to ignore emergency warning devices.

Traffic waiting to make a left-hand turn may pull to the right or left, depending upon the driver. In situations where all lanes of traffic in the same direction as the responding apparatus are blocked, a departmental SOP may allow the apparatus to be driven into the opposing lane of traffic and to proceed through the intersection at an extremely reduced speed. In some departments, an SOP requires that the street or road be blocked by law enforcement before fire apparatus is driven in an opposing traffic lane. Oncoming traffic must be able to see the approaching apparatus. Full use of warning devices is essential.

Driving in the opposing lane of traffic is not recommended when oncoming traffic is unable to see the apparatus for any reason. Be alert for traffic approaching from a side street/road or a driveway. Traffic coming over the crest of a hill, slow-moving traffic in both directions, and other responding emergency vehicles must be closely monitored.

Fire apparatus driver/operators should realize that emergency signals may be obscured or muffled by other warning devices and street noises. This is especially true when two or more emergency vehicles are approaching the same intersection from different directions. Serious collisions and fatalities have been caused by overreliance on warning devices.

Anticipating other drivers' actions. Never assume what another driver's actions will be. Instead, expect the unexpected. Anticipation is the key to safe driving. Always remember the following control factors:

- Aim high in steering — Find a safe path well ahead.
- Get the big picture — Stay back and see it all.
- Keep your eyes moving — Scan; do not stare.
- Leave yourself an escape route way out — Do not expect other drivers to leave you an exit.
- Make sure others can see and hear you — Use lights, horn, and signals in combination.

Visual Lead Time — The concept that driver/operators should scan far enough ahead of the apparatus, for the speed it is being driven, to assure that evasive action can be taken if it becomes necessary.

Visual lead time. The distance of attention that a driver must heed is known as **visual lead time**. By scanning far enough ahead of the apparatus, the driver/operator assures that evasive action can be taken, if necessary. For example, if a driver/operator is concentrating on vehicles that are 100 feet (30 m) in front of the apparatus, but driving at a speed that would take 200 feet (60 m) to stop or perform an evasive maneuver, a collision is likely to occur. The driver/operator needs to learn to match the speed of travel with the distance being surveyed ahead of the vehicle. Visual lead time interacts directly with reaction time and stopping distances. By aiming high in steering and getting the big picture, it is possible to become more keenly aware of conditions that may require slowing or stopping. Some departments train their driver/operators to practice looking 12 seconds ahead on city streets and 20 seconds ahead on highways, estimating where they will be in that time.

Figure 6.19 Driver/operators are tested on their skills.

Driver/Operator Testing

After driver/operators have been selected and trained, their performance should be evaluated by a standard method. These evaluations should occur before a driver/operator is allowed to operate the apparatus under emergency conditions. NFPA 1002 provides some specific directions on how driver/operator candidates should be tested. Various aspects of driving, pumping, raising aerial devices and trailering are requisite knowledge, and the abilities to perform duties are requisite skills. These directions need to be followed by agencies that certify their personnel to the standard. Other agencies should follow the standard to avoid civil law liabilities if the driver/operator is involved in a collision.

Most agencies use a combination of written and practical testing for driver/operator candidates **(Figure 6.19)**. For volunteer fire departments, successful completion of this testing may simply result in making the person qualified to drive the fire apparatus. For career fire departments, this may be a promotional exam.

Driver Training Programs

In addition to driver training programs developed by the fire and emergency services organization, nonprofit and for profit organizations provide additional training opportunities. Nonprofit organizations with course offerings include:

- **National Safety Council** —The NSC offers state-certified and nationally recognized courses in defensive driving, including Coaching the Emergency Vehicle Operator (CEVO).
- **VFIS** —VFIS offers educational materials, training classes, and online training options for firefighters and emergency service personnel, including an Emergency Vehicle Driver Training Instructor course.
- **Federal Motor Carrier Safety Administration** — The FMCSA's compliance and enforcement program offers resource materials to educate drivers about their safety responsibilities.
- **National Highway Traffic Safety Administration** — The NHTSA's Office of EMS coordinates a system of EMS education and provides free webinars on federal EMS efforts and programs.

Additional driver training programs are available through for-profit driver training companies that offer road training, classroom seminars, E-learning, training certifications, and on-site consulting. Training in The Smith 5 Keys® To Space Cushion Driving focuses on driving fundamentals of space, visibility, and time.

Vehicle Operation

Everyone who responds to emergencies or incidents knows it is one part of the many risks faced by emergency providers. Driver/operators must make safe and efficient operation of the vehicle a priority, whether it is a crew of two, company of four, or a bus loaded with fire control laborers.

It is generally required that all occupants riding in vehicles equipped with seat belts are seated and using the belts prior to the vehicle moving. Yet, through negligence, lack of communication, or lack of experience, emergency service personnel die in motor vehicle accidents every year. Standing on an apparatus while the vehicle is moving is still tolerated in some organizations, even when it violates the law and organizational policy. This highly hazardous behavior has led to numerous injuries and fatalities. Individual, supervisory, or organizational tolerance of willful violations of any safety policy indicates tacit approval, which will likely result in financial costs, lost time, and loss of respect from the general public.

In Motion

To ensure the safety of all personnel, certain actions must be taken when the vehicle is in motion. These actions must occur whether responding to, or returning from, an emergency; during training; or during any nonemergency function. Responsibility for these actions falls first on the individual, then on the driver/operator, and finally on the supervisor of the unit.

Riding in Apparatus

Since 1991, the use of seat belts or three-point restraints has been mandatory in all fire apparatus. Personnel must be secured prior to the movement of the vehicle and until it comes to a full stop. Seat belts prevent personnel from being ejected if the vehicle is involved in a collision. Being strapped into an SCBA while it is still in the mounting bracket is not considered to be wearing a seat belt or safety restraint.

NFPA 1500 prohibits standing on the tailboard, on the running boards, or within the crew or passenger compartment of a moving apparatus. Only two potential exceptions are allowed in NFPA 1500, A6.3.4 1-7: hose-loading activities and tiller operator training. In both cases, the emergency services organization is required to develop a policy that addresses the safety requirements for these exceptions. A possible alternative is the use of a safety harness that secures the firefighter to the vehicle while performing the task. NFPA 1500 also requires that personnel in the patient compartment of ambulances be secured with a seat belt or safety harness while providing patient care.

Company Officer

The company officer has ultimate authority over the safe operation of the emergency vehicle and sets a tone that safe behavior is expected. While the driver/operator has physical control of the vehicle, the company officer must advocate and enforce safe operation. If a driver/operator is exhibiting unsafe behaviors, it is the officer's responsibility to correct those behaviors immediately. Likewise, under the crew resource management (CRM) concept discussed in Chapter 3, other crew members are responsible for alerting the company officer and driver/operator of potential unsafe acts or conditions.

Figure 6.20 The use of spotters whenever an apparatus is backing should be a part of all risk management plans. *Courtesy of Ron Jeffers, Union City, NJ.*

Backing Apparatus

Backing emergency apparatus is hazardous and should be avoided. If backing is unavoidable, spotters should safely direct the operation from outside the vehicle. When backing the apparatus, all members (excluding the driver/operator and tiller operator) should dismount and act as spotters for the backing operation. Spotters should not be permitted to ride on the apparatus from an exposed position such as the tailboard.

The spotters should discuss with the driver/operator a backing plan (that is consistent with the SOP). They should also agree on the communication or warning process prior to backing. Communication can be verbal, or by a remote electronic signaling device, hand signals, or radio. The vehicle should not be moved until the spotters are in position and have communicated their approval to start backing. Once it begins, the spotters should remain visible to the driver **(Figure 6.20)**. Spotters not wearing PPE should always wear high-visibility reflective vests. At least one spotter should be positioned at the left rear corner to operate as the primary spotter. When the driver cannot see the primary spotter, the backing operation should stop immediately.

To reduce the risk associated with operating a vehicle in reverse, additional mechanical or electronic aids can be added. Fire apparatus may be equipped or retrofitted with mirrors, commercially available video cameras, or sensing devices. However, these devices cannot take the place of spotters.

Water Tender Issues

Also known as *mobile water supplies*, tankers, or water tenders **(Figure 6.21)**, these vehicles account for many firefighter response-related fatalities. *Safe Operations of Fire Tankers* (FA-248), available through the United States Fire Administration (USFA) website, analyzes the causes of water tender accidents and suggests solutions.

Figure 6.21 Water tenders should meet the dump requirements of NFPA 1901.

Several factors contribute to water tender accidents. They include physical characteristics unique to the water tender apparatus:

- **Weight** — Water tenders are heavier than most other types of apparatus. Straight-frame water tenders may weigh as much as 25 tons (22.5 metric tons), while tractor-trailer water tenders can weigh much more. The additional weight makes the vehicle more difficult to control and stop.
- **High center of gravity** — Water tenders have a high center of gravity, making them more unstable than other vehicles. The instability can result in the vehicles being difficult to control on curves and uneven road surfaces.
- **Improper design or alteration** — Improper design of new water tenders or modification to vehicles can create unsafe conditions. New apparatus chassis must be designed to carry the total weight of the tank when filled with water. Consideration must also be given to the added weight of tools, equipment, and personnel plus a safety factor. Modifications to vehicles used for transporting other liquids can also cause water tender accidents. For instance, trucks used to carry liquid food products lack the interior tank baffles to prevent movement of the liquid, which can cause rollovers. Cargo

tank trucks, used for carrying fuel, are designed for a lighter load because gasoline weighs less than water, and this weight difference must be taken into consideration. Finally, the braking systems of all modified vehicles must be capable of handling the added weight of the water in the tank.

- **Liquid surge** — The movement of water in an unbaffled tank causes many water tender accidents. Liquid surge can add to the forward momentum of the vehicle at the same time a driver/operator is attempting to stop. Tanks must be designed or modified to conform to the requirements detailed in NFPA 1901 for mobile water supplies.

NOTE: It is common practice that water tenders are driven with tanks either full or empty to reduce the possibility of surges.

Besides the physical characteristics of water tenders, driver/operator behavior has been shown to contribute to their accidents. Driver/operators who lack training, are unfamiliar with driving water tenders, exceed the safest speed, ignore operation SOPs, and disregard state/provincial motor vehicle laws cause accidents, as well.

Reserve Apparatus

Reserve apparatus may be unsafe for a variety of reasons. Driver/operators, especially personnel new to the position, may not be familiar with the operation and design of older reserve vehicles.

Additional factors that contribute to accidents involving reserve vehicles include the following:

- **Age** — Older apparatus show signs of hard service and wear. Metal fatigue, reduced braking capacity, and electrical problems affect the operation of reserve vehicles.
- **Maintenance** — Since reserve vehicles are not in continuous service, maintenance may not be as frequent or thorough.
- **Infrequent use** — Lack of use can cause a deterioration of parts and mechanical systems.
- **Design** — Reserve vehicles may not meet the most recent safety requirements. In addition, older vehicles may have manual transmissions and lack power steering or air or power brakes.
- **Lack of familiarity** — Driver/operators, especially personnel new to the position, may not be familiar with the operation and design of older reserve vehicles.

Privately Owned Vehicles (POVs)

Firefighters killed in POVs during the course of their duties account for the largest percentage of vehicle-related deaths. From 2005 to 2013, 52 fatalities occurred, with 46 of those being volunteers in their POVs. Seventy-three percent of these fatalities occurred while responding to calls.

Most volunteer departments do not require personnel to stand by in the fire station. Members may respond directly to incidents from their homes or work places, and they often drive POVs to improve response times. Some organizations provide emergency lights for personnel to use when they respond to incidents. State/provincial motor vehicle laws and local policies determine the type, location, and use of emergency lights on POVs.

However, driver attitude may be the most significant factor in accidents involving POVs. If the firefighter believes the emergency response is more important than safety, chances for an accident increase. Speeding and violating traffic laws are the results of that attitude. This is not only unsafe, it makes an individual and department legally liable. Emergency warning lights on a POV on the way to an emergency may not exempt an emergency responder from all traffic laws or liability. Some automobile insurance carriers may deny damage claims if the responder's POV is not insured for that type of use. Check with your state motor vehicle agency and insurance carrier to determine emergency response requirements.

Training programs for apparatus driver/operators are important, and so are training programs for personnel who respond in POVs. The chief officer is responsible for providing adequate training in defensive driving skills to personnel who drive their own vehicles to an incident. Many programs teach defensive driving skills to emergency service workers, and most state law enforcement agencies offer an Emergency Vehicle Operations Course (EVOC). Topics of instruction include anticipation of other drivers, effective braking techniques, confidence courses, and skidding-recovery techniques.

Nonemergency Operations

The majority of operations conducted by fire and emergency services vehicles are nonemergency in nature **(Figure 6.22 a and b, P. 206)**. Staff vehicles are used to transport personnel to fire and life safety education programs, make building inspections, deliver documents, and attend meetings. Support services deliver fuel and supplies, provide facility and apparatus maintenance in the field, and move reserve apparatus around the jurisdiction. Ambulances are sometimes used to perform nonemergency patient transfers and pick up medical supplies from hospitals and supply facilities.

Other times when fire apparatus should be operated in a nonemergency manner include:

- Returning from an incident
- Service training
- Preincident inspections and surveys
- Driver/operator familiarization
- Driving to/from shop for repairs
- Driving to pick up fuel and supplies.

All of these activities, though nonemergency, are essential to the operation of fire and emergency services organizations. Driver/operators must adhere to all state/provincial and local motor vehicle laws when performing these activities. The safety of other crew members and the general public depends on the ability of the driver/operator to maneuver the vehicle through traffic and constricted streets. In the case of oversized vehicles, such as tractor-trailer aerial devices, care is essential. Aerial devices equipped with a tiller require coordination and communication between the driver and the tiller operator.

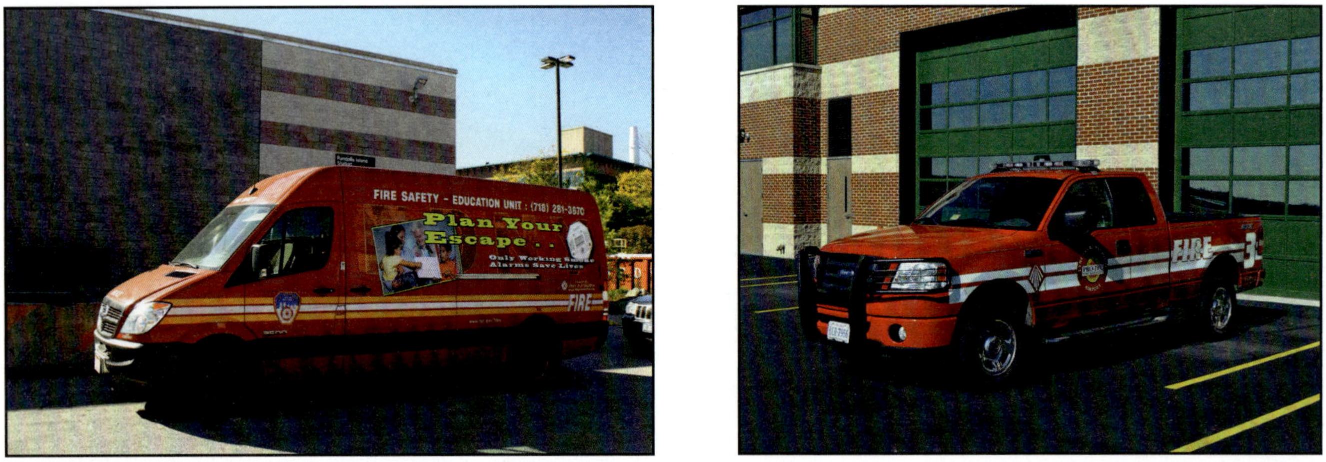

Figure 6.22 a and b a) A non-emergency vehicle used for fire and life safety education programs. *Courtesy of Ron Jeffers, Union City, NJ.* b) A non-emergency vehicle used to transport personnel. *Courtesy of Edwin W. Hall, Lynchburg Regional Airport, Lynchburg, VA.*

Emergency Response

During an emergency response, the driver/operator's responsibility is to safely deliver the apparatus, its equipment, and its crew to the emergency scene. Making this transit must be done within the parameters of the department's response policies and the state/province's motor vehicle laws. A driver must remember that no emergency is worth the lives of members of the unit or of the public that may be endangered by unsafe acts.

Some training and operational concerns for safe emergency responses include the following:

- Right-of-way, one-way streets
- Appropriate speeds
- Intersection safety
- Highway access ramps
- Limitations of optical and audible warning devices
- Driver/operator multitasking
- Tillering issues **(Figure 6.23)**
- Skid recovery
- Towing trailers **(Figure 6.24)**
- Nighttime
- Inclement weather
- Public and private bridges

Total Stopping Distance — Sum of the driver reaction distance and the vehicle braking distance.

NOTE: Refer to the IFSTA **Pumping Apparatus Driver/Operator Handbook** for more information about driving techniques.

Driver Reaction Distance — The distance a vehicle travels while a driver is transferring the foot from the accelerator to the brake pedal after perceiving the need for stopping.

Braking and Reaction Time

Driver/operators should know the braking characteristics for the apparatus they operate, including total stopping distance. The **total stopping distance** is the sum of the driver/operator's reaction distance and the vehicle's braking distance. **Driver reaction distance** is the distance a vehicle travels while

Figure 6.23 The tiller operator steers the rear wheels of a tractor-drawn apparatus. *Courtesy of Ron Jeffers, Union City, NJ.*

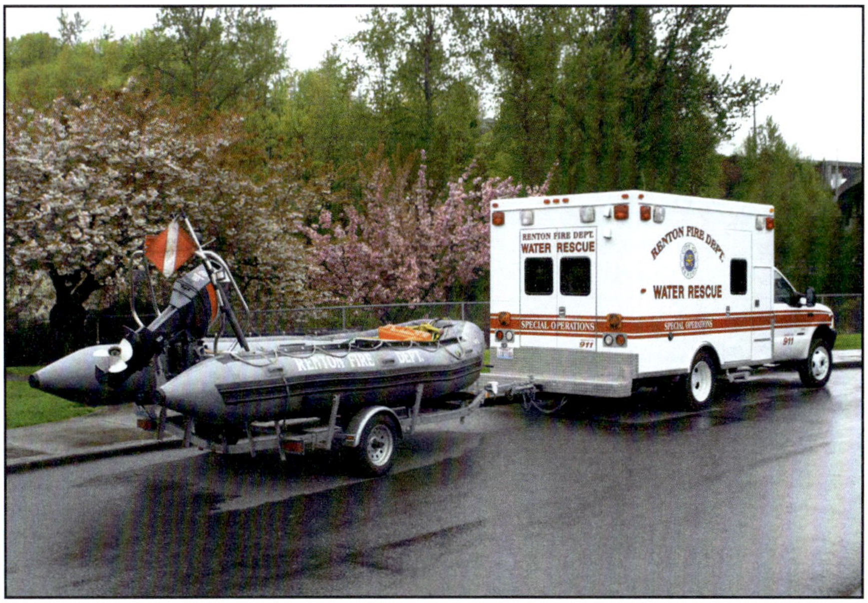

Figure 6.24 Equipment trailers, air compressors, lighting units, and boat trailers are often towed behind fire apparatus.

a driver is transferring the foot from the accelerator to the brake pedal after perceiving the need for stopping. Reaction distance can be affected by the driver/operator's age, aptitude, emotions, level of fatigue, and use of any intoxicating substances. The **braking distance** is the distance the vehicle travels from the time the brakes are applied until the apparatus comes to a complete stop **(Tables 6.1 and 6.2, p. 208)**. As statistics may vary for different emergency vehicles, each department should conduct braking distance tests with its apparatus. Manufacturers of specific apparatus may be able to provide this information.

Braking Distance — The distance a vehicle travels from the time the brakes are applied until it comes to a complete stop.

There are factors that influence a driver/operator's ability to stop an apparatus, such as the following:

- Condition and slope of the driving surface
- Speed being traveled
- Weight of the vehicle
- Type and condition of the vehicle's braking system
- Weather conditions
- Type and condition of tires
- uxiliary traction devices
- Weight transfer
- Passing of other vehicles

A flat, dry, paved road provides the optimal stopping surface. A driver/operator's ability to stop the apparatus is impaired by steep, wet, snowy, icy, or unpaved roads. Driver/operators must compensate for these conditions by reducing their speeds accordingly.

Table 6.1 (Customary)
Braking and Stopping Distances (dry, level pavement)

Speed (mph)	Average Driver Reaction Distance (feet)	Braking Distance (feet)				Total Stopping Distance (feet)			
		Vehicle A	Vehicle B	Vehicle C	Vehicle D	Vehicle A	Vehicle B	Vehicle C	Vehicle D
10	11		7	10	13		18	21	24
15	17		17	22	29		34	39	46
20	22	22	30	40	50	44	52	62	72
25	28	31	46	64	80	59	74	92	108
30	33	45	67	92	115	78	100	125	148
35	39	58	92	125	160	97	131	164	199
40	44	80	125	165	205	124	169	209	249
45	50	103	165	210	260	153	215	260	310
50	55	131	225	255	320	186	280	310	375
55	61	165	275	310	390	226	336	371	451
60	66	202	350	370	465	268	426	436	531

Typical Brake Performance

A–Average automobile
B–Light two-axle trucks
C–Heavy two-axle trucks
D–Three-axle trucks and trailers

Table 6.2 (Metric)
Braking and Stopping Distances (dry, level pavement)

Speed (km/h)	Average Driver Reaction Distance (meters)	Braking Distance (meters)				Total Stopping Distance (meters)			
		Vehicle A	Vehicle B	Vehicle C	Vehicle D	Vehicle A	Vehicle B	Vehicle C	Vehicle D
16	3.4		2.1	3.0	4.0		5.5	6.4	7.3
24	5.2		5.2	6.7	8.8		10.4	11.9	14.0
32	6.7	6.7	9.1	12.2	15.2	13.4	15.8	18.9	21.9
40	8.5	9.4	14.0	19.5	24.4	18.0	22.6	28.0	32.9
48	10.1	13.7	20.4	28.0	35.1	23.8	30.5	38.1	45.1
56	11.9	17.7	28.0	38.1	48.8	29.6	39.9	50.0	60.7
64	13.4	24.4	38.1	50.3	62.5	37.8	51.5	63.7	75.9
72	15.2	31.4	50.3	64.0	79.2	46.6	65.5	79.2	94.5
80	16.8	39.9	68.6	77.7	97.5	56.7	85.3	94.5	114.3
88	18.6	50.3	83.8	94.5	118.9	68.9	102.4	113.1	137.5
96	20.1	61.6	106.7	112.8	141.7	81.7	129.8	133.0	161.8

Typical Brake Performance

A–Average automobile
B–Light two-axle trucks
C–Heavy two-axle trucks
D–Three-axle trucks and trailers

The correlation between vehicle weight and speed and stopping distance should be obvious. At the same speed, it will take a greater distance to stop a fully-loaded, three-axle water tender than a lighter vehicle. It will also take a greater distance to stop a vehicle that is going 50 mph (80 km/h) than the same vehicle traveling at 30 mph (48 km/h).

The ability to stop a fire apparatus depends greatly on the type and condition of the braking system. Poor maintenance of the braking system can contribute to serious accidents. Obviously, a vehicle with a properly maintained braking system will stop faster than one with a system in disrepair.

Tire condition can be a significant factor in stopping. Ones with worn tread will have less road grip than tires with full tread, especially on wet roads. Tires with treads designed for snow conditions will perform better in those conditions than general use tires. However, some snow tires are at a disadvantage in dry road conditions.

Auxiliary traction devices (cable chains, link chains, and automatic traction devices) all require special driving and handling skills. If road conditions require the use of these devices, slower speeds and increased stopping distances must be used.

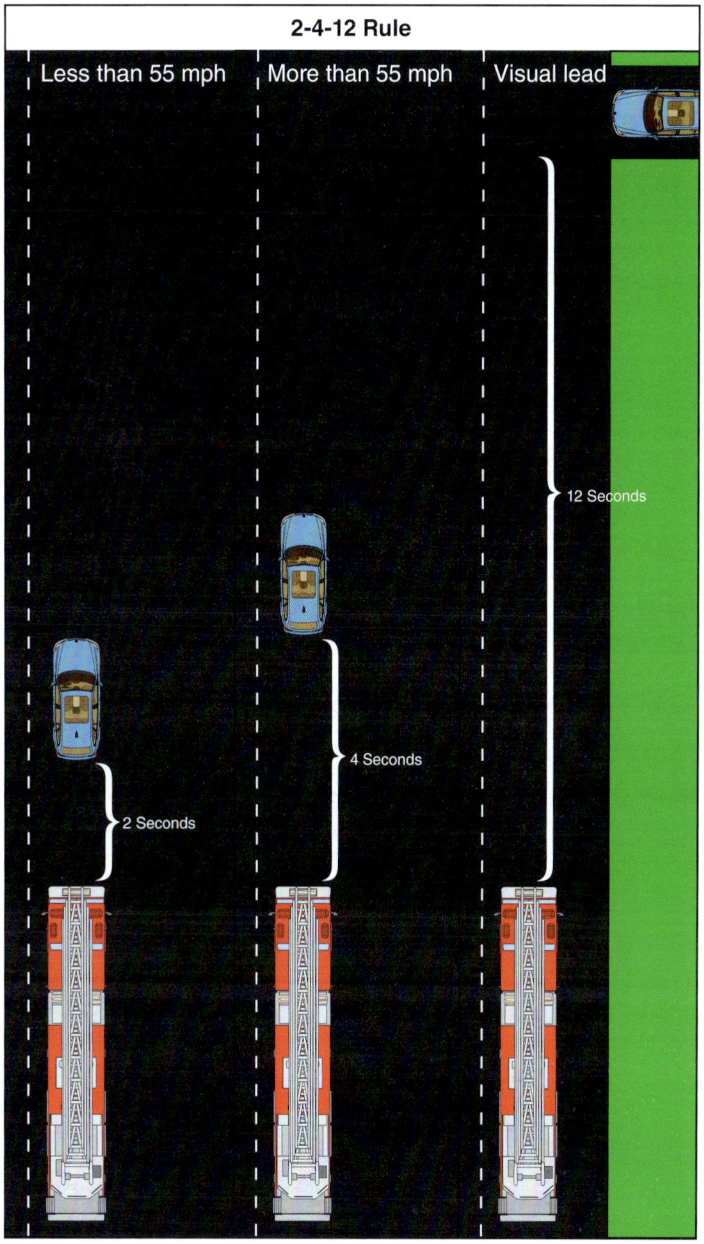

Figure 6.25 The 2-4-12 rule helps drivers maintain a safe following distance regardless of speed.

Following Distance

Whether responding to an emergency or simply driving a personal car on an errand, maintain a safe following distance at any speed. To determine the proper following distance use the 2-4-12 rule **(Figure 6.25)**. This rule helps drivers to see conditions well ahead of their apparatus and to take action early. The 2-4-12 rule states the following:

- Maintain a two-second interval between your vehicle and the vehicle ahead for speeds below 55 mph (88 km/h).
- Maintain a four-second interval between your vehicle and the vehicle ahead for speeds above 55 mph (88 km/h). This allows for increased stopping distances required by higher speeds.
- Maintain a visual lead time of 12 seconds; that is, look ahead for possible hazards and potential alternate paths of travel in case an emergency occurs. The 12-second visual lead time permits the driver to make continual, smooth adjustments to speed and position.

VFIS recommends a slightly different means of calculating safe following distance. It suggests that driver/operators allow one second of following distance for every 10 feet (3 m) of vehicle length for speeds under 40 mph (64 km) and add one additional second for each 10 mph (16 km) for speeds over 40 mph (64 km). That would translate into 4 seconds for speeds under 40 mph (64 km) and 5 seconds for 50 mph (80 km).

Adverse Weather

Weather is another factor to consider in terms of safe driving. Rain, snow, ice, and mud make roads slippery. Driver/operators should decrease speed gradually, slow down while approaching curves, keep off low or soft shoulders, and avoid sudden turns. Driver/operators should recognize areas that first become slippery, such as bridge surfaces, northern slopes of hills, shaded spots, and areas where snow is blowing across the roadway.

Because the stopping distance is greatly increased on wet and icy surfaces, it is prudent to apply the brakes carefully while in an area free of traffic to find out how slippery the road is. Speed and the use of auxiliary braking systems must be adjusted to road and weather conditions so that the apparatus can be stopped or maneuvered safely.

NOTE: Departments may have additional equipment such as tire chains or auxiliary traction control devices. Keep abreast of manufacturer and department recommendations for their safe and effective operation.

Braking bad

Auxiliary braking devices such as engine brakes, transmission retarders, or electromagnetic driveline brakes may cause the rear wheels to lock on slick roadways with very little pressure applied to the brake pedal. Based upon the manufacturer's policy, the performance of the vehicle, and the experience of the driver/ operator, the auxiliary brake may be turned off at the dash-mounted switch.

Multiple Unit Responses

There have been numerous collisions involving fire apparatus and other emergency vehicles. It is not always possible to hear the warning devices of other emergency vehicles when the audible warning devices on both vehicles are sounded. When more than one emergency vehicle is responding along the same route, units should travel at least 300 to 500 feet (90 m to 150 m) apart. Some fire departments rely upon designated response routes. This practice can be hazardous if a company is delayed or detoured. SOPs may call for radio reports of location and status, particularly when vehicles might be approaching the same intersection. Regardless of the system or pattern used, always take precautions to ensure a safe, collision-free response. This includes coming to a complete stop before proceeding at any intersection that has a stop sign or a red signal light.

Apparatus Placement at the Incident Scene

Apparatus, crew, and driver/operator safety continue as the unit arrives at the emergency scene. Safety can be enhanced by apparatus placement, control of traffic, and other activities.

Roadways are dangerous work environments, especially at night or in rainy or icy conditions. Aerial apparatus driver/operators must be aware of wind and overhead power lines when positioning the apparatus. In addition, the slope of the grade and condition of the soil surface will affect the position of the aerial stabilizer jacks.

At the emergency scene, the driver/operator becomes vulnerable to the actions of other drivers. While standing at the pump panel, attaching hoselines, or removing equipment along the roadside, the driver/operator and other personnel become targets for inattentive drivers. Crew members are also exposed as they gain access to apparatus compartments. Place traffic cones and other control and warning devices at the rear of the apparatus to direct traffic away from the work area.

The driver/operator, under the supervision of the company officer, is responsible for placing or positioning the apparatus at the emergency incident scene. The placement of the vehicle depends on the type of incident and location. The sections that follow suggest where to place apparatus in the following situations:

- Structural incidents
- Wildland incidents
- Hazardous materials incidents
- High-rise incidents
- Technical rescue incidents
- Medical incidents
- Highway incidents

Structural Fire Scenes

At a structural fire scene, vehicles directly involved in fire suppression (engines, aerial apparatus) should be positioned to facilitate fire suppression operations **(Figure 6.26)**. These vehicles provide the tools and equipment needed to gain access, perform search and rescue, apply extinguishing agents, and protect property through ventilation and loss-control activities. Additional apparatus that provide power for electric or hydraulic tools, or incident scene lighting, may also be positioned in this area.

Figure 6.26 Once a pumper is positioned and connected to a water supply, it is very difficult to relocate. *Courtesy of Ron Moore, McKinney, TX, Fire Department.*

Aerial devices should be strategically positioned to provide access to upper stories or to place elevated fire streams into operation. When positioning these devices, consider the length and type of aerial device, as well as overhead obstructions like power lines, street lights, or traffic control devices. The aerial device must be properly stabilized before the aerial device is operated. Narrow, crowded streets or uneven terrain can affect the stability of these vehicles. See the IFSTA Aerial Apparatus Driver/Operator Handbook for additional information.

Wildland Fire Scenes

At a wildland scene, driver/operators can expect access and egress difficulties. Place apparatus with primary consideration for safety, mobility, and efficiency. Special considerations are fuel sources, terrain, and potential smoke **(Figure 6.27, p. 212)**.

Figure 6.27 Wildland fire trucks must be able to traverse rugged off-road terrain. *Courtesy of Bob Esposito, Pennsburg, PA.*

Some structural fire suppression apparatus are not designed for off-road use. The angle of approach and departure, and the ground clearance on the underside of the apparatus are inadequate to clear some obstacles. Due to a higher center of gravity, the height of the apparatus can cause it to overturn on steep inclines. Even though all-wheel- or four-wheel-drive vehicles have superior climbing ability, they typically have a high center of gravity that makes them susceptible to rollover. In addition, the weight of an apparatus can cause it to become stuck in soft, loose, or wet soil.

NOTE: For more information, see the IFSTA **Wildland Fire Fighting for Structural Firefighters and Pumping Apparatus Driver/Operator Handbook** manuals.

Hazardous Materials Incidents

Hazardous materials incidents can pose very complex, rapidly changing problems. General-placement and positioning guidelines for apparatus at hazardous materials incidents include the following:

- Park apparatus upwind of the incident scene and uphill whenever possible.
- Position apparatus well away from downed power lines, damaged transformers, or escaping flammable gas.
- Place apparatus so that it can be repositioned easily and quickly.
- Place or position apparatus so that its hoselines can protect the route of egress through the spill when rescue is required.

Figure 6.28 High-rise structures create challenges to apparatus placement.

High-Rise Incidents

Incidents in multistory or high-rise structures create special requirements for apparatus placement and positioning **(Figure 6.28)**. The structures may be located in a series or row of similar structures with only the front and rear of the structure accessible. They may also be independent of other structures and located on individual parcels of land, surrounded by parking lots or connected to parking structures.

High-rise structures are generally considered to be at least seven stories in height. Some of the following characteristics influence apparatus placement and positioning at high-rise structures:

- Limited ground-level access with personnel access doors located only in the front and rear walls (when the building is in a series of structures).
- Intermediate parking garage areas between grade level and upper occupied floors, which increase the distances for transporting personnel and equipment. Parking levels will also be filled with vehicles that can be a source of ignition or fire.

- Fences, planting beds, decorative walls, or other barriers near the base of independently located structures.
- Limited street frontage.
- Building setbacks from the street, parking areas, or driveways, which may be obstructed by barriers in the setback area.

If possible, apparatus must be located away from any potential collapse zones. A collapse zone is described as 1.5 times the height of the wall or structure. Collapse concerns apply to all types of emergency scenes.

NOTE: For more information, see the IFSTA **Structural Fire Fighting: High-Rise Fire Fighting** manual.

Technical Rescue Incidents

At a technical rescue scene (water, ice, confined-space, structural collapse, trench, or high-angle), rescue vehicles (or apparatus carrying rescue tools and equipment) should be positioned nearest the incident. Place apparatus used for fire suppression operations (engines, ladders, and trucks) in the staging area.

Place apparatus close enough to unload or support powered tools and equipment, and consider the weight of the apparatus. For example, the weight of the apparatus and engine vibration can cause soil in a trench-collapse area to cave in. Vehicles that do not need to be close to the scene, including Command vehicles and POVs, should be parked in a staging area at least 150 feet (50 m) from the incident. All vehicle roadway traffic should be stopped if it is within 300 feet (100 m) of a trench collapse rescue.

Medical Incidents

Many medical emergencies are time-critical, requiring that a specific level of care be provided as quickly as possible. Cardiac arrest, asphyxiation, and massive blood loss are just three medical emergencies that demand immediate care. Guidelines for apparatus and ambulance placement and positioning during medical incidents include the following:

- Place apparatus and ambulances as close to the incident as possible.
- Park units that are designed for patient transport with clear access to the loading door.
- Position vehicles far enough from entry to or exit from the emergency site to make moving a patient gurney simpler and easier.
- Position apparatus so that victims or emergency responders are not exposed to vehicle exhaust, vibration, or noise.
- For active-shooter incidents, place apparatus between rescuers and the area of the shooter. Contact with law enforcement prior to entering the scene area is essential to the safety of emergency medical providers.

Highway Incidents

Highway incidents pose special problems for emergency vehicles. Traffic congestion, higher speeds, and inattention on the part of other drivers add to the risk posed to emergency services personnel. There should be a safe response followed by protection of personnel, patients, and other drivers at an incident. This should be followed by an attempt to allow continued traffic flow at a slower, safer speed around the incident site. Some states require that law enforcement officers (highway patrol) serve as the designated Incident Commands (ICs) during highway incidents. Operations should be coordinated with them in a spirit of cooperation to resolve the incident.

Firefighters and other emergency responders come in contact with other vehicles at a roadway incident scene due to the following factors:

- **Lack of training** — Responders are not trained in the hazards associated with roadway incidents and the proper ways to minimize their occurrences.
- **Lack of situational awareness** — Due to insufficient training or lack of experience,responders fail to recognize the dangers associated with the particular roadway situation they are facing.
- **Failure to establish a proper temporary traffic-control zone** — Many fire departments do not have sufficient training, equipment, or SOPs to set up a properly marked work area when operating at a roadway incident scene. In some cases, responders did have good training, equipment, and SOPs, but failed to use or follow them.
- **Improper positioning of apparatus** — Numerous cases have been cited where apparatus were not positioned in a manner that protected the work area. In other cases, apparatus were unnecessarily positioned in the roadway.
- **Inappropriate use of or lack of scene lighting** — Inappropriate use of vehicle headlights, warning lights, and floodlights can confuse or blind approaching motorists. This may cause the drivers to strike an emergency vehicle, responder, or other vehicle in the incident area.
- **Failure to use safety equipment** — Responders working in the roadway must wear appropriate protective garments and use all available traffic control devices in order to prevent being struck by oncoming traffic.
- **Careless, inattentive, or impaired drivers** — Even when everything is done correctly, there are drivers who will not react appropriately to the altered traffic pattern that occurs at a roadway incident and may drive directly into work zones.
- **Reduced vision driving conditions** — Although firefighters may be struck by vehicles at any scene, the chances are greater during obscured vision conditions, including darkness, fog, rain, snow, and blinding sunshine.
- **Altered traffic patterns** — Drivers may be confused by the traffic-control measures used at an incident scene or those being employed in a construction zone.

In response to this information, the USFA and the Department of Transportation/Federal Highway Administration (DOT/FHWA) formed a partnership with the International Fire Service Training Association (IFSTA) to research and identify effective technical guidance and training programs for fire and emergency service providers in Traffic Incident Management Systems (TIMS).

The purpose of the project was to enhance responder safety and provide guidance to local level fire departments on compliance with the Manual on *Uniform Traffic Control Devices* (MUTCD) from DOT/FHWA, which is available on the MUTCD website. *Model Procedures Guide for Highway Incidents* by the National Incident Management System Consortium recommends the following:

- Always establish an Incident Command organization at the scene. In more complex operations, a formal unified command structure may be necessary to ensure close cooperation among the involved agencies
- Establish a temporary traffic control (TTC) zone or traffic incident management area (TIMA) **(Figure 6.29)**. This zone includes:
  - Advance Warning Area: Folding signs or lighting are used to get the attention of drivers.
  - Transition Area: Cones are used to redirect traffic into an alternate lane.
  - Activity Area: The apparatus is positioned as a barrier to protect workers and patients.

Figure 6.29 Roadway incidents are marked to warn drivers, redirect traffic, and protect workers and patients.

- If necessary, assign a member of the unit to direct traffic into an alternate lane. This person must wear a retroreflective vest, be equipped with a hand light at night and a flag during the day, and have been trained as a flagger.
- Turn off white spot lights or direct them into the activity area rather than at oncoming traffic. Spotlights can blind drivers.
- Use directional arrows to indicate the required lane change. Generally, these arrows are amber in color.
- Make sure all personnel wear reflective vests or PPE with reflective trim.
- Position the apparatus to provide an effective barrier to the work area.
- For incidents located in the inner lanes of a divided highway or in the median, it may be necessary to position apparatus in the inner lanes of both directions of travel to protect the incident scene. In some areas, such as highways with very wide dividing medians or dividing walls, placing apparatus on both sides of the median would be counterproductive.

General Safety Concerns

There are other operational safety concerns that may not fall into a specific category. These include the securing of tools and equipment, the addition of cabinets or lockers designed for medical equipment, and the refueling procedures for apparatus. These concerns overlap design requirements and crew activities.

Securing Equipment

In a motor vehicle accident involving a fire or emergency vehicle, tools and equipment can become deadly projectiles. NFPA 1901 requires that all tools be securely mounted when carried in the crew compartment. SCBAs must be firmly clamped into the passenger seat back or into a quick-release bracket. Spare air cylinders must also be secured in crash-resistant brackets. Hoses must be secured.

In the body of the apparatus, tools and equipment should be secured to compartment walls, slide-out panels, or trays. In these areas, securing the tools is less of a safety issue and more of an issue of efficiency and damage control. Secure heavy equipment such as generators or ventilation fans to the cabinet floor or to roll-out brackets to improve access and prevent shifting during travel.

The gurney must be secured to the floor in the patient compartment of ambulances, and the passenger and attendant seats must have seat belts. All cabinet doors must have positive latches to prevent them from opening in an accident.

Refueling

Apparatus can be refueled at fixed fueling sites, from fuel trucks, or at temporary dispensing sites during emergency operations. In any case, care must be taken in transferring fuel into the apparatus. Never assign refueling to anyone not trained to perform the task. Make sure a Class B extinguisher is available.

Fixed Fueling Site

The person filling the apparatus remains at the pump nozzle during the filling operation. As the possibility of static electricity exists, the person filling the tank should not reenter the apparatus until the operation is complete. Smoking must be prohibited during fueling operations. Spilled fuel should be cleaned up immediately.

Fuel Truck

Transferring fuel from the fuel tank truck to an apparatus may take place at the station, emergency scene, or any other location. When refueling at the station, the operation should take place outside the apparatus bay to prevent vapors from building up in the structure. The fuel truck and apparatus should be connected with a grounding wire to prevent a static charge. Spills should be cleaned up as soon as possible.

Temporary Dispensing Sites

During long-duration emergency operations, it may be necessary to establish temporary dispensing sites. Fuel may be transferred from fuel trucks, 55-gallon (220 L) drums with hand pumps, or from 5-gallon (20 L) fuel cans. Safety precautions include no smoking during the operation, spill or leak prevention, grounding or bonding to eliminate static electricity, and cleaning up all spills immediately.

Chapter Review

1. What are five common causes of apparatus accidents?
2. What types of audible and visual warning devices should be included on all emergency response vehicles and apparatus?
3. Why should driver/operators be familiar with basic defensive driving concepts?
4. How can hazards hinder the placement of an apparatus at different types of local emergency incidents?
5. What are four safety precautions that should be followed during refueling?

Discussion Questions

1. Have you had any experience with preventive or corrective vehicle maintenance? Why are these types of maintenance activities crucial to safely operating an emergency apparatus and other types of vehicles?

Tools and Equipment Safety

Chapter Contents

chapter 7

Key Terms

FESHE Learning Outcomes

Occupational Safety and Health for Emergency Services

4. List and describe the components of risk identification, risk evaluation, and incident management.
5. Describe the relevance for safety in the workplace, including the importance of PPE.

Tools and Equipment Safety

Learning Objectives

After reading this chapter, students will be able to:

1. Describe how data collected from tools and equipment accidents is used in causation-based investigations.
2. Explain how regulations and manufacturer responsibilities influence tool and equipment safety features and design.
3. List general and specific inspection points for common fire service tools and equipment.
4. Describe design features and policies and procedures that can reduce potential hazards in a maintenance area.
5. Review safety-centered maintenance methods for batteries, unpowered and powered tools and equipment, and portable fire extinguishers.
6. Differentiate between cleaning and disinfecting rooms.
7. Determine appropriate storage and risk management methods for common hazardous materials.
8. Explain the importance of testing and documentation following tool repairs.
9. Describe how a firefighter can use situational awareness to select, carry, and use tools and equipment.

Chapter 7
Tools and Equipment Safety

Fire fighting tools and equipment are vital to efficient, effective, and safe operations at emergency and nonemergency incidents. Powered tools have improved both efficiency and effectiveness of emergency and nonemergency operations. When used properly, they can minimize hazards to the firefighter. However, if not used and maintained properly, tools can increase the risk of injury during operations. See **Table 7.1** for a list of hand tools and **Table 7.2 (both tables on page 222)** for a list of power-operated tools commonly used in the fire service.

This chapter describes recommended tool and equipment design criteria, operations, inspections, and maintenance practices. It also addresses the consequences of unsafe practices. Injuries and death often result from improper use and maintenance of tools and equipment; lack of knowledge; insufficient skills; and careless attitudes.

Equipment Accidents

Unsafe behaviors while using tools and equipment can result in injuries; permanent disabilities; fatalities; damage to or loss of tools and equipment; and delayed completion of assigned tasks. Tool or equipment failure and accidents can also be caused by poor tool choice, inadequate training, and a lack of periodic inspections, preventive maintenance, or repairs. Recognizing the results and causes of tool and equipment accidents is the first step in reducing or eliminating them.

Data Collection

Cuts, abrasions, contusions, and broken bones occur in the fire and emergency services. It is reasonable to assume that some of these injuries are due to the unsafe use of tools and equipment, or tools and equipment that are in poor condition and/or infrequently serviced. Some injuries, such as hearing loss, eye injuries, or cuts, are caused by a lack of personal protective equipment (PPE) or failure to use it correctly.

It is difficult to statistically link firefighter injuries to the use of tools and equipment. Sharing near misses can help identify root causes, prevent future accidents, and improve safety. Near Miss, managed by the IAFC, collects and shares near-miss experiences of firefighters. The program was launched in 2005 to capture, share, and analyze near-miss reports. Reports may be submitted anonymously on the Firefighter Near Miss website or through a department-level online reporting tool. Stories range from the failure of a coupling caused

Table 7.1
Hand Tools

Cutting	Striking	Prying	Pushing/Pulling	Wildland Tools	Misc. Tools
Pick-head axe	Sledgehammer	Crowbar	Standard pike pole	Pick-head axe	Shovel, scoop
Flat-head axe	Maul	Halligan bar	Clemens hook	Flat-head axe	Broom, push
Bold cutters	Battering ram	Pry (pinch) bar	Plaster hook	McLeods	Mop
Carpenter's saw	Pick	Hux bar	Drywall hook	Pulaskis	Squeegee
Keyhole saw	Flat-head axe	Claw tool	San Francisco hook	Brush hook	Mop bucket
Drywall saw	Mallet	Kelly tool	Multipurpose hook	Pick	Water key
Hack saw	Hammer	Pry axe	Roofman's hook	Fire rakes	
Wire cutters	Punch	Flat bar (nail puller)	Rubbish hook	Hoes	
Burning bars	Chisel	Ram bar		Wire Brooms	
Cutting flares				Flails	
				Combination tools	
				Shovel, pointed tip	

Table 7.2
Power-Operated Tools

Saws	Cutting	Prying	Ventilation	Wildland Tools	Power Plants
Circular saw	Rebar cutter	Hydraulic door opener	Fans, electric, gas, hydraulic	Chain saw	Gasoline generator
Reciprocating saw	Plasma cutter	Extrication or rescue tools		Power (leaf) blower	Wire reels
Chain saw	Oxy-gasoline cutting torch			Portable pumps	
Ventilation saw	Oxyacetylene cutting torch				
Rotary saw					

by a design flaw to the result of a water hammer in a hoseline caused by unsafe behavior. Any fire department can create its local near-miss database and use it to help identify issues (tool selection, PPE, training, maintenance), or address funding needs that should be considered by HSOs or safety committees to avert accidents.

Firefighter fatalities are easier to associate with tool or equipment usage since incident-related fatalities are more thoroughly investigated and reports are made public. However, when the relationship between a tool or piece of equipment and a fatality is indirect, a tool or piece of equipment may not be reported as a cause. For instance, when the failure of a forcible entry tool prevents a firefighter from a successful self-escape, the cause of the fatality may be listed as asphyxiation.

Appropriate data collection of tool and equipment accidents and/or failures can provide the basis for safety training and preventive maintenance. Maintenance and repair staffers are good sources of information regarding frontline tool care and use issues. On a department level, the employee/management safety committee should examine injuries for their relationship to tool failure and use.

Causation

When causation is the main focus of reviews or investigations, the need to place blame on one person or a crew of individuals is not the primary function. As a result, participants are more likely to explain the reasons for their actions. The organization can then work to prevent future occurrences rather than take precautions to protect themselves from reprisal. This is especially true if organizational SOPs or management policies support protection of the individual, unless it can be determined that his/her actions were deliberate or careless. The discovery findings should be turned over to safety committees for the development of an organizational improvement recommendation report.

Every tool requires user training and knowledge of appropriate use. Tools also require routine maintenance, inspection, and repair. They lose reliability and usefulness if they are poorly cared for; if users are inadequately trained in their use; or if they are misused due to the pressures of an event. Accidents that result from tool failure can usually be attributed to one or more of these issues.

Figure 7.1 Material fatigue to hose jackets or couplings can lead to a loss of water pressure at the nozzle.

Equipment Failure

Tools and equipment can fail for many reasons. Some of the reasons may be directly or indirectly linked to poor choices on the part of the tool operator, maintenance personnel, administration, or manufacturer. Tool and equipment failures may be caused by the following:

- Inappropriate specifications
- Design flaws
- Unreported/unrepaired damage
- Use beyond design, life, or safety limits
- Material fatigue **(Figure 7.1)**
- Lack of preventive maintenance
- Lack of periodic or scheduled inspections

When an accident review indicates that causation is related to manufacturer error or inappropriate design, the manufacturer should be notified immediately. If the manufacturer agrees with the report, the company will make changes to the tool or equipment or will issue a product safety warning to all organizations that own similar tools or equipment.

Some accident reporting will result in a product recall. The fire and emergency services administration must monitor product recalls, regularly checking information trade journals and websites.

Safe Tool and Equipment Design

Most tools and equipment used by fire and emergency services organizations are designed for emergency operations. Many are required to meet NFPA design standards, while others must meet federal standards. Firefighters who are assigned the task of selecting or purchasing tools and equipment for an organization must be familiar with applicable standards and regulations.

Intrinsically Safe — Describes equipment approved for use in flammable atmospheres; must be incapable of releasing enough electrical energy to ignite the flammable atmosphere. Formerly known as *Explosion Proof.*

Because tool and equipment manufacturers must meet design standards and regulations, as well as industry standards for materials, they are responsible for recommending or requiring how their products are to be used, stored, cleaned, and maintained. Organizations must follow the manufacturer's recommendations when using their tools and equipment. The organization is also responsible for training employees in the proper use, care, and maintenance of tools and equipment.

Tools and equipment that may be used in an atmosphere containing flammable vapors must be **intrinsically safe**. Intrinsic safety (IS) is a protection technique for safe operation of electronic equipment in explosive atmospheres. Theoretically, IS ensures the available electrical and thermal energy in the system is always low enough to prevent ignition of the hazardous atmosphere. This is achieved by ensuring that only low voltages enter the hazardous area. Hand tools such as scoop shovels must be constructed of materials that will not cause a spark. Communications equipment, such as radios and ventilation fans, must also be intrinsically safe **(Figure 7.2)**.

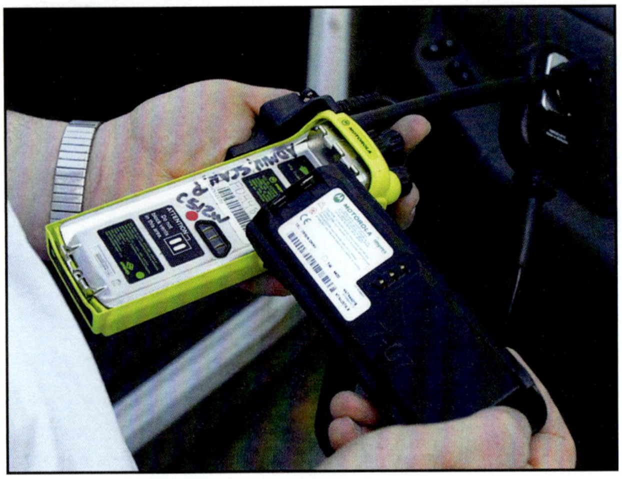

Figure 7.2 Intrinsically safe radios are designed for use where flammable vapors may be present.

Apparatus and portable power supply units may have ground fault circuit interrupter (GFCI) receptacles installed to reduce the risk of electrocution. These receptacles shut off the flow of electricity through the system if a short circuit occurs. GFCIs on equipment are similar to those found in building areas that contain water or moisture, such as kitchens and bathrooms.

Most powered tools are designed with safety guards and features. Safety guards are intended to protect the operator's hands and fingers from contact with blades or chains, depending on the tool.

Automatic power shutoff switches, known as *dead man's switches*, are required on all powered tools to prevent them from working if the operator releases the switch. Eye guards are also provided on tools that produce sparks or metal shavings, such as grinders. These required safety guards and switches must not be tampered with or removed.

Tool and Equipment Inspection

When tools and equipment are purchased, they must be inspected and, in some cases, tested to determine if they meet the purchasing specifications. Hand tools are generally inspected when they are received and placed into service. Powered equipment, fire hose, respiratory protection equipment, and generators must be acceptance tested when received. Requirements for acceptance testing are usually included in the purchase contract or agreement.

Following the acceptance tests, periodic inspections and tests are performed to ensure that tools and equipment remain in good condition. The local jurisdiction establishes inspection schedules based on federal regulations, NFPA standards, and manufacturer's recommendations. These mandated inspections, which may include testing by third-party organizations, apply to equipment such as respiratory protection and breathing air equipment, powered rescue tools, ground ladders, fire hose, and generators.

In addition, the individual fire and emergency services organizations will require frequent inspections performed by fire and emergency responders. These inspections occur at the beginning of each work shift, weekly, monthly, or after each use of the tool or equipment. The following sections present general inspection points for all tools and equipment and specific inspection points for fire fighting tools and equipment.

Figure 7.3 A firefighter conducts a thorough tool inspection after use.

General Inspection Points

Firefighters must inspect tools and equipment at the beginning of each work shift, weekly, monthly, and after each use **(Figure 7.3)**. A thorough safety inspection should check for the following:

NOTE: Remember to wear proper PPE when performing maintenance or inspections.

- Damage
  - Broken parts
  - Chipped blades
  - Missing parts
  - Dents, cracks, discoloration or splits
- Wear
  - Worn or dull surface
  - Chipped paint
  - Metal fatigue; cracks or splits
  - Worn belts
  - Corrosion
- Lack of cleanliness
  - Dirt, grime, or debris
  - Contamination; medical or hazardous material
- Proper operation of powered equipment
  - Unit starts and runs per normal operation
  - Unit opens or closes per normal operation
- Effects of heat or direct flame contact
  - Blackened surface
  - Singed surface, bubbled paint or material
  - Charring
- Cylinder hydrostatic testing date
  - Includes all self-contained breathing apparatus (SCBA) cylinders
  - Includes fuel and fire extinguisher cylinders

Specific Inspection Points

By nature of their design and use, some tools and equipment require inspections that exceed the general points mentioned previously. The following sections discuss inspection points used to more closely examine specific tools and equipment used in the fire service.

Ground Ladders

All fire service ground ladders must meet the inspection requirements of NFPA 1932, *Standard on Use, Maintenance, and Service Testing of In-Service Fire Department Ground Ladders.* It states that ground ladders must be service tested annually; inspected monthly and after each use; and inspected according to the manufacturer's recommendations.

When performing a visual inspection monthly or after each use, confirm that:

- Heat-sensor labels on ladders have not changed color, indicating heat exposure.
- All rungs are snug where they connect to the beams.
- All bolts and rivets are tight; bolts on wood ladders are snug without crushing the wood.
- Welds are free of cracks or apparent defects.
- Beams and rungs are free of cracks, splintering, breaks, gouges, checks, wavy conditions, or deformation.
- Butt spurs are free of excessive wear or other defects, and firmly attached to the ladder.
- Halyards are free of fraying or kinking **(Figure 7.4)**.
- Roof hooks are sharp and operate properly without binding.
- Rungs are free of punctures, wavy conditions, worn serrations, or deformation.
- Surfaces are free of corrosion.

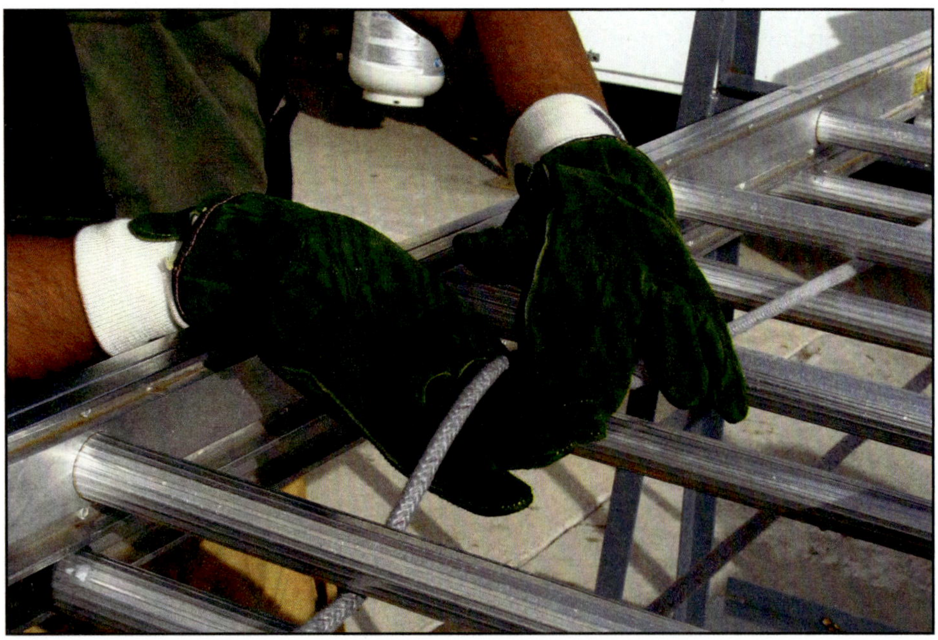

Figure 7.4 All components of an extension ladder must be carefully inspected, including the halyard cable.

Ropes

Fire service rope falls into two classifications: **escape rope** and **life safety rope**. Escape rope is solely used as a lowering rope for firefighters who need to leave a hostile environment. Life safety rope is used to support rescuers and/or victims during emergency incidents or training exercises.

Escape Rope – A single-purpose, emergency self-escape (self-rescue) rope; not classified as a life safety rope.

NOTE: Common utility rope is not classified.

All rope, life safety harnesses, belts, and other auxiliary equipment must conform to NFPA 1983, *Standard on Life Safety Rope and Equipment for Emergency Services.* This standard specifies that only rope of block creel construction (without knots or splices in fibers) using continuous filament virgin fiber for load-bearing elements is suitable for life safety and escape rope applications. Rope made of any other material or construction must not be used to support firefighters or victims.

Life Safety Rope — Rope dedicated to supporting people during rescue, fire fighting, other emergency operations, or training evolutions.

All types of rope can be damaged by high stresses, impact loading, abrasions, kinking, and exposure to heat and chemicals **(Figure 7.5)**. All rope should be inspected by qualified personnel, according to manufacturer recommendations. A record-keeping system must be established for all life safety ropes. Information regarding inspections, after-incident reviews, and in-service dates are kept in the record.

Figure 7.5 Rope must be inspected regularly to ensure safe use.

- Inspect all types of rope after each use.
- Inspect unused ropes at least annually. Inspect rope visually and by touch.
- Inspect rope for embedded shards of glass, metal shavings, wood splinters, or other foreign objects that could cause damage. If any of these are found, take the rope out of service.
- Document all inspections in a rope log.

Any inspection or other test for certification or recertification purposes must be conducted by a certification organization accredited under International Standards Organization (ISO). There are performance requirements and numerous test methods for each rope classification and all rope-related auxiliary equipment. Reference NFPA 1983 for these performance and test methods if an organization carries rope for any type of rescue work.

Fire Hose

Annual service tests are required for all fire hose according to NFPA 1962, *Standard for the Care, Use, Inspection, Service Testing, and Replacement of Fire Hose, Couplings, Nozzles, and Fire Hose Appliances.* The tests ensure a hose is able to maintain pressure without leaking or breaking at a coupling. Hose is inspected monthly, usually when the apparatus hose load is removed and repacked, or after use as it is reloaded onto the apparatus. After each use, all hose must be cleaned, stored out of direct sunlight in a well-ventilated area, and not stored until it is completely dry.

Inspect fire hose for the following conditions:

- Dented or damaged couplings
- Damaged coupling threads
- Missing or damaged coupling gaskets
- Abraded or damaged exterior hose jacket
- Discoloration of exterior hose jacket due to heat exposure **(Figure 7.6)**
- Slippage during testing
- Presence of mildew or mold on hose jacket or lining
- Lining detached from jacket
- Lining detached from coupling
- Unclean or contaminated jacket

Figure 7.6 A firefighter inspects hose for any signs of damage.

Portable Fire Extinguishers

Annual inspections and tests of extinguishers are listed in NFPA 10, *Standard for Portable Fire Extinguishers*. Persons conducting maintenance and recharging of extinguishers must be certified. Each extinguisher class (A-K) has maintenance and inspection frequency requirements as set out by NFPA 10. At a minimum, all extinguishers must be inspected annually. Items to look for when inspecting extinguishers include the following **(Figure 7.7)**:

- Damage to the extinguisher shell or cylinder
- Damage to discharge nozzle, horn, or hose
- Obstructions in nozzle, horn, or hose
- Locking pin or tamper seal to ensure they are in place
- Correct pressure on gauge of stored-pressure extinguisher
- Hydrostatic test date on cylinder to ensure that it is current
- Evidence that the extinguisher has been discharged and not refilled

Figure 7.7 Portable fire extinguishers must be inspected according to NFPA standards.

Electronic Equipment

New electronic equipment becomes available annually for the fire and emergency service. More sophisticated technology means that equipment may be more sensitive to abuse, damage, or the environment. Electronic equipment must be inspected at the start of each work shift and after each use. In some cases, it may also be necessary to recalibrate the equipment or charge batteries. A trained and certified technician is required to calibrate some equipment. However, with proper training, most personnel can calibrate the electronic equipment they use. Examples of electronic equipment include thermal imagers (TIs), dosimeters, and gas meters.

Tool and Equipment Maintenance

Tool and equipment maintenance is the responsibility of fire and emergency responders, as well as maintenance technicians employed by, or under contract to, the jurisdiction. When an inspection reveals the need for maintenance, it should occur as soon as possible. Local policy and procedures will dictate how tools and equipment are removed from service and repaired or discarded.

Maintenance Area Safety

While some tool and equipment maintenance, such as respiratory protection equipment and radio maintenance, must be performed by certified technicians, much of the work can be done by fire and emergency responders. However, training is necessary to use powered maintenance equipment safely and to repair tools and equipment according to the manufacturer's recommendations. All required PPE — including hand, eye, respiratory, and hearing protection — must be worn while performing maintenance.

The potential for injuries in maintenance areas can be reduced through the design of the space and by keeping it clean and orderly. The preferred method for ensuring safety is through risk reduction during the design of the shop or maintenance area. Other safety hazards can be controlled by adding safeguards to the individual equipment. Some of these design features include the following:

Figure 7.8 Designated maintenance areas for self-contained breathing apparatus (SCBA) contain spare parts, repair tools, testing equipment, and cleaning areas. *Courtesy Cherry Hill (NJ) Fire Department.*

- Eye shields and mechanical safeguards installed on all powered tools
- Large tools such as grinders and drill presses secured to the floor, work-bench, or wall
- Noise-reducing materials installed on walls, floors, and ceilings where possible
- General room ventilation provided to filter and remove particulates, vapors, and odors from the area
- Spray-paint booths provided in any area where spray-painting takes place
- Designated rooms for the maintenance of SCBA regulators and oxygen regulators **(Figure 7.8)**
- Adequate lighting; fixtures equipped with screens to prevent or contain light bulb breakage
- Electrical receptacles equipped with ground fault circuit interrupters
- GFCIs to reduce shock hazards in the work area

Potential hazards can be controlled in the maintenance area through the following policies and procedures that promote cleanliness in the area:

- Keep maintenance areas clean and neat.
- Store tools in a designated area when not in use.
- Provide waste and trash containers and mark them according to their contents to ensure that solid waste is not contaminated with hazardous waste.
- Remove all waste from the maintenance area daily and from the facility at least weekly.
- Remove hazardous waste, such as oils, batteries, and grease, and dispose of it properly in accordance with OSHA and EPA requirements.
- Keep walkways clear to prevent tripping hazards.
- Mark floors to indicate walkways and segregate equipment areas.
- Clean up oil, grease, and liquid spills when they occur to prevent slipping and contaminating other areas.

Battery Charging and Changing

Battery-operated tools and equipment allow increased flexibility in areas where electrical power may not be available. The following items may be powered by a battery:

- Radios
- Thermal imagers (TIs)

- Medical monitors/automatic external defibrillators (AEDs)
- Flashlights
- Gas/radiation detectors
- SCBA Personal Distress Locators/PASS devices
- Drills
- Saws
- Computers/mobile data terminals (MDTs)
- Global Positioning Systems (GPS)

Batteries contain lead-acid, nickel-cadmium, alkaline, lithium-ion, or nickel zinc. All batteries can become dangerous when overheated or exposed to the elements. Batteries may be rechargeable or designed for single use. Some rechargeable batteries must be completely depleted before recharging. If the battery is not completely discharged, it will establish a memory based on the recharged amount, thereby reducing its ability to hold a complete charge.

Figure 7.9 Radio batteries are recharged as needed or when the battery must be recycled by draining the charge completely.

Battery inspection and replacement periods vary. Batteries used in portable radios, SCBA personal locator devices, and TIs should be inspected at the beginning of each work shift **(Figure 7.9)**. Batteries should be changed or charged if they appear to be weak or if the unit fails to operate properly.

Charging batteries can be hazardous. Older batteries can leak contents because of damage to the casing, repeated use, and exposure to heat. Read the manufacturer's recommendations for charging and changing batteries. Verify that the battery matches the charging system. Case configurations may look the same, but contacts may not match the battery bank's contacts. Department policy may dictate when and how to charge a battery safely. Facilities that house electric forklifts and golf carts will contain charging stations.

Battery charging should take place in areas devoted exclusively to this type of activity and isolated from other employees. Wood-slat floorboards or rubber mats should be utilized in the area to minimize slips and falls and the danger of electric shock. The metal alloy composition in lead-acid batteries can generate two highly toxic byproducts: arsine and stibine. Batteries also release oxygen and hydrogen gases when they are charging and can create a potentially explosive atmosphere. Locations for charging must be well-ventilated to prevent concentrations of hydrogen and oxygen from reaching volatile levels.

Water must be available in the charging area for fire protection and for flushing spills of electrolytes on the floor. Flammable liquids must not be located in the charging area. Other safety precautions include:

- Do not permit smoking in charging areas.
- Use proper lifting techniques or mechanical hoists to lift batteries.
- Allow only trained personnel to charge or change batteries.

- Do not place metal tools on top of batteries.
- Prevent metal tools from contacting both battery terminals and causing a short circuit.
- Check the fluid levels of wet cell batteries regularly.

Sulfuric acid is a common component in batteries. In the event of a battery acid spill, neutralizing agents should be spread on the spill. After the spill is neutralized, it can be safely cleaned up and disposed of by trained personnel in accordance with local ordinances.

Maintenance of Unpowered Tools, Powered Tools, and Extinguishers

Crew members will be assigned the task of caring for tools that are kept in the station or maintenance area. Most of these tools are unpowered hand tools. However, powered tools and equipment will also require frequent care and maintenance. Crew members are also responsible for maintaining the portable fire extinguishers carried on the apparatus and located in the fire station. This section includes general information on maintaining most types of unpowered and powered tools, equipment, and portable extinguishers. Remember that you must always follow the manufacturer's instructions for cleaning and maintaining any tool or equipment.

Unpowered Tools

Unpowered tools, also known as hand tools, are the most common type used in the fire and emergency services. Most are designed to be carried and used by one person, but some require two people to operate. Examples of unpowered tools include, but are not limited to:

- Pike poles
- Sledgehammers
- Axes, flat-head and pick-head
- Claw hammers and roofing hammers
- Picks
- Shovels
- Bolt and wire cutters
- Pry bars
- Halligans
- Battering rams

Unpowered tools may be constructed from a single material such as metal, or a combination of materials such as metal and wood. The working end is usually made from a very hard grade of steel, while the handles are generally wood or fiberglass. Metal parts may be natural (unpainted) or painted.

Wooden handles. Most wooden handles are found on hammers, sledgehammers, axes, pike poles, and wildland tools. Care and maintenance procedures for wooden handles include the following:

- Inspect the handle for cracks, blisters, or splinters.
- Sand the handle, if necessary, to eliminate splinters **(Figure 7.10)**.
- Wash the handle with mild detergent, rinse, and wipe dry. Do not soak the handle in water because it will cause the wood to swell.
- Apply a coat of boiled linseed oil to the handle to preserve it and to prevent roughness and warping. Do not paint or varnish the handle.

- Check the tightness of the tool head.
- Limit the amount of surface area covered with paint for tool marking. Some departments use self-adhesive barcodes on the handle for identification.

Fiberglass handles. Fiberglass handles have become popular because they are sturdy and require less maintenance than wooden handles do. When caring for fiberglass handles, use the following procedures:

- Wash the handle with mild detergent, rinse, and wipe dry.
- Check for damage or cracks.
- Check the tightness of the tool head.
- Do not paint or varnish fiberglass handles.

Cutting edges. Cutting edges are found on hand tools such as axes, bolt and wire cutters, wildland tools, and some tools used for opening walls. Inspect the tool's cutting edge for chips, cracks, or burrs.

If the edge is damaged, replace it. While hand-sharpening is preferred, some cutting edges may require the use of a grinding wheel or stone. Take care when using a grinding wheel, because it will generate enough heat to weaken the metal of the cutting tool.

Figure 7.10 Tools with wooden handles, such as axes, must be sanded and refinished to remove splinters and prevent permanent damage to the wood surface.

The choice of sharpening tools depends on preference and may be regulated by local policy. Firefighters should follow the manufacturer's instructions or departmental procedures for sharpening the cutting edge of a tool. Sharpening tools include the following:

- Whetstones — Most common sharpening tool; available in a variety of surfaces and sizes **(Figure 7.11)**
- Diamond-coated flat file — Durable and long lasting; requires only water for lubrication and remains flat while sharpening
- Sharpening steel — Useful for finishing or for repairing minor dullness or damage

Figure 7.11 A file or rough whetstone can be used to sharpen the cutting edge of an axe blade.

NOTE: Consult the IFSTA **Essentials of Fire Fighting** manual for more information on tool care.

Plated surfaces. Plated surfaces are protected by chromium or other metal applied by an electroplating process. General maintenance includes inspecting for damage, wiping plated surfaces clean, or washing with mild detergent and water. Repairing a plated surface may release cadmium fumes that require the use of respiratory protection equipment. Plating materials that are flaking can produce a sharp edge and result in cuts or abrasions.

Unprotected metal surfaces. Unprotected metal surfaces are blades and other tool components that are not electroplated to protect them from rust or corrosion. General maintenance includes the following:

- Keep the surface clean, dry, and free of rust.
- Oil the metal surface lightly. Light machine oil works best. Avoid using any metal protectant that contains 1,1,1-trichloroethane, because this chemical may damage and weaken the handle.

- Do not paint metal surfaces, because paint hides defects.
- Inspect the metal for chips, cracks, or sharp edges, and file them off when found.

Powered Tools

Follow the manufacturer's recommendations for care and maintenance of powered equipment. Maintenance for powered equipment includes the following **(Figure 7.12)**:

- Check all fluid levels to be sure they are full.
- Use only lubricants, hydraulic fluids, and grades of fuels recommended by the manufacturer.
- Be sure that battery packs are fully charged.
- Inspect powered tools at the beginning of the work shift and verify that they will start manually.
- Check blades for damage or wear.
- Replace blades that are damaged or worn.
- Check all electrical components, including cords and receptacles, for cuts or other damage.
- Verify that all guards are functional and in place.
- Verify that fuel is fresh. A fuel mixture may separate or degrade over time.
- Inspect hydraulic and fuel supply hoses for damage.
- Inspect hydraulic hose couplings (quick-disconnect fittings) to ensure that they are clean and functional.
- Ensure that all parts and support items are available for immediate access.

Figure 7.12 Powered tools such as generators and fans are inspected regularly.

Extinguishers

General fire extinguisher maintenance may be performed by trained and qualified personnel. Firefighters should not underestimate the dangers of working with stored-pressure containers. Injuries and fatalities have occurred with extinguishers that malfunctioned or were not maintained.

NOTE: Consult the IFSTA **Essentials of Fire Fighting** and **Fire Protection, Detection, and Suppression Systems** manuals for more information about fire extinguishers.

> **CAUTION**
>
> **When pressurizing a cartridge-type extinguisher, do not have your head or any other part of the body above the top of the extinguisher. If the fill cap is not properly screwed on, the cap and/or a cloud of agent can forcibly discharge.**

> **CAUTION**
>
> **As with handheld units, the top of the extinguisher should be pointed away from the operator and from any other nearby personnel when pressurizing the unit. Because of the size of the nozzle, the operator should be prepared for a significant nozzle reaction when it is opened.**

Cleaning and Disinfecting Tools and Equipment

NFPA 1581, *Standard on Fire Department Infection Control Program*, requires fire stations to have separate areas designated for cleaning and disinfecting. One area is designated for cleaning equipment and protective clothing, and the other area is for disinfecting contaminated emergency medical equipment. The health and safety officer (HSO) or the infection control officer is responsible for ensuring these areas comply with local standards and meet NFPA standards.

Local health department codes, as well as federal and state/provincial environmental protection regulations, determine the appropriate filtration, storage, or disposal of contaminated wastewater from the cleaning and disinfecting areas. Cleaning materials used in these two areas (scrub brushes, disinfectant solutions, soaps, and sponges) should be stored within easy reach of the sink. To prevent further exposure or contamination, the fire and emergency services organization must provide chemical goggles, aprons, and vinyl medical gloves. A written protocol for disinfecting and cleaning protective clothing, medical equipment, and fire fighting tools and apparatus must be implemented. This protocol should be posted near the disinfecting and cleaning areas and be included in health and safety training for all personnel. In addition, Safety Data Sheets (SDS) must be located so they are readily available to employees.

The HSO or the infection control officer must ensure that all personnel wear proper clothing when involved in cleaning or disinfecting activities. Equipment that has been cleaned or disinfected should be drained into the sink and not allowed to spill onto the floor. This prevents the possibility of tracking contaminated water into the living quarters and prevents a slipping hazard. Provide well-ventilated drying areas for clothing and equipment.

The disinfecting, cleaning, and drying of SCBA facepieces is extremely important **(Figure 7.13)**. If they are not properly cleaned, SCBA facepieces can spread infection through the eyes, nose, and mouth of the wearer. Disinfecting and cleaning procedures must follow the manufacturer's recommendation and the guidelines found in the current edition of NFPA 1981, *Standard on Open-Circuit Self-Contained Breathing Apparatus (SCBA) for Emergency Services*, or other applicable respiratory protection standards. The written protocol for cleaning and disinfecting should designate the location for and method of maintaining the SCBA facepieces.

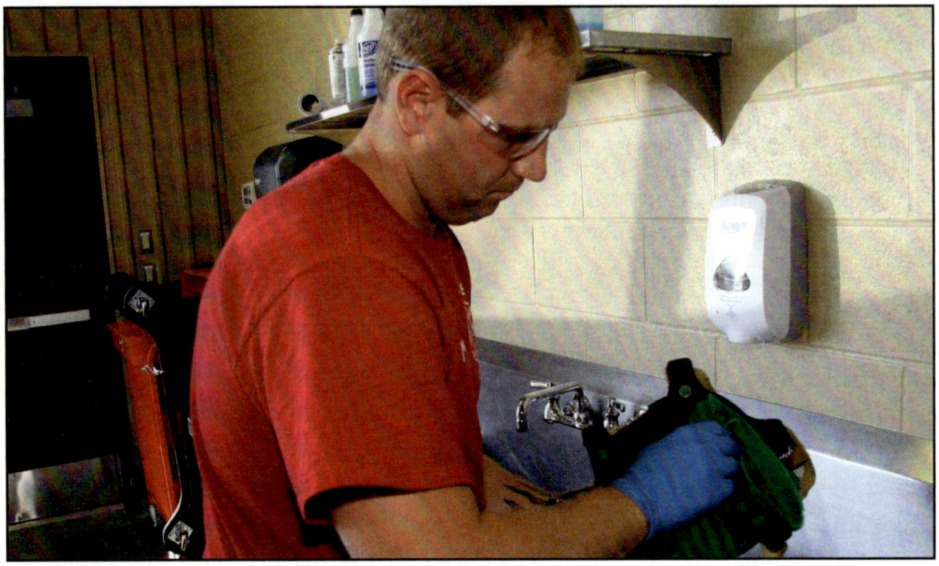

Figure 7.13 An SCBA facepiece must be disinfected, cleaned, and dried properly.

Figure 7.14 Designated equipment cleaning areas are located in the apparatus bay and equipped with sinks for washing equipment, as well as laundry equipment.

Designated Cleaning Area

The designated cleaning area must be equipped with proper ventilation, lighting, and drainage facilities **(Figure 7.14)**. Drainage must be connected to the sanitary sewer system or septic tank and located away from the kitchen, sleeping and living quarters, bathroom, and designated disinfecting area. Personal protective clothing, fire fighting equipment, and portable equipment are cleaned in this area. The immediate cleaning of contaminated clothing and equipment prevents the potential contamination of other personnel or other portions of the facility.

Facility cleaning equipment may also be cleaned in this area. Check with the local health department, water/sewer department, and state/local EPA to ensure there are no regulations for using the public sewer system.

Designated Disinfecting Area

Immediate disinfection of contaminated medical equipment reduces the possibility of exposure to bloodborne pathogens by other personnel. A designated disinfecting area is primarily intended for the disinfection of medical equipment exposed to bloodborne pathogens. The disinfecting area should not be used for cleaning mops, equipment, or clothing items that have not been contaminated by blood or other body fluids. The area should be equipped with a stainless steel, two-bay utility sink with both hot and cold water service, non-porous rack shelving, adequate drainage, proper lighting, and proper ventilation. The disinfecting area should be located in or adjacent to the apparatus bay to encourage use immediately upon return to quarters. Administrators should check with the local health department, water/sewer department, and state/local environmental agencies to ensure there are no regulations against dumping contaminated water into the public sewer system.

Biological waste should be disposed of in properly marked containers. Used needles and other disposable sharp objects should be placed in sharps containers and disposed of according to local protocol.

Hazardous Materials Storage and Use

Fire and emergency services facilities store and use limited quantities of hazardous materials. The following sections describe the types of storage containers and safe practices needed for flammable/combustible liquids, corrosives, and compressed/liquefied gases.

Flammable/Combustible Liquids

Flammable and combustible liquids may be found in a variety of containers:

- Glass containers
- Metal containers
- Polyethylene containers
- Safety cans

Because flammable and combustible liquids have ignition temperatures of varying degrees, they are divided into classes **(Tables 7.3 and 7.4, p. 238)**. The classes are used to determine the type and capacity of storage containers that these liquids require. Flammable/combustible and corrosive liquids should be stored in proper containers and kept in approved storage cabinets or rooms. These cabinets and rooms are designed to contain any spill or fire involving the materials or substances stored in them.

Glass containers. Glass containers or bottles may be found in sizes up to 1 gallon (4 L). Because glass will not react with acids and bases, it is suitable for storing corrosives like cleaning or housekeeping solutions, antiseptic or medical solutions, and pesticides.

Due to the potential for breakage, the quantity of flammable and combustible liquids that can be stored in glass containers is restricted by NFPA 30, *Flammable and Combustible Liquids Code*. The maximum quantities by class of liquids, are as follows:

- ***Class IA*** — 1 pint (0.5 L)
- ***Class IB*** — 1.05 quarts (1 L)
- ***Class IC*** — 1.3 gallons (5 L)
- ***Class II*** — 1.3 gallons (5 L)
- ***Classes IIIA*** — 5.3 gallons (20 L)
- ***Class IIIB*** — 5.3 gallons (20 L)

Safety cans. An inspector most often encounters flammable or combustible liquids stored in containers that are 5 gallons (20 L) or less in size. Although there are other acceptable containers for storing small amounts, the safest ones are approved safety cans. Safety cans are constructed to reduce the chance of leakage or container failure. Under normal conditions, they are also designed to virtually eliminate vapor release. Safety cans use self-closing lids

Table 7.3
ICC and NFPA Flammable Liquids Classes

Class	Flash Point	Boiling Point	Examples
Class IA	Below 73°F (23°C)	Below 100°F (38°C)	Acetaldehyde Cyclohexane Ethel ether
Class IB	Below 73°F (23°C)	At or Above 100°F (38°C)	Acetone Benzene Gasoline
Class IC	At or Above 73°F (23°C)	Below 100°F (38°C)	Hydrazine Styrene Turpentine

Table 7.4
ICC and NFPA Combustible Liquids Classes

Class	Flash Point Minimum	Flash Point Maximum	Examples
Class II	At or Above 100°F (38°C)	Below 140°F (60°C)	Acetic acid Diesel fuel Naphtha Stoddard solvent
Class IIIA	At or Above 140°F (60°C)	Below 200°F (93°C)	Cyclohexanol Formic Acid Nitrobenzene
Class IIIB	At or Above 200°F (93°C)	N/A	Formalin Motor oil Picric acid

with vapor seals and contain a flame arrester in the dispenser opening. The self-closing lid also acts as a pressure-relief device when the can is heated. Metal bulk, plastic and metal consumer-use, nonmetallic, and fiber drums all have restrictions regarding quantities and contents. NFPA 30 has lists and tables that offer direction for the storage of flammable and combustible liquids.

Storage cabinets. OSHA and the model fire codes require the use of storage cabinets and lockers that safely contain hazardous materials **(Figure 7.15).** Designed primarily for flammable and combustible liquids, these storage cabinets have the following safety features:

- A fire-resistant steel enclosure that permits materials to be stored near the point of use, and reduces the risk of fire
- A leak-proof sill under the cabinet doors and welded seams at all joints to contain spills
- Vents with flame arrestors to provide ventilation and to prevent fire from extending into the cabinet

- High-visibility reflective warning labels and yellow painted exteriors to improve visibility and recognition of the cabinet under reduced light conditions
- Built-in grounding connectors to prevent the accumulation of static electric charges

Storage rooms. Facilities storing flammable or combustible liquids in containers that exceed the allowable capacity of storage cabinets must be stored in storage rooms (also called *control areas* by the model fire codes). Storage rooms are constructed of fire-resistive materials in order to protect the containers from any nearby fire exposures. These rooms must have appropriate secondary containment, such as closed bottles or safety cans, as required by the locally adopted code. Storage rooms must also have some form of natural or mechanical ventilation.

Figure 7.15 Flammable and combustible liquids must be properly stored in approved storage cabinets.

Corrosives

Corrosives include acids and bases that can cause significant injuries to human flesh. They must be stored and used carefully. Generally, storage requirements for corrosives are the same as those for flammable/combustible liquids. Corrosives are found in glass and plastic containers and should be kept in cabinets or storage rooms that will retain the material if it is spilled. Emergency responders must wear proper PPE when dispensing or using corrosives, including eye, respiratory, and hand protection.

Compressed/Liquefied Gases

Compressed/liquefied gases are stored in cylinders approved by the U.S. Department of Transportation (DOT). These cylinders must be secured in racks or chained to the structure to prevent them from falling or being struck. Cylinders must be hydrostatically tested with the test date stamped on the surface of the cylinder. Remember that SCBA cylinders are also compressed gas cylinders and must meet the same safety requirements. To ensure that compressed/liquefied gases are stored safely, perform the following procedures:

- Verify that the hydrostatic test date is current.
- Inspect cylinders carefully for any external physical damage or discoloration.
- Store and use gas cylinders/gases according to the manufacturer's recommendations.

General Safe Usage Practices

A number of important safety practices can prevent a fire or explosion. Take the following preventive measures where flammable/combustible liquids and compressed/liquefied gases are stored and used:

- Eliminate all sources of ignition, such as the following:
  - — Electrical, mechanical, or frictional sparks
  - — Static electricity (Ground containers when appropriate.)
  - — Open flames
  - — Hot surfaces
  - — Incompatible materials

- Store flammable/combustible liquids and compressed/liquefied gases in approved, closed containers.
- Verify that all storage areas are ventilated to prevent the accumulation of vapors within flammable ranges or above flammable limits.
- Verify that all storage containers or areas are properly marked for the type of material stored in them.
- Wear the correct personal protective equipment (PPE), as indicated by the manufacturer in the material safety data sheet/safety data sheet (MSDS/SDS) or by department policy when using or handling hazardous materials.
- Provide proper ventilation when using these materials, including exhaust-hoods, fans, and openings in the structure.
- Separate products containing chlorine from ammonia or acid-based cleaners. Mixing these incompatible chemicals can produce a dangerous toxic gas.

Figure 7.16 Compressed gas cylinders should be secured in place.

- Keep hazardous materials in their original containers or in approved and marked secondary containers. All containers must be tightly sealed and stored in an upright position.
- Ensure that different products are never combined or mixed.
- Keep storage and work areas clean and free of trash.
- Ensure that numbers or marks stamped on compressed/liquefied gas cylinders remain in place and unchanged, with the exception of those made by hydrostatic testing organizations.
- Protect compressed/liquefied gas cylinders from cuts or other physical damage.
- Do not tamper with safety devices on the valves of cylinders.
- Mark empty compressed/liquefied gas cylinders with the word *Empty* or the letters *MT* (the phonetic equivalent of *empty*), and secure their valve caps in place.
- Cover compressed/liquefied gas cylinder valves with safety caps when cylinders are not in use.
- Anchor stored compressed/liquefied gas cylinders to a wall or other secure object with chains around the tanks so that they cannot be accidentally knocked over **(Figure 7.16)**.
- Always disconnect a cutting torch head or instrument from the supply hose before storing it in a tool box or cabinet.
- Ensure that oxygen cylinders, hoses, and connections are free of oil, grease, or other hydrocarbon (petroleum-based) lubricants.
- Store incompatible gases in separate areas.
- Store compressed/liquefied gas cylinders in an area that is clean and free of accumulated combustible materials.

- Ensure that abused, dented, arc-struck, damaged, or severely corroded cylinders are removed from service and replaced.
- Report compressed/liquefied gas cylinders that have expired hydrostatic test dates and remove them from service.
- Ensure that portable fire extinguishers of the proper type, size, and number are located in the storage and work areas, and that they are visible and mounted to the structure.

Testing Following All Major Repairs

A tool must be inspected and, in some cases, tested before it is returned to service. Simple hand tools and ladders may only need to be inspected to verify that the repairs have been properly made. Fire hose should be inspected, pressure-tested, and checked for leaks, especially if couplings have been replaced. Powered equipment should be tested using the initial acceptance or annual test criteria. Those requirements may include a minimum operating time, pressure, or other criteria.

Records should be kept on tool and equipment repairs and replacement. These records are essential for providing an accurate history of the tool or equipment and may be required by law or ordinance. They also help the department to keep track of valuable property and to develop a budget based on the actual cost of repair and replacement.

Disposal of Damaged or Obsolete Tools and Equipment

Departments must properly dispose of tools and equipment that are obsolete, have reached the end of their life cycle, are damaged beyond repair, or have accumulated a high operating cost. Disposals should be made in accordance with local, state/provincial, and federal laws. Most jurisdictions have policies regulating equipment disposals. In some cases, the equipment must be destroyed due to the liability potential in the event of an accident by a future owner.

Tool and Equipment Operation

Emergency responders are responsible for safely using tools and equipment at the emergency scene. Follow the organization's policies, procedures, and manufacturer's recommendations for the use of all tools and equipment. Safety must be the first consideration. This section addresses selecting the proper tool for the task, the proper way to carry tools and equipment, and the proper way to use them.

Selecting the Proper Tool

Most tools are designed to perform specific functions. Very few universal tools can perform multiple tasks. Practice helps firefighters remember what they learned in training; experience helps crew members quickly select the correct tools.

Always select the tool based on the task. Remember, pry tools are designed and constructed to allow force to be applied to one end in order to move something on the other end. Cutting tools are designed to cut or sever an object. Utility rope is designed to lift light loads. It is not for lifting humans or making rescues. Specific types of materials, such as steel and concrete, require specialized cutting blades on powered equipment.

Carrying Tools

Firefighters must carry tools and equipment as safely as possible to protect the carrier, other firefighters, and bystanders. Some recommended safety practices for carrying specific tools are as follows:

- ***Axes*** — If not in a scabbard, carry the axe with the blade away from the body. With pick-head axes, grasp the pick with a hand to cover it. Never carry an axe on the shoulder. A best practice is to carry an axe in a scabbard when ascending a ladder. This practice frees both hands to hold onto the ladder.
- ***Prying tools*** —Carry these tools with any pointed or sharp edges away from the body. With multiple surfaces, this will be somewhat difficult.
- ***Combinations of tools*** — Strap tool combinations together. Halligan bars and flat-head axes can be nested together and strapped. Short sections of old hose can be slipped over the handles of some tools and smaller prying tools inserted into the hose **(Figure 7.17)**.

Figure 7.17 A prying tool and an axe can be easily carried together.

- ***Pike poles and hooks*** —When outside a structure, carry these tools with the tool head down, close to the ground and ahead of the body. When entering a building, carefully reposition the tool. Carry it with the head upright and close to the body to facilitate prompt use. These tools are especially dangerous, because they are somewhat unwieldy and can severely injure anyone accidentally jabbed with the working end of the tool.
- ***Striking tools*** — Keep the heads of these tools close to the ground. Maintain a firm grip. Mauls and sledgehammers are heavy and may slip.
- ***Powered tools*** — The weight of the tool will dictate the number of firefighters needed to carry it **(Figure 7.18)**. Regardless of the number, always lift with the legs by crouching to grasp and raise the tool. Never bend over to lift an object, as this leads to back strains and injuries. Back or shoulder injuries can result from lifting heavy objects stored above shoulder height from apparatus compartments. These items should be carried in lower compartments or come equipped with some form of mechanical assistance to lower them to the ground.
  - Refer to department policy as to how a running powered tool should be carried. In most cases, the safest method is to transport the tool to the work area and then safely start it.
  - Prior to starting any powered equipment, determine if the atmosphere is free of flammable vapors that might ignite.

Figure 7.18 Lifting and carrying heavy powered tools, such as a cutting saw kit, requires two persons.

- ***Ground ladders*** — Ground ladders may be carried and raised by one or more firefighters. To prevent injuries, firefighters must employ proper lifting and carrying techniques. Wear PPE, including gloves, when handling ladders. Take special care to avoid electrical wires when using metal ladders. Information on ladder lifts and carries is contained in the IFSTA **Essentials of Fire Fighting** manual.
- ***Portable Fire Extinguishers*** — Carry the extinguisher by its handle. Do not pull the operating pin until you are ready to use the extinguisher. Protect the operating mechanism from impact or damage.

Safe Tool Usage

Use all tools and equipment with safety in mind. Improper use of tools can result in injuries, unnecessary damage, and loss. Improper use can also result in an inefficient use of energy and time that may endanger the people who need help.

Hand Tools

A key to hand tool safety is situational awareness. Because each situation is different, take time to evaluate it before using the tool. For instance, cutting a roof made of wet plywood with an axe can result in the axe bouncing off the surface instead of penetrating it. Extremely dry wood may splinter, causing pieces of wood to fly in many directions. In both cases, the axe or the splinters could strike the operator or other firefighters.

Cheater Bar — Piece of pipe added to a prying tool to lengthen the handle and provide additional leverage.

As with all tools, using prying tools incorrectly creates a safety hazard. For example, it is not acceptable to use a **cheater bar** or to strike the handle of a pry bar with other tools. Use of a cheater bar can put forces on the tool that are greater than the tool was designed to withstand. This can cause serious injury to the operator if the tool slips, breaks, or shatters. Furthermore, such action can damage the tool. If a job cannot be done with a particular tool, do not strike the handle of the tool. Use a larger tool. Also, do not use a prying tool as a striking tool unless it has been designed for that purpose. Firefighters must communicate clearly to avoid injuries when working as a team.

Powered Tools

When operating powered equipment, crew members must be even more aware of the situation, the potential of the equipment, and the environment. Powered equipment can cause harm to operators, and to anyone nearby, if the operator loses control of it. Wearing the appropriate PPE is mandatory when operating any powered equipment. Firefighters must work to maintain their balance when using powered saws, chisels, cutters, or spreaders.

Rotary saws must be used with extreme care because the rapidly rotating blade continues to spin after the throttle has been released. Another hazard associated with the rotary saw is the twisting (gyroscopic or torsion effect) caused by the spinning blade. When using a rotary saw, it is important to start all cuts at full rpm to prevent the blades from binding into the material.

Remember that blades from different manufacturers may look alike, but they may not be interchangeable. When replacing blades, be sure they are mounted with the correct direction of rotation. Store blades in a clean, dry environment. Do not store composite blades in any compartment where gasoline fumes accumulate, such as where spare saw fuel is stored, because the hydrocarbons will attack the bonding material in the blades and make them subject to sudden and violent disintegration during use.

Use the following safety rules when operating a powered saw **(Figure 7.19)**:

- Match the saw to the task and to the material. Never force a saw beyond its design limitations.
- Always wear proper protective equipment, including gloves, hearing protection, and eye protection.
- Fully inspect the saw before and after use.
- Do not use any powered saw when working in a flammable atmosphere or near flammable liquids.
- Maintain situational awareness.
- Keep unprotected and nonessential people out of the work area.
- Follow the manufacturer's guidelines for proper saw operation.
- Keep blades and chains well sharpened. A dull saw is more likely to cause an accident than a sharp one.
- Be aware of hidden hazards such as electrical wires, gas lines, and water lines.
- Start all cuts with the saw running at full rpm.

Figure 7.19 Complete personal protective equipment (PPE) must be worn when operating a powered tool such as a rotary saw.

Extrication Tools

When using a powered extrication tool, such as a spreader, cutter, or ram, use the following safety precautions **(Figure 7.20)**:

- Check the atmosphere for the presence of flammable vapor.
- Confirm that a charged hoseline is in place.
- Wear full PPE.
- Determine if the vehicle has any passive restraints that might be activated by the cutter blade.
- Do not put hands or fingers between the blades or spreader tips.
- When using chains with a spreader to pull a vehicle apart, anchor the chain securely and work slowly.
- Keep all nonessential personnel away from the work area.

Figure 7.20 When performing a vehicle extrication with powered tools, safety precautions must be used.

Providing lighting at an emergency scene improves the visibility and safety of all personnel working in areas that are dark, whether during the day or night. Depending on the organization's size and resources, the auxiliary light units available may vary from handheld battery-powered portable lights and apparatus-mounted lights to portable light units powered by generators. Lights can become very hot and cause severe burns. Gloves must be worn when handling or touching light units. Care must also be taken when handling energized lighting equipment. Extension or power cords must be of sufficient size and capacity to carry the required wattage and should have twist-locking devices to prevent disconnection. Cords must be kept out of the water to prevent electrocution and should be replaced if frayed or damaged. Grounding of all electrical equipment is important, as well.

Nonemergency Tools and Equipment

Nonemergency tools and equipment must also be used safely **(Figure 7.21)**. Wearing appropriate PPE when operating a lawnmower is as important as wearing respiratory protection when working at a structure fire. Follow these safety procedures to operate nonemergency tools and equipment:

- Wear work boots with safety toe caps when operating lawn mowers and edgers.
- Wear hearing and eye protection when operating powered equipment.
- Wear respiratory protection (dust masks or respirators) when spray painting or sanding.
- Wear eye, face, and hand protection when using a cutting or welding torch.
- Never remove or alter guards on equipment.

Figure 7.21 Safety procedures can prevent injury when operating nonemergency tools and equipment.

Chapter Review

1. What features make a tool intrinsically safe?
2. What are the general tool/equipment conditions that should be checked during a thorough safety inspection?
3. What are the general procedures for unpowered and powered tool/equipment maintenance?
4. What are some general safe practices that can prevent a fire or an explosion in a hazardous materials storage room?
5. What are common procedures for the safe operation of tools/equipment on the fireground?

Discussion Questions

1. How can accident reporting, data collection, and accident investigations focused on causation strengthen a department's health and safety committee?
2. Are there any specific tools that departments have that are not listed in the text? What are they and what are the inspection procedures for each?
3. Describe how cleanliness and organization in maintenance areas promotes risk management.

Protective Clothing and Equipment

Chapter Contents

chapter 8

Key Terms

FESHE Learning Outcomes

Occupational Safety and Health for Emergency Services

5. Describe the relevance for safety in the work place, including the importance of PPE.
6. Apply the knowledge of an effective safety plan to pre-incident planning, response, and training activities.

Protective Clothing and Equipment

Learning Objectives

After reading this chapter, students will be able to:

1. Explain the importance of using PPE properly.
2. Select PPE based on an incident's hazards.
3. Identify the functions of PPE accessories.
4. Summarize how to properly select, don, and doff PPE.
5. Describe how to inspect various pieces of PPE.
6. Explain post-incident PPE care and maintenance procedures.
7. Explain the various disposal procedures for PPE.
8. Differentiate between the types of PPE records and the information they contain.

Chapter 8
Protective Clothing and Equipment

There are three ways to protect from hazards:

- Eliminate the hazard through engineering
- Provide administrative controls that prevent exposure to the hazard
- Wear personal protective equipment (PPE) to minimize exposure to the hazard

Along with training and good judgment, personal protective clothing and equipment shield against the hazards faced in emergency operations. PPE enables firefighters to:

- Work more safely around most types of fires
- Make entry into Immediately Dangerous to Life and Health (IDLH) and hazardous environments to perform essential tasks
- Treat traumatically injured and medically ill patients

As the types and variety of hazards increase, the protective clothing and equipment industry responds with greater levels of protection in its products. Standards and regulations have become more rigorous and comprehensive. Technological advances have resulted in more sophisticated equipment and safety systems.

This chapter provides information on the components of personal protective clothing and equipment, its selection, use, and inspection. Additional information is provided on maintenance and care, disposal of damaged, contaminated, or aged equipment, and the record keeping required by NFPA standards.

PPE-Related Injuries and Fatalities

PPE, when used correctly, helps to prevent injuries and fatalities. Unfortunately, when PPE is inappropriately used because a hazard is not understood or is not communicated in a timely manner, injuries and fatalities may result. Selecting the correct type and level of protection is critical.

Firefighters must be cautious about placing too much confidence in PPE and avoid exposing themselves to dangers that the garments are not designed to withstand. For instance, structural fire fighting PPE provides thermal protection by absorbing heat and providing significant protection to firefighters in the initial stage of the fire. However, as turnout gear has improved, today's firefighters often move closer to the interior of a structure during a fire.

When fire fighting teams enter structures, their turnouts absorb heat and smoke, which brings turnouts to their maximum heat-absorbing potential. This situation becomes especially dangerous because modern fire behavior results in more heat in the initial stages of the fire, and then repeats the heat-releasing process when additional oxygen is introduced into the structure **(Figure 8.1)**. Firefighters face increased risks in this situation because the heat-absorbing potential of the turnout is quickly reached, just as the fire begins to rebuild with additional oxygen.

Figure 8.1 PPE provides protection for firefighters at a structure fire, but heat-absorbing potential can be reached quickly as the fire rebuilds with additional oxygen. *Courtesy of Bob Esposito, Pennsburg, PA.*

Not providing the correct fit, type, or NFPA-compliant PPE, or continuing to use outdated, contaminated, or damaged PPE can also contribute to firefighter injuries and fatalities. Emergency service administrators must continually evaluate the provided PPE to ensure its compliance and adequacy.

Types of PPE Based on Hazard

Personal protective clothing must be designed to meet NFPA standards. Then it must be tested and certified through third-party testing organizations to verify that it will protect firefighters and emergency responders from specific types of hazards. Certification may be provided for a single garment, such as a pair of fire fighting boots, or for an entire ensemble, such as a proximity fire fighting entry suit. Those responsible for purchasing PPE must be aware of the specific use for which a garment or ensemble is certified. Likewise, the person using the PPE must be aware that not all ensemble pieces are designed or rated the same way. For example, most self-contained breathing apparatus (SCBA) masks are rated to fail at temperatures far below that of structural fire fighting gear. Personal protective clothing should be specified and purchased based on the department's intended use.

While NFPA develops criteria for PPE design, construction, and performance, the use of the equipment may be mandated by the federal, state/provincial, or local government. In the U.S., the Occupational Safety and Health Administration (OSHA), or a state equivalent, mandates the use of PPE that meets NFPA standards. These agencies also require that the employer, whether public or private, provide employees with the correct type of PPE.

Figure 8.2 Turnout coats and pants have evolved through many types of materials, resulting in higher levels of thermal protection.

NFPA standards establish the minimum protection requirements to meet each type of hazard. Some garments, however, may have dual certification for more than one function. An example of a dual-certified garment would be one that protects the wearer from the hazards of a structural fire incident and also from biological hazards encountered in an emergency medical services (EMS) incident. When dual-certified garments are specified, purchased, and issued, they must meet the requirements of each NFPA standard that applies to them **(Figure 8.2)**.

All PPE has limitations. For instance, personal protective clothing certified for structural fire fighting, while protecting the wearer from high temperatures encountered in a fire, may also contribute to heat stress during warm weather. Firefighters must undergo training and education to understand how to select and use PPE **(Figure 8.3)**.

Figure 8.3 Selecting the correct PPE will help firefighters work safely in different types of emergencies.

The types of responses that PPE are designed to meet, specific hazards, and the applicable NFPA standard(s) are found in **Table 8.1, p. 254-255**. Each response type and the PPE it is designed to address are discussed in the following sections.

Medical Hazards

With an increasing awareness of the variety and virulence of infectious diseases - such as staph, noroviruses, and hepatitis — emergency responders should be very concerned with personal protection during emergency medical responses **(Figure 8.4)**. Traditionally, personnel who performed both EMS and structural fire fighting duties depended on their structural clothing for protection, believing that the multiple layers of fabric would protect against biological contamination. However, unless the ensemble is dual-certified, it will not provide the required level of protection. Additionally, the cost of decontaminating structural PPE from biological exposures can outweigh the cost of disposable medical PPE.

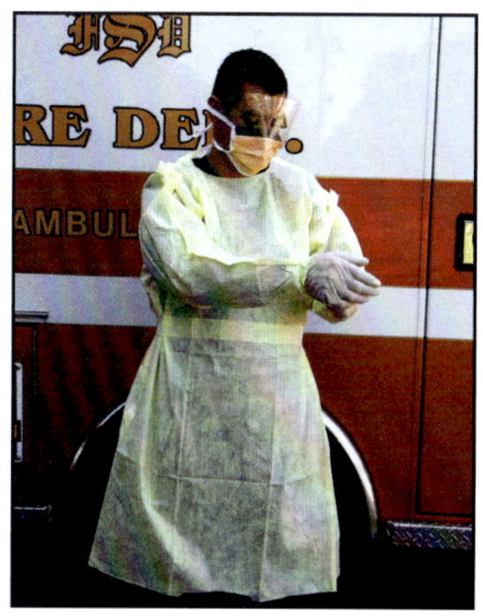

Figure 8.4 EMS-certified protective clothing is designed to provide a barrier against liquid-borne pathogens and other bodily fluids.

EMS-certified ensembles that provide protection against liquid-borne pathogens and other bodily fluids consist of garments, gloves, and face protection that meet the requirements of NFPA 1999, Standard on Protective Clothing and Ensembles for Emergency Medical Operations. This standard is compatible with the other protective clothing standards in its requirements for testing, labeling, and third-party certification during manufacture.

Table 8.1
NFPA Standards To Meet Specific Hazards and Responses

Response Type	Specific Hazard	NFPA Standard
Medical	Contaminated or low-oxygen atmospheres	NFPA 1981, *Standard on Open-Circuit Self-Contained Breathing Apparatus (SCBA) for Emergency Services*
	Chemical, Biological, Radiological, or Nuclear	NFPA 1994, *Standard on Protective Ensembles for First Responders to Hazardous Materials Emergencies and CBRN Terrorism Incidents*
	Biological contamination	NFPA 1999, *Standard on Protective Clothing for Emergency Medical Operations*
Structural	Heat, flame, sharp objects	NFPA 1971, *Standard on Protective Ensembles for Structural Fire Fighting and Proximity Fire Fighting*
	Contaminated or low-oxygen atmospheres	NFPA 1981, *Standard on Open-Circuit Self-Contained Breathing Apparatus (SCBA) for Emergency Services*
	Disorientation and unconsciousness	NFPA 1982, *Standard on Personal Alert Safety Systems (PASS)*
	Contaminated or low-oxygen atmospheres	NFPA 1989, *Standard on Breathing Air Quality for Emergency Services Respiratory Protection*
	Chemical, Biological, Radiological, or Nuclear	NFPA 1994, *Standard on Protective Ensembles for First Responders to CBRN Terrorism Incidents*
Wildland	Heat, flame, embers, and sharp objects	NFPA 1977, *Standard on Protective Clothing and Equipment for Wildland Fire Fighting*
Hazardous Materials	Contaminated or low-oxygen atmospheres	NFPA 1981, *Standard on Open-Circuit Self-Contained Breathing Apparatus (SCBA) for Emergency Services*
	Contaminated or low-oxygen atmospheres	NFPA 1989, *Standard on Breathing Air Quality for Emergency Services Respiratory Protection*
	Contact with dangerous or hazardous vapors or gases	NFPA 1991, *Standard on Vapor-Protective Ensembles for Hazardous Materials Emergencies CBRN Terrorism Incidents*

Continued

Table 8.1
NFPA Standards To Meet Specific Hazards and Responses

Response Type	Specific Hazard	NFPA Standard
Medical	Contaminated or low-oxygen atmospheres	NFPA 1981, *Standard on Open-Circuit Self-Contained Breathing Apparatus (SCBA) for Emergency Services*
	Chemical, Biological, Radiological, or Nuclear	NFPA 1994, *Standard on Protective Ensembles for First Responders to Hazardous Materials Emergencies and CBRN Terrorism Incidents*
	Biological contamination	NFPA 1999, *Standard on Protective Clothing for Emergency Medical Operations*
Structural	Heat, flame, sharp objects	NFPA 1971, *Standard on Protective Ensembles for Structural Fire Fighting and Proximity Fire Fighting*
	Contaminated or low-oxygen atmospheres	NFPA 1981, *Standard on Open-Circuit Self-Contained Breathing Apparatus (SCBA) for Emergency Services*
	Disorientation and unconsciousness	NFPA 1982, *Standard on Personal Alert Safety Systems (PASS)*
	Contaminated or low-oxygen atmospheres	NFPA 1989, *Standard on Breathing Air Quality for Emergency Services Respiratory Protection*
	Chemical, Biological, Radiological, or Nuclear	NFPA 1994, *Standard on Protective Ensembles for First Responders to CBRN Terrorism Incidents*
Wildland	Heat, flame, embers, and sharp objects	NFPA 1977, *Standard on Protective Clothing and Equipment for Wildland Fire Fighting*
Hazardous Materials	Contaminated or low-oxygen atmospheres	NFPA 1981, *Standard on Open-Circuit Self-Contained Breathing Apparatus (SCBA) for Emergency Services*
	Contaminated or low-oxygen atmospheres	NFPA 1989, *Standard on Breathing Air Quality for Emergency Services Respiratory Protection*
	Contact with dangerous or hazardous vapors or gases	NFPA 1991, *Standard on Vapor-Protective Ensembles for Hazardous Materials Emergencies CBRN Terrorism Incidents*

Continued

Figure 8.5 Fire fighting PPE includes a helmet with face shield and ear cover, protective hood, eye protection, turnout or bunker coat, turnout or bunker pants, boots, and gloves.

Budget constraints may tempt emergency response agencies, including EMS, to depend solely on structural fire fighting clothing for all types of responses. However, this practice places firefighters and patients at risk. Fire fighting clothing exposes firefighters and patients to contamination and risk of infection from hydrocarbons, soot, and chemicals on the clothing. The agency's health and safety officer (HSO) should provide agency administrators with supporting documentation, including NFPA and OSHA requirements, that justify the funding to provide the proper PPE for medical responses.

Violence

Violence directed at fire and emergency services personnel justifies the need for body armor similar to the types worn by law enforcement and military personnel. Body armor is designed to protect the torso from sharp objects and some ammunition. It is usually donned when there is an indication that the responders may be in danger.

Emergency responders can experience violent attacks from bystanders and from patients, particularly patients with mental health and/or substance abuse issues. A balanced approach that includes increased situational awareness, as well as protective equipment, is a consideration for responder protection.

Structure Fires

A covered structural firefighter should be equipped with the following **(Figure 8.5)**:

- Helmet with face shield and ear cover — The helmet protects the head from falling debris and water. The face shield provides eye protection and can be used even while wearing SCBA. The ear cover, also referred to as neck protection, is intended to protect the area between the helmet and coat.
- Protective hood — Worn over the head, the hood protects the exposed portions of the face and neck and interfaces with the SCBA facepiece and coat collar.
- Eye protection — Provided by the SCBA facepiece, helmet face shield, or goggles.
- Turnout or bunker coat — Multilayered garment consisting of an outer shell of fire-resistant fabric such as polybenzimidazole (PBI), Kevlar®, Nomex®, or cotton and an inner liner or thermal barrier. The liner may or may not be removable from the shell. The coat provides protection from heat, debris, and momentary flame impingement.
- Turnout or bunker pants — Pants are the same as the turnout coat in construction and design.
- Boots — Waterproof boots are made of rubber or leather with reinforced shanks to protect the feet from puncture wounds and safety toe caps to protect the toes from crushing. Boots must be a minimum of 8 inches (200 mm) high and fit beneath the trouser cuffs.
- Gloves — Gloves protect the hands from steam burns and injury due to contact with sharp objects. It is important to select the proper glove for the job.

Although not required, ear plugs can be worn under the protective hood or ear cover. When operating power tools, ear muffs may be used.

The HSO, and anyone responsible for purchasing protective clothing, should be familiar with the requirements of NFPA 1971, Standard on Protective Ensembles for Structural Fire Fighting and Proximity Fire Fighting. The most recent changes to the standard include requirements for aircraft rescue and fire fighting, including proximity clothing.

CAUTION

Structural fire fighting clothing is tested as a complete unit. Removing the liner from the coat or trouser shell and wearing the shell alone will void any manufacturer's warranty, as well as place a legal liability on the wearer and the organization. It can also result in injuries to the wearer due to steam and heat burns.

Wildland

Wildland fire fighting presents conditions that require specially designed PPE. The design of PPE for wildland fire fighting must balance the need for adequate protection from the hazards created by the fire with the need to wear the clothing for long periods in a very hot and hostile environment. The design and testing requirements are in NFPA 1977, *Standard on Protective Clothing and Equipment for Wildland Fire Fighting.*

A fully equipped wildland firefighter is protected by the following equipment **(Figure 8.6)**:

- Protective helmet with shroud — Protects the wearer from injury due to contact with low-hanging vegetation or falling debris. Most wildland protective helmets resemble a typical construction-type hard hat, although some lightweight structural fire fighting helmets are available. One neck shroud is attached to the helmet and protects the ears and neck of the wearer from flying embers.
- Eye protection — Protects eyes from smoke, soot, and cinders. Goggles can also help the wearer to see in smoky conditions. Helmet-mounted face shields provide some protection but do not keep smoke away from the eyes.
- Protective clothing — Jackets provide upper-torso protection from embers and cinders, while trousers protect lower torso and legs. Wildland PPE must have fire-resistant qualities and be waterproof, yet lightweight. A second layer of clothing, including long-sleeved shirts, is worn under the wildland PPE.
- Gloves — Protect the hands from cuts and abrasions and permit the wearer to pick up or move warm objects in the burn area.

Figure 8.6 Wildland-certified personnel protective equipment (PPE) is a lighter weight than structural fire fighting clothing.

- Protective footwear — Protects and supports the feet and ankles. Boots designed for wildland fire fighting are similar to hiking boots with ankle supports, steel soles, and toe caps, and have minimum heights of 8 inches (200 mm). Boots should be waterproof and designed for comfort during long periods of wear.
- Respiratory protection — Protects the wearer from particles in the air. Respiratory protection consists of dust masks, respirators, and bandanas. Structural SCBA are not used in wildland situations due to the weight and limited air capacity. Some new designs of respiratory protection provide increased face and lung protection. These devices are made of multiple layers of fire-resistant material and support an activated carbon filter inside the mask to capture smoke and ash **particulates**. Some of these masks only filter particulates.

Particulates — Very small particles of solid material, such as dust, suspended in the atmosphere.

- Fire shelter — Provides some protection for the user from heat and direct flame contact when caught in a fire. This type of equipment is a last resort and is used when timely evacuation is not possible.
- Hearing protection — Required when operating any noise-producing power-operated equipment, such as chain saws. Styles may include earmuffs or earplugs.

Hazardous Materials

Many hazardous materials can cause a variety of injuries, including chemical burns, allergic reactions and rashes, diseases, and absorption of toxic materials into the body through contact with the skin. PPE is designed to prevent these injuries. In most incidents, the first type of PPE available is structural fire fighting clothing with SCBA. With proper monitoring, this equipment may provide sufficient protection to enable a rescue or an initial evaluation of the incident before other types of equipment are available or needed.

When structural fire fighting protective clothing does not provide sufficient protection against the hazards present, responders may need to use an ensemble of **chemical protective clothing (CPC)** and respiratory equipment. The purpose of this ensemble is to shield or isolate individuals from the chemical, physical, and biological hazards that may be present at hazardous materials incidents.

Chemical Protective Clothing (CPC) — Clothing designed to shield or isolate individuals from the chemical, physical, and biological hazards that may be encountered during operations involving hazardous materials.

Design and testing standards generally recognize two types of CPC: liquid-splash protective clothing and vapor-protective clothing. CPC is made from a variety of materials, none of which protects against all types of chemicals. The manufacturer of an ensemble must provide a list of chemicals for which the suit is designed to offer protection. Selection of appropriate CPC depends on the specific chemical and on the specific tasks to be performed by the wearer.

Liquid-Splash Protective Clothing

NFPA 1992, *Standard on Liquid Splash-Protective Ensembles and Clothing for Hazardous Materials Emergencies*, classifies and addresses performance criteria for splash-protective ensembles. Liquid-splash protective clothing is designed to protect users from chemical liquid splashes, not from chemical vapors or gases **(Figure 8.7)**. Liquid-splash protective clothing can be encapsulating or nonencapsulating.

Neither encapsulating nor nonencapsulating liquid-splash protective clothing resists heat or flame exposure unless it is specifically designed for working in these environments. It does not protect against projectiles or shrapnel. Liquid-splash protective clothing is made from the same types of material used for vapor-protective suits.

Vapor-Protective Clothing

NFPA 1991, *Standard on Vapor-Protective Ensembles for Hazardous Materials Emergencies and CBRN Terrorism Incidents*, classifies and addresses performance criteria for vapor-protective ensembles. Vapor-protective clothing is designed to protect the wearer against chemical vapors or gases and offers a greater level of protection than liquid-splash protective clothing. Vapor-protective ensembles must be worn with positive-pressure SCBA, be used in combination SCBA/ SAR on the inside of the suit, or have special attachments for Supplied Air Respirators (SAR). These suits provide the greatest degree of protection against respiratory hazards from vapors, gases, particulates, sudden splashes, immersion, or contact with hazardous materials.

Figure 8.7 Vapor-protective clothing also protects against liquid splashes.

The approach in selecting PPE must encompass clothing and equipment that are easily integrated to provide an appropriate level of protection, yet still allow responders to perform work. For example, simple protective clothing such as gloves and a work uniform, in combination with a face shield (or safety goggles), may prevent exposure to such etiological agents as bloodborne pathogens. The use of vapor-protective encapsulating suits, combined with positive-pressure SCBA, is considered the minimum level of protection necessary for dealing with vapors, gases, or particulates that are harmful to skin or capable of being absorbed through the skin.

EPA Ensembles

Hazardous material PPE ensembles are classified by OSHA and the EPA. Levels of PPE ensembles, in descending levels of protection, are Level A, B, C, and D **(Figure 8.8 a-d, p. 260)**.

Level A ensembles. The Level A ensemble provides the highest level of protection against vapors, gases, mists, and particles for the respiratory tract, eyes, and skin. Level A protection provides very little protection against fire. Operations Level responders do not typically operate in situations requiring **Level A protection**. However, you must be appropriately trained to wear Level A PPE if you are required to wear it.

Level A Protection — Highest level of skin, respiratory, and eye protection afforded by personal protective equipment. Consists of positive-pressure, self-contained breathing apparatus, totally encapsulating chemical-protective suit, inner and outer gloves, and chemical-resistant boots.

- **Required components of Level A ensembles:**
  - Positive-pressure, full facepiece, SCBA, or positive-pressure airline respirator with escape SCBA, approved by NIOSH
  - Vapor-protective suits: Totally-Encapsulated Chemical Protective (TECP) suits constructed of protective-clothing materials that meet the following criteria:
    - Cover the wearer's torso, head, arms, and legs
    - Include boots and gloves that may either be an integral part of the suit or separate and tightly attached
    - Enclose the wearer completely by itself or in combination with the wearer's respiratory equipment, gloves, and boots

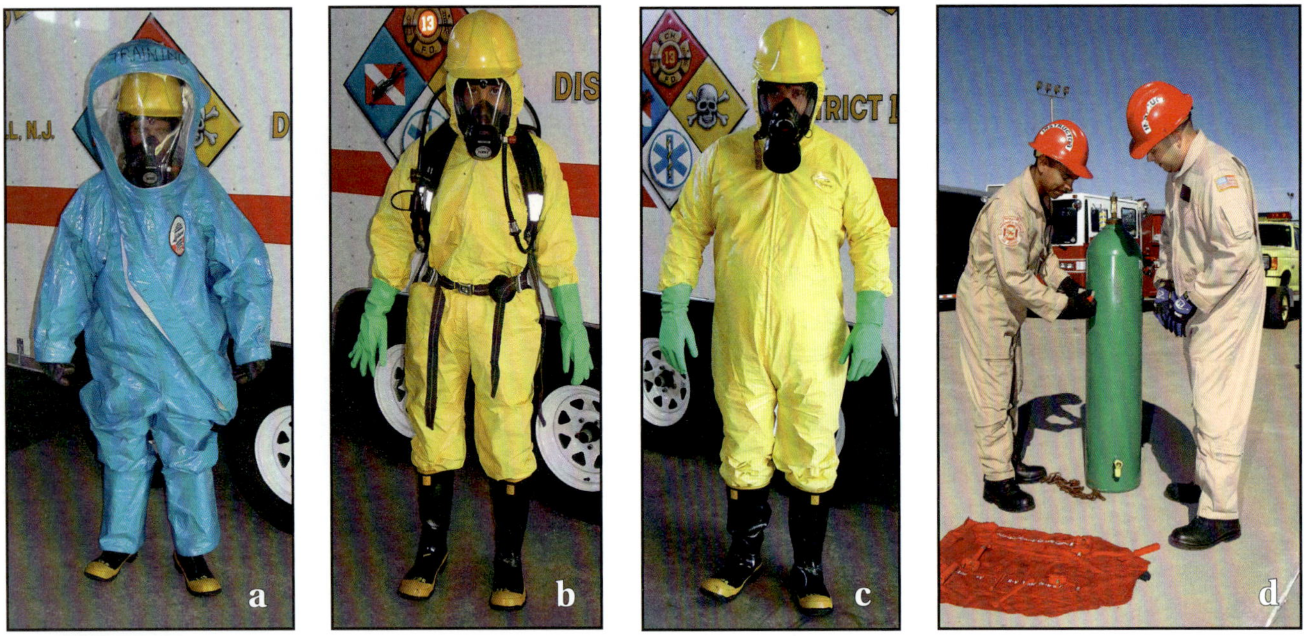

Figure 8.8 a-d The U.S. Environmental Protection Agency (EPA) has established four levels of hazardous materials protection; a) Level A; b) Level B; c) Level C; and d) Level D. Photos a-c *Courtesy of Cherry Hill (NJ) Fire Department.*

  - Provide equivalent chemical-resistance protection for all components of a TECP suit (such as relief valves, seams, and closure assemblies)
  - Meet the requirements in NFPA 1991

— Coveralls (optional)

— Long underwear (optional)

— Chemical-resistant outer gloves

— Chemical-resistant inner gloves

— Chemical-resistant boots with steel toe and shank

— Hardhat (under suit) (optional)

— Disposable protective suit, gloves, and boots (can be worn over totally encapsulating suit, depending on suit construction)

— Two-way radios (worn inside encapsulating suit)

- **Protection provided** — Highest available level of respiratory, skin, and eye protection from solid, liquid, and gaseous chemicals.
- **Use** — Level A ensembles are used when risk analysis indicates it is appropriate. For example, Level A protection may be selected when site operations and work functions involve a high potential for splash, immersion, or other exposure to unexpected vapors, gases, or particulates of material that are harmful to skin, or capable of damaging or being absorbed through intact skin.

Level B Protection — Personal protective equipment that affords the highest level of respiratory protection but a lesser level of skin protection. Consists of positive-pressure self-contained breathing apparatus, hooded chemical-resistant suit, inner and outer gloves, and chemical-resistant boots.

Level B ensembles. **Level B protection** requires a garment that includes an SCBA or a SAR and provides protection against splashes from a hazardous chemical. This ensemble is worn when the highest level of respiratory protection is necessary, but a lesser level of skin protection is needed. Level B protection provides very little protection against fire. The Level B CPC ensemble may be encapsulating or nonencapsulating.

- **Required components of Level B ensembles:**
  - Positive-pressure, full facepiece, SCBA, or positive-pressure airline respirator with escape SCBA, approved by NIOSH
  - Hooded chemical-resistant clothing that meets the requirements of NFPA 1992 (overalls and long-sleeved jacket, coveralls, one- or two-piece [encapsulating or nonencapsulating] chemical splash suit, and disposable chemical-resistant overalls)
  - Coveralls (optional)
  - Chemical-resistant outer gloves
  - Chemical-resistant inner gloves
  - Chemical-resistant boots with steel toe and shank
  - Disposable, chemical-resistant outer boot covers (optional)
  - Hardhat (outside or on top of nonencapsulating suits or under encapsulating suits)
  - Two-way radios (worn inside encapsulating suit or outside nonencapsulating suit)
  - Face shield (optional)
- **Protection provided** — Level B ensembles provide the same level of respiratory protection as Level A but have less skin protection. These ensembles provide liquid splash-protection, but they do not protect against chemical vapors or gases.
- **Use** — Level B ensembles may be used in the following situations:
  - Type and atmospheric concentration of substances have been identified and require a high level of respiratory protection, but less skin protection.
  - Atmosphere contains less than 19.5 percent oxygen or more than 23.5 percent oxygen.
  - Presence of incompletely identified vapors or gases is indicated by a direct-reading organic vapor detection instrument, but the vapors and gases are known not to contain high levels of chemicals harmful to skin or capable of being absorbed through intact skin.
  - Presence of liquids or particulates is indicated, but they are known not to contain high levels of chemicals harmful to skin or capable of being absorbed through intact skin.

Level C ensembles. **Level C protection** differs from Level B in the area of equipment needed for respiratory protection. Level C is composed of a splash-protecting garment and an Air-Purifying Respirator or Powered Air-Purifying Respirator (APR or PAPR). Level C protection includes any of the various types of APRs. Periodic air monitoring is required when using this level of PPE. Level C protection provides very little protection against fire.

Level C Protection — Personal protective equipment that affords a lesser level of respiratory and skin protection than Levels A or B. Consists of full-face or half-mask APR, hooded chemical-resistant suit, inner and outer gloves, and chemical-resistant boots.

- **Required components of Level C ensembles:**
  - Full-face or half-mask APRs, NIOSH approved
  - Hooded chemical-resistant clothing (overalls, two-piece chemical-splash suit, and disposable chemical-resistant overalls)

- Coveralls (optional)
- Chemical-resistant outer gloves
- Chemical-resistant inner gloves
- Chemical-resistant boots with steel toe and shank
- Disposable, chemical-resistant outer boot covers (optional)
- Hardhat
- Escape mask (optional)
- Two-way radios (worn under outside protective clothing)
- Face shield (optional)

- **Protection provided** — Level C ensembles provide the same level of skin protection as Level B, but they have a lower level of respiratory protection. Level C ensembles provide liquid splash-protection, but they do not protect against chemical vapors or gases on the skin.
- **Use** — Level C ensembles may be used in the following situations:
  - Atmospheric contaminants, liquid splashes, or other direct contact will not adversely affect exposed skin or be absorbed through any exposed skin.
  - Types of air contaminants have been identified, concentrations have been measured, and an APR is available that can remove the contaminants.
  - All criteria for the use of APRs are met.
  - Atmospheric concentration of chemicals does not exceed IDLH levels. The atmosphere must contain between 19.5 and 23.5 percent oxygen.

Level C equipment is only used by emergency response personnel under the following conditions:

- The specific material is known.
- The specific material has been measured.
- This protection level is approved by the IC after all qualifying conditions for APRs and PAPRs have been met:
  - The product is known.
  - An appropriate filter is available.
  - The atmospheric oxygen concentration is between 19.5 and 23.5 percent.
  - The atmosphere is not IDLH.

Level D Protection — Personal protective equipment that affords the lowest level of respiratory and skin protection. Consists of coveralls, gloves, and chemical-resistant boots or shoes.

Level D ensembles. Level D ensembles consist of typical work uniforms, street clothing, or coveralls. **Level D protection** cannot be worn when atmospheric hazards exist.

- **Required components of Level D ensembles:**
  - Coveralls
  - Gloves (optional)
  - Chemical-resistant boots/shoes with steel toe and shank
  - Disposable, chemical-resistant outer boot covers (optional)
  - Safety glasses or chemical splash goggles

— Hard hat

— Escape device in case of accidental release and the need to immediately escape the area (optional)

— Face shield (optional)

- **Protection provided** — Level D ensembles provide no respiratory protection and minimal skin protection.
- **Use** — Level D ensembles are not typically worn in the hot zone and are not acceptable for hazmat emergency response above the Awareness Level. Level D ensembles are used when both of the following conditions exist:

— Atmosphere contains no hazard.

— Work functions preclude splashes, immersion, or the potential for unexpected inhalation of or contact with hazardous levels of any chemicals.

Chemical, Biological, Radiological, Nuclear, or Explosive Devices (CBRNE)

Both NFPA 1971, *Standard on Protective Ensembles for Structural Fire Fighting and Proximity Fire Fighting*, and NFPA 1994, *Standard on Protective Ensembles for First Responders to Hazardous Materials Emergencies and CBRN Terrorism Incidents*, provide requirements for the design and use of CBRNE protective ensembles. NFPA 1994 stipulates design requirements for Class 2, Class 3, and Class 4 levels of ensembles and ensemble elements to meet different scenarios of risk analysis:

- Class 2 ensembles protect against incidents involving vapor or liquid chemical where concentrations exceed IDLH thresholds.
- Class 3 ensembles protect against incidents of terrorism involving low levels of vapors or liquid chemicals where concentrations are below IDLH thresholds.
- Class 4 ensembles protect against incidents of terrorism involving biological hazards or radiological particulate hazards where concentrations are below IDLH thresholds and the use of APRs and PAPRs is permitted.

Terrorism agents include chemical warfare agents, industrial toxic chemicals, biological terrorism agents, and radiological particulate terrorism agents. NFPA 1971 provides additional design requirements for SCBAs to provide firefighters with a minimal level of protection against CBRNE events that are unknown to responders at the onset of the event.

Water Rescue

PPE used in water rescue should be designed for the intended purpose and for the qualifications of the rescuer. Structural fire fighting clothing and SCBA should never be used for water rescues because they will retain water; even with a personal flotation device (PFD) the rescuer may drown. Generally, all personnel working within 10 feet (3 m) of the water's edge must wear water rescue PPE. The rescuer must train with the PPE regularly to ensure a safe outcome in the event of a water rescue deployment **(Figure 8.9)**. Minimum water rescue PPE should include:

Figure 8.9 Proper water rescue PPE includes helmets and wetsuits that provide thermal protection.

- **Personal Flotation Device (PFD)** — PFDs are assigned types by the U.S. Coast Guard. Type 3 and Type 5 PFDs are the standard for rescue. A Type 3 (Flotation Aid) PFD is considered a general-use PFD. The Type 5 (Wearable Special Use Device) PFD is designed for water rescue operations, and the wearer must be trained in its use. Type 5 PFDs generally allow the trained rescuer to make contact rescues with a releasable tether system. NFPA standards require that the rescuer carry a whistle and cutting device.
- **Water rescue helmets** — Holes in the helmet shell allow water to pass through the helmet.
- **Footwear** —Rescuers may wear specifically designed shoes/booties or tennis/basketball high-top shoes. Footwear must allow for the use of swim fins if the rescuer is going to deploy them. Rescuers should not wear structural fire fighting PPE or boots during a water rescue.
- **Thermal protection** —The amount and type of thermal protection must be matched to the water hazard. The most common types of thermal protection ensembles are wet suits and dry suits with a liner. Wet suits come in many thicknesses to provide more or less thermal retention for the rescuer. Wet suits put rescuers in contact with water which, in floods or sewage, exposes them to contamination. Dry suits can be worn for many water hazards and water temperatures. By varying the liner material or layers, rescuers can increase or decrease the amount of thermal retention. Dry suits encapsulate most of the rescuer, except for hands and head, and are preferred for cold water and floodwater. Dry suits also provide some measure of protection over wet suits in contaminated water.
- **Gloves** —The main issue with gloves is the tradeoff between thermal protection and dexterity. The rescuer will have to match the gloves to the intended use. Gloves that permit the handling of debris, ropes, and equipment during the search process may not provide enough thermal protection, thus rendering the rescuer ineffective. Diving-type gloves and work gloves can be used interchangeably, as long as rescuers are aware of the greatest hazard to their hands.
- **Eye protection** — Operating in extreme weather or around aircraft may necessitate the use of eye protection. Water craft operations also dictate the use of eye protection.
- **Clothing** — Generally, non-cotton shorts, T-shirts or swimming trunks/suits are worn with the water rescue equipment.

Personal protective clothing and equipment used for rescues in ice-covered rivers, ponds, lakes, and coastal areas are designed for use in extremely cold temperatures. Ice rescue PPE is similar to a dry suit and protects the wearer from hypothermia during rescue operations. Many states/provinces and local jurisdictions require that personnel performing ice rescue be certified Ice Rescue Technicians (IRTs).

WARNING

Ice-rescue suits are not designed to be used in moving water.

CAUTION

Throughout the U.S. and Canada, companies and organizations offer IRT certificates. Before contacting a private or nonprofit training agency, department HSOs and agency risk management personnel should evaluate where fire department or emergency response personnel should go for certification.

WARNING

Structural self-contained breathing apparatus (SCBA) and self-contained underwater breathing apparatus (SCUBA) are not the same. SCBA must never be worn in place of SCUBA.

Industrial

There are a variety of levels of training and response capabilities for industrial fire fighting. Referenced standards particular to industrial fire fighting include:

- OSHA 1910.156: *Fire Brigades*
- OSHA 1910.120: *Hazardous waste operations and emergency response*
- NFPA 600: *Standard on Facility Fire Brigades*
- NFPA 1081: *Standard for Facility Fire Brigade Member Professional Qualifications*

The type of PPE required for industrial fire fighting is based on the level of training and type of incident **(Table 8.2)**.

Table 8.2
Industrial Training Level and Appropriate PPE

Industrial Training Level	Appropriate PPE
Incipient	Normal work clothing
Advanced Exterior	Structural PPE with respiratory protection
Interior Structural	Structural PPE with respiratory protection
Advanced Exterior and Interior Structural	Structural PPE with respiratory protection

Technical Rescue

NFPA 1951, *Standard on Protective Ensembles for Technical Rescue Incidents*, outlines three categories of technical rescue ensembles:

- Utility — No liquid hazards in structural collapse and confined space settings
- Rescue and Recovery — May include some body fluids or liquid hazards in structural rescue operations
- CBRN — After terrorist events

These ensembles are lighter weight and more breathable than structural turnouts, which makes them more comfortable to wear for longer durations and in smaller spaces.

Wildland fire PPE is commonly used instead of specialized search and rescue PPE, although it lacks some of the desired performance and special needs for search and rescue. Search and rescue ensembles include the following **(Figure 8.10)**:

Figure 8.10 Technical responders wear PPE specific to search and rescue operations.

- **Head protection** — Protects the head from falling or protruding objects. Although structural fire fighting helmets can be worn, specially designed rescue helmets that meet NFPA 1951 standards are lighter and smaller and are intended for working in confined spaces. Brims on structural fire fighting helmets can be difficult to wear in confined spaces. Additionally, in the event of fall or head impact, the brim strikes first, thus causing neck trauma.
- **Eye protection** — Protects the eyes from debris, liquids, airborne particles, and harmful types of light. Safety goggles and glasses should meet ANSI Z87.1 design criteria for use in confined-space operations.
- **Hearing** — Earplugs or earmuffs should be worn to protect the hearing from the sound of power-operated equipment or noise that is amplified in confined spaces.
- **Footwear** — Boots with steel or composite insoles and toe caps should be worn to protect the feet from puncture wounds.
- **Hand wear** —Comfortable, close-fitting leather or synthetic gloves allow dexterity but protect the hands. When performing emergency medical aspects of rescue, latex or other impermeable exam gloves must always be worn. In some situations, they should be worn inside protective leather or synthetic outer gloves.
- **Knee and elbow pads** — Knee and elbow protection is essential, as rescuers may be required to crawl into confined, debris-strewn spaces. Strap-on pads, specifically designed for these situations or for use in the construction trades, are commercially available

- **Body protection** — Structural fire fighting PPE may be used in search and rescue operations, although the clothing is heavy and bulky. Work uniforms may be more appropriate, providing complete coverage and protection while allowing the wearer more flexibility. Coveralls made from fire-resistant materials, such as Nomex®, meet NFPA requirements.
- **Respiratory protection** — Respiratory protection and atmospheric monitoring is mandatory and in some cases legally required. The type of respiratory protection will depend on the hazard assessment. Following a structural collapse, fine particles of concrete and other materials are released into the atmosphere. In addition, confined spaces may have toxic gases or low oxygen levels in them.

Nonemergency Work Apparel

Fire fighting, search and rescue, response to hazardous materials incidents, and medical activities are not the only tasks that emergency responders perform. They spend most of their time in station wear or work uniforms. A department may also require the use of physical fitness uniforms and clothing that is specific to inspections or investigations. This apparel is not considered PPE, but it may become part of the emergency response ensemble if worn beneath wildland or structural PPE.

Work Apparel

Work apparel should meet the requirements of NFPA 1975, Standard on Emergency Services Work Clothing Elements. It is generally recommended that undergarments be made solely from cotton.

While the structural PPE will protect firefighters during emergency situations, work apparel is the only protection around the station or in the repair shop **(Figure 8.11)**. Approved work uniforms should be worn to protect from sharp objects, chemical spills, or flash fires.

Figure 8.11 Station wear is intended to provide a minimum level of protection against hazards encountered during nonemergency activities.

Physical Exercise Uniforms

Physical fitness clothing usually consists of shorts, T-shirts, sweat suits, socks, and athletic shoes. Items to be worn under PPE or as sleepwear must also be 100 percent cotton.

Inspections and Investigations Clothing

Inspectors and investigators should wear approved work apparel when performing their duties. However, some conditions may warrant the use of additional PPE. Inspectors should be provided with fire-resistant coveralls, gloves, hard hats, and hearing and eye protection when working at construction or industrial sites. Investigators must wear structural PPE and respiratory protection when investigating fire scenes or chemical spills.

PPE Accessories

In addition to the task-specific PPE ensembles, there are other items essential to firefighter health and safety. They are addressed in this section and include:

- Hand protection
- Eye protection
- Hearing protection
- Personal alert safety system (PASS) devices
- Respiratory protection
- Traffic control vests and PPE reflective striping
- Life safety rope, life belts, harnesses, and personal escape/rescue systems
- Infection Control PPE
- Personal lighting
- SCBA Heads up display
- Thermal imager
- Lighted rope
- Rescue loop on PPE coat
- Truck man belt for tools and equipment

Hand Protection

Gloves designed for fire fighting and medical responses have already been described in previous sections. Station or shop activities and apparatus operations also require hand protection. Stripping paint, using welding or cutting equipment, and working with hazardous materials, all require specific types of protection. Driving apparatus gloves enable the driver/operator to grip the steering wheel without slipping. Moisture, such as perspiration, is absorbed into the liner of the glove to prevent a slippery steering wheel or gearshift knob.

In order to afford the correct level of protection, the type of glove used should meet the type of activity performed. Work gloves may be cotton, leather, or vinyl depending on the task. Organizations should be diligent about not allowing members to bring personal hand protection products for work activities.

Eye Protection

During emergency operations, the helmet face shield, safety goggles, and SCBA facepiece provide eye protection. Safety goggles or glasses should also be worn when operating any tools or equipment that create a risk of debris coming in contact with eyes. Due to the risk of **airborne pathogens**, most departments mandate the use of eye protection to prevent disease transmission at all medical calls.

Airborne Pathogens — Disease-causing microorganisms (viruses, bacteria, or fungi) suspended in the air.

Hearing Protection

Hearing protection is required by NFPA 1500 and OSHA for noise levels greater than 90 dB. There is no single best type of hearing protection for all users or situations. General types of hearing protection include:

- Expandable foam plugs —Designed to expand and conform to the user's ear canal. They are inexpensive and disposable.
- Molded, reusable plugs —Made of silicone, rubber or plastic material and available in several sizes. They are inexpensive and can be reused.
- Canal caps —Flexible tips on a flexible plastic or metal headband that act as caps to close the ear opening. They can be taken off and put back on relatively easily.
- Ear muffs — Many models and sizes of ear cups designed to fit most people and cover the ear. Easy to put on and take off. May include extra materials or electronic components. People with heavy beards or glasses may have difficulty blocking out noise due to a poor seal around the ear.

The best type of hearing protection is the protection that is worn. HSOs need to train firefighters how to know when hearing protection is needed; how to insert, adjust, or use it; and how to select protection that is comfortable and convenient to use.

Personal Alert Safety System (PASS) Devices

OSHA or equivalent state agencies mandate the use of PASS devices when operating in IDLH areas. Designed to alert other emergency responders when someone is incapacitated or lost, these devices work when they are activated manually or automatically.

Inspections of the units at the beginning of each work shift and prior to entering the IDLH area, along with constant training and supervision, are essential to the successful operation of a PASS device. Due to strong support from regulatory agencies and manufacturers, most PASS devices are integrated into SCBA units and turn on automatically when the air supply valve is activated.

Respiratory Protection

Many emergency response sites have the potential to become IDLH atmospheres. Respiratory protection should also be used in non-IDLH atmospheres, such as wildland fires. To protect their personnel, fire and emergency services organizations should provide a wide variety of respiratory protection choices for both emergency and nonemergency use. Respiratory protection ranges from simple particle or dust masks to positive-pressure SCBA units. Selection of the appropriate type of protection is based on the identity and physical state of the hazard or potential hazard.

Figure 8.12 NFPA standards require that each person who is required to wear SCBA be fit-tested annually and issued their own facepiece.

Personnel must be fit-tested according to specific criteria, and trained in the use and limitations of all types of respiratory protection equipment **(Figure 8.12)**. Training includes the selection of the proper type of mask for the type of hazard and the use, care, cleaning, and disposal of masks.

Often overlooked as a form of respiratory protection, the N95 particle respirator and/or dust mask should be used when exposure to airborne particulates is possible. These masks are generally considered a one-time-use item and should be disposed of properly. Firefighters are responsible for selecting the appropriate type of particle mask for the task they are performing. Uses of particle masks include:

- Medical responses — To protect against airborne germs and some body fluids.
- Lawn mowing and yard work — To protect against dust and allergens, such as pollen.
- Painting — To protect against paint fumes and spray paint particulates.
- Tool and equipment repair — To protect against dust caused by sanding and grinding.
- Building maintenance — To protect against dust from demolition, sanding, or cleaning.
- Fire investigations — To protect against particulate matter where other atmospheric hazards have been ruled out.

APRs provide the next level of protection. APRs may be full or half-face masks that have replaceable canister filters. They are more effective than particle masks and can be used for the same types of activities. APRs may only be used in environments where contaminants are known and stable, and oxygen levels are normal.

Positive pressure SCBAs and SARs offer the highest degree of respiratory protection. Both are designed for IDLH atmospheres. However, due to the potential for damage to the air supply hose, SAR units are not intended for situations that involve fire.

Traffic Control Vests and PPE Reflective Striping

In 2008, the Federal Highway Administration (FHWA)/DOT established Rule 23 CFR 634, *Worker Visibility (High-Visibility Vests) Compliance Guidance for Fire and Emergency Services Personnel* to require the use of high-visibility safety apparel and provide guidance on its application.

Besides structural fire fighting ensembles with reflective trim, safety vests are also available and recommended. These lightweight vests have reflective trim and colors that are intended to alert drivers to the presence of responders. These vests must meet the requirements of ANSI/ISEA 207-2011, *American National Standard for High Visibility Public Safety Vests*, and state department of transportation regulations **(Figure 8.13)**.

Figure 8.13 Reflective vests provide higher visibility during roadway incidents.

The rule requires emergency workers to wear a minimum of an ANSI Class 2 high visibility vest during operations in and alongside a public roadway. ANSI Class 2 vests have either a high visibility green or high visibility orange/red background with reflective striping. ANSI Class 3 garments provide more visibility than Class 2 and are acceptable alternatives.

Life Safety Rope, Life Belts, Harnesses, and Personal Escape/Rescue Systems

Life safety rope, life belts, harnesses, and similar equipment are only issued to and used by personnel who are trained and certified as Rescue Technicians. A certified third party designs, labels, tests, and inspects this equipment. Some life safety rope is carried and used for self-rescue by fire departments that have determined a need for a lowering/raising or belay system **(Figure 8.14)**. A structural fire fighting pants ensemble may also have an integrated Class II harness.

Figure 8.14 Personnel that are trained and certified as Rescue Technicians use life safety rope, life belts, and harnesses.

Personal Lighting

Personal lighting identifies the location of personnel and makes them visible in the dark. These units are battery-operated flashing lights that can be color-coded for rapid identification. They are worn on the coat, belt, or SCBA harness and are visible up to 100 feet (30 m). If approved by the U.S. Coast Guard, personal lighting may also be used with flotation devices. Make sure all equipment and batteries are operational and intended for the atmosphere in which they will be used.

Emerging Technology

Emerging electronic technology has provided many innovations to the fire service. Among these are SCBA heads-up displays (HUDs), electronic accountability systems, built-in back support systems, and rescue harness loops built into the structural turnout jackets.

Heads-Up Display (HUD)

Designed into the SCBA face piece, the HUD provides wearers with a visual indication of the amount of air remaining in the SCBA cylinder. Four lights of different colors indicate when the cylinder is at 75 percent, 50 percent, 25 percent, and 10 percent capacity.

Electronic Accountability System

Electronic accountability systems are designed to offer improved accountability and monitoring. These systems provide notification if a firefighter is trapped, injured, or low on air. Based on radio transmissions, the master unit located at the Incident Command post tracks the units worn by firefighters. If a wearer stops moving for an extended period or activates the remote unit, an alarm is transmitted to the control panel.

Rescue Loop

Mandated by NFPA 1971, rescue harnesses are required in all new protective coats. The harness has an attachment loop located at the back of the coat near the collar. If a firefighter is incapacitated, a rescue rope can be attached to this loop, permitting the wearer to be pulled or lifted to safety.

Cooling Vests

Cooling vests are designed to offset the stress-inducing heat and humidity created by thermal protective clothing and some health conditions. Cooling vests are made of fabrics and other materials that absorb and remove perspiration. Some hold chemical packs or contain water cells that can be cooled prior to use. Others use a personal cooling system with a remote cooler and pump. Some vests are designed with retroreflective strips for use as outerwear at night.

Selection and Use

Because no single type of PPE will work for multiple types of emergencies, it is important that equipment be carefully selected. This section will provide a general overview of the purchasing process, including a risk-based model for selecting the proper PPE and the criteria that must be met.

Sources for PPE design standards include NFPA, OSHA (federal and state), and the National Institute for Occupational Safety and Health/Mine Safety and Health Administration (NIOSH/MSHA). The following sources for design criteria include both certification organizations and professional organizations:

- American National Standards Institute (ANSI)
- UL and Underwriters Laboratories of Canada (ULC)
- FM Global
- Southern Area Fire Equipment Research (SAFER)
- Fire Industry Equipment Research Organization (FIERO)
- Safety Equipment Institute (SEI)

Even the best PPE cannot protect unless it is worn properly and under the recommended set of circumstances. Emergency responders determine when to wear the equipment. They must also understand that PPE has limitations and that using it beyond those limitations increases the risk of being injured. Other safety considerations include:

- Understanding the hazard
- Donning and doffing techniques for PPE
- Equipment distribution and load carrying
- Heat, hydration, and stress issues caused by the weight of the PPE
- Heat and moisture retention caused by moisture barriers

Determining when to wear PPE is based on a number of factors. The first is the hazard at the scene. Firefighters should be equipped with PPE for each type of emergency response. Education and experience with similar situations and training scenarios will also help in making the correct selection.

After selecting the proper PPE, firefighters must put on or don the equipment. Follow the donning and doffing procedures of the department and manufacturer. Procedures for donning the SCBA face mask are very important for a good facial seal **(Figure 8.15)**.

Figure 8.15 A firefighter tests the complete seal of his facepiece.

Firefighters should don medical response PPE before the incident or upon arrival at the scene. Responders should also don respiratory protection before approaching a patient or entering a patient's room. If the patient is known to have a communicable disease such as tuberculosis, respiratory protection is required. In most cases, eyes, nose, mouth, and hands should be protected when working with a patient.

Disposable PPE, such as particle masks and medical gloves, should be removed carefully and placed in a waste container without contaminating the skin. Remove contaminated medical gloves by pulling the wrist edge down over the glove, trapping the contamination inside.

When removing fire fighting PPE, be careful to keep the contaminated area away from exposed skin. Use of disposable wipes on temple areas, ears, neck, armpits, wrists, hands, and groin is recommended immediately after any fire fighting efforts. Responders should also remove work uniforms and shower as soon as possible. Bag all contaminated PPE for transport back to the stations, then clean it as soon as possible, according to NFPA, organizational, and manufacturer standards. The department should consider supplying an additional set of turnout gear while the first set is being cleaned.

Finally, when selecting the PPE, err on the side of safety. Select the highest level of protection based on the potential danger created by the hazard **(Figure 8.16)**. Entering a structure in a work uniform to investigate a reported smoke odor is dangerous. When respiratory and structural PPE were not worn into unknown hazardous conditions, firefighters have been exposed to harmful chemicals from burning electrical devices, pesticides, and chemical mixtures.

Figure 8.16 When selecting PPE, choose the highest level of protection based on the potential danger of the hazard.

Figure 8.17 Regular PPE inspections should ensure equipment is safe for use.

Figure 8.18 The organization's health and safety officer is responsible for inspecting PPE on a regular basis.

Another consideration is equipment distribution and load-carrying ability in both structural and wildland fire fighting. PPE can weigh as much as 50 pounds (23 kg). Additional items like hose bundles, forcible entry tools, and extinguishers can increase that load considerably. Emergency responders must distribute the load evenly across the body or share the load with another firefighter or emergency responder.

The design of the PPE will add to physical stress. Structural PPE is designed with a moisture barrier as part of the inner liner. This barrier is intended to provide protection from external water and heat. However, it also helps to contain body heat and perspiration. The barrier can create an artificial environment within the PPE that can increase body temperature and cause dehydration. Heat, dehydration, and stress will result in rapid fatigue. Firefighters must be aware of this potential and the symptoms to respond appropriately.

Inspections

Like all apparatus, tools, and equipment, PPE must be inspected and tested in accordance with NFPA 1851, *Standard on Selection, Care, and Maintenance of Protective Ensembles for Structural Fire Fighting and Proximity Fire Fighting,* or the appropriate NFPA standard for a specific type of PPE. Inspections should be conducted regularly, following a formal schedule, as well as when the PPE is first accepted by the department and after any repair.

When new PPE is purchased, it is inspected when it is delivered to the department. This initial inspection is necessary to ensure that the item is complete and meets the original specifications. Some items, like SCBA and PASS devices, should be tested to ensure that they perform properly. At this time, new equipment is assigned an inventory control number and an individual record is started.

In-service PPE is inspected at the beginning of each work shift. Regular inspections ensure that the PPE is ready for immediate use and all items work properly **(Figure 8.17)**. Any required repairs are reported through a formal notification process. PASS devices and SCBAs are tested according to manufacturer's directions. Following the inspection, the PPE is placed on or next to the apparatus.

Formal, annual, or preventive maintenance inspections are more thorough than regular inspections. They are usually performed by trained mechanics, repair technicians, or the department HSO **(Figure 8.18)**. These inspections should take place in accordance with NFPA 1851 at regularly scheduled intervals and coincide with a periodic cleaning and preventive maintenance program. A PPE inspection by the health and safety officer should:

- Check the date the PPE was manufactured
- Check the condition of the PPE:
  - Stitching
  - Seams
  - Buttons
  - Clasps
  - Hook and pile fasteners
  - Zippers
  - Retroreflective trim
  - Leather/vinyl trim
- Look for unauthorized alterations or modifications
- Look for cleanliness
- Check for correct size on the wearer

SCBAs are inspected by respiratory protection technicians who perform operational tests and conduct repairs. Facepiece fit testing must occur annually to ensure that the facepiece seals correctly to the wearer's face. Fit testing may occur more often if the wearer has lost more than 15 pounds or has had recent facial surgery.

Care and Maintenance

The importance of proper care and maintenance of PPE cannot be overstressed. From a safety standpoint, equipment that is clean and in serviceable condition continues to perform as it was designed. From an economic standpoint, proper care and maintenance ensures that the original cost/benefit ratio used to select the equipment is met. Care and maintenance is the cooperative responsibility of both the firefighter and the administration. The administration provides materials, establishes procedures for care and maintenance, and may contract with vendors to provide maintenance. The firefighter cleans the equipment and reports any maintenance needs.

Care

In the past, soot-covered PPE was considered a mark of a "real firefighter." The appearance was worn with pride: the dirtier, the better. Today, firefighters are more aware that the byproducts of combustion (smoke, soot, and other contaminants) are hazardous to their health and increase their risks of cardiovascular disease and cancer. PPE that is covered in soot increases the potential for both short and long-term illnesses.

PPE should be cleaned, according to department policies, following each exposure to smoke or contaminated atmospheres **(Figure 8.19)**. Use water and a mild deter-

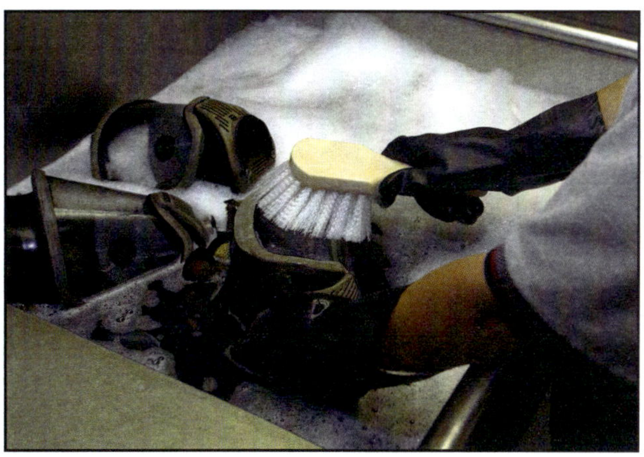

Figure 8.19 Fire fighters are responsible for cleaning and maintaining their own PPE and reporting any repair needs.

gent to clean the outer shell of the coat and trousers, including the cuffs and collar, boots, hood, gloves, helmet shell, and neck shroud. Clean the helmet headband and chinstrap, as needed, in a similar fashion. Advanced or special cleaning will be required to decontaminate PPE that has been exposed to hazardous materials, such as chemical or biological waste. General cleaning and care procedures include:

- Never use bleach or a product containing bleach on PPE.
- Clean PPE in the designated cleaning area.
- Remove liners from shells and wash separately.
- Clean contaminated PPE in a designated decon sink. Keep PPE in the assigned storage area and away from diesel exhaust, if possible.
- Do not store contaminated PPE in a personal vehicle or in the passenger cabs of apparatus.
- Wash PPE in a designated washer, not one used for sheets and towels.
- Do not clean PPE at home.
- Dry PPE in a designated dryer or by hanging it on a drying rack.
- Do not hang PPE in direct sunlight to dry, as ultraviolet light will degrade the fabric.

To follow these procedures properly, each fire and emergency responder needs two sets of PPE. For some departments, this is not economically feasible. In such cases, SOPs must specify procedures for replacing or decontaminating available PPE.

All PPE must be stored properly in well-ventilated, secure lockers in or adjacent to the apparatus room. PPE should never be stored or taken into the living quarters of the station. Ventilation is necessary to keep damp clothing from molding, and to allow any contamination to be diluted by the atmosphere.

Maintenance

Simple maintenance, such as the replacement of a helmet chin strap, can easily be performed by the wearer. Most maintenance and repairs, however, must be performed by certified or qualified repair technicians with the proper tools and repair parts. This is especially true of respiratory protection equipment. Only trained and certified technicians should repair SCBAs and SARs. These units must be tested following maintenance and before returning to service.

Some departments and personnel have a tendency to make modifications to PPE to fit their local needs or personal preferences. However, each piece of PPE has been tested and certified by the manufacturer to perform in its manufactured condition. Modifications nullify any warranty and can alter the performance of the item. Never modify any piece of PPE. If additional pockets or attachment points are needed, they should be specified when PPE is ordered. Even the use of decals on helmets can affect how the helmet reacts in a fire situation.

Disposal

At some point, all PPE requires disposal. Some equipment becomes so contaminated that it cannot be properly cleaned. Badly damaged equipment may be impossible or too costly to repair. When equipment reaches its end of service life (ESL), state/provincial or local laws determine the proper means of disposal **(Figure 8.20)**.

Figure 8.20 Personal protective equipment contaminated with chemical, radiological, or biological materials must be disposed of properly.

End of Service Life (ESL)

There are very few established guidelines for determining the service life of a piece of PPE. Some manufacturers do not publish any guidelines. However, others suggest turnout ESL is 10 years or less. Criteria to determine the ESL of PPE include:

- Amount of use
- Cumulative maintenance costs
- Changes in regulations
- Obsolescence
- Changes in technology
- Changes in function or mission

As a general guideline, some departments keep equipment in first-line service for 7 years. After that time, the item is inspected and placed in reserve, used for training, or disposed of. Records for each item must be maintained to monitor condition, assignment, and eventual disposal. If items are used for training, they should be permanently marked as *No Live Fire (NLF)* or *Training Only* so they do not get mishandled or find their way back into use on the front line.

Contaminated Equipment

PPE may become contaminated over time through repeated exposures or all at once through a single exposure to chemical, radiological, or biological contaminants. Equipment that cannot be thoroughly cleaned and decontaminated must be disposed of properly.

Equipment contaminated by chemical, radiological, or biological material must be disposed of in accordance with the federal OSHA and EPA hazardous materials procedures. While wearing the proper PPE, place the contaminated items in a plastic trash bag, secure the opening, and place the bag in an overpack drum. The drum is then sealed. A company that is approved for handling hazardous materials waste is given the drum for disposal.

Damaged Equipment

The decision to dispose of damaged equipment is usually based on a cost/benefit analysis. The item should be disposed of if the cost of repair, or cumulative cost of maintenance, meets or exceeds its value. Damaged equipment may be stripped of usable parts and then destroyed or placed in a trash container. Due to liability issues, it is not a good policy to donate damaged equipment to other organizations.

Records

Records must be kept on all PPE. The records serve many purposes, including inventory control, budget justification, and liability mitigation. PPE records include purchasing, inspection, maintenance, and disposal information. Records help plan for replacement costs and determine when preventive maintenance or disposal is required. They can also be reviewed by accident investigators.

Types of PPE Records

Modern record-keeping systems permit various types of files to be cross-referenced for easy access. In some cases, the information is located in remote sites that may be under the control of a different division or department within the jurisdiction. A common identification number or bar code for each item links these records.

Purchase Information

Purchasing information is usually maintained by the jurisdiction's purchasing department, although it may be retained by the fire and emergency services department. The information is used to verify that products have been received and that they meet the original specifications. Purchase information also provides background data that may help with future purchases. This includes:

- Product specifications
- Purchase order
- Purchase contract
- Acceptance inspection date and inspector name
- Warranty
- Shipping/delivery papers
- Manufacturer/vendor information

Inspection Information

Local jurisdictions usually establish periodic inspection requirements for PPE. Some of these requirements are based on legal mandates, such as the inspection cycle for SCBA. Inspection information provides a basis for other types of records, such as maintenance and disposal dates, tracking inventory, and determining liability in the event of a PPE-related accident. Inspection information usually includes:

- Inventory tracking number
- Inspector's name
- Item assignment(s) (to a person or location)
- Condition
- Age
- Repairs needed
- Reason(s) for removal from service

Maintenance Information

Maintenance records include information about cleaning, preventive maintenance, and corrective work. Cleaning records are maintained on protective clothing, ropes, and facepieces. Preventive maintenance includes scheduled repairs, such as periodic battery or parts replacements, calibrations, and any repairs recommended by the manufacturer. Corrective maintenance is the result of inspection findings or of any reported equipment failure or damage. The PPE is removed from service and a replacement is issued.

Maintenance records generally include the following:

- Inventory tracking number
- Name of person requesting maintenance
- Name of person performing maintenance
- Type of maintenance
- Reason for maintenance
- Parts used
- Labor (in hours/minutes)
- Cost of parts
- Cost of labor
- Recommendation for disposal

Disposal Information

All state/provincial and local governments have laws concerning the disposal of property. When PPE reaches its normal ESL or repairs cannot be justified, the item must be disposed of according to those laws. Generally, there are only two ways to dispose of PPE: surplus sale or destruction.

Records must indicate how the PPE was disposed of. Property that was purchased with tax money must be tracked all the way from receipt to disposal. When PPE is disposed of in a surplus sale, any revenue received from its sale must be documented.

The second way to dispose of PPE is to destroy it. This is sometimes mandated if the jurisdiction continues to be liable for the equipment after it is transferred to another owner. Remember that protective clothing that is no longer acceptable for use in one department is not acceptable for use in another department.

Disposal records include the inventory tracking number, reason for disposal, and final means and date of disposal. If the final disposal is the result of damage or loss that can be reimbursed, additional information may be included to indicate how and when the item was replaced.

Budget Justification

PPE is purchased with revenue from taxes or other sources. The process generally involves assigning an inventory control number to the item, sometimes in the form of a bar code. As indicated previously, this number is the basis of the record-keeping system and is used to determine the location, condition, maintenance, and disposal of the item throughout its service life.

Another use for PPE records is to develop and justify the budget requests of the department. Records provide the initial cost of equipment; its expected service life; the cost of repairs for parts and labor; how well various types of equipment have worked; and the cost/benefit the PPE has provided. These records can be used to justify purchasing backup or replacement equipment, changing to another type of equipment, or requesting an increase in budget allotment.

Liability Mitigation

The final reason to keep records on PPE is to reduce the risk of liability. Detailed records help to determine the role of the PPE in an accident, injury, or fatality.

Chapter Review

1. What factors involving PPE contribute to firefighter injuries or fatalities?
2. What pieces of PPE comprise a full structural fire fighting ensemble?
3. What are the components of EPA-Level A, B, C, and D ensembles?
4. Why should fire and emergency services provide a wide variety of respiratory protection to firefighters?
5. What are the general PPE donning and doffing procedures?
6. What should the HSO or firefighters check for when tasked with conducting PPE inspections?
7. What are the general procedures for cleaning and drying PPE used in an incident?
8. Why might a department need to dispose of PPE?
9. How are records used other than documenting the condition of PPE?

Discussion Question

1. What types of hazards, whether it be from manufacturing, residential, or geographic features, are in your jurisdiction, and what types of PPE may be employed to help mitigate them?

Emergency Operations

Chapter Contents

chapter 9

Key Terms

FESHE Learning Outcomes

Occupational Safety and Health for Emergency Services

4. List and describe the components of risk identification, risk evaluation, and incident management.
6. Apply the knowledge of an effective safety plan to pre-incident planning, response, and training activities.
7. Explain the components of an accountability system in emergency service operations.
8. Discuss the need for and the process used for post-incident analysis.
9. Describe the components and value of critical incident management programs.
12. Identify and analyze the major causes involved in line-of-duty firefighter deaths related to health, wellness, fitness, and vehicle operations.

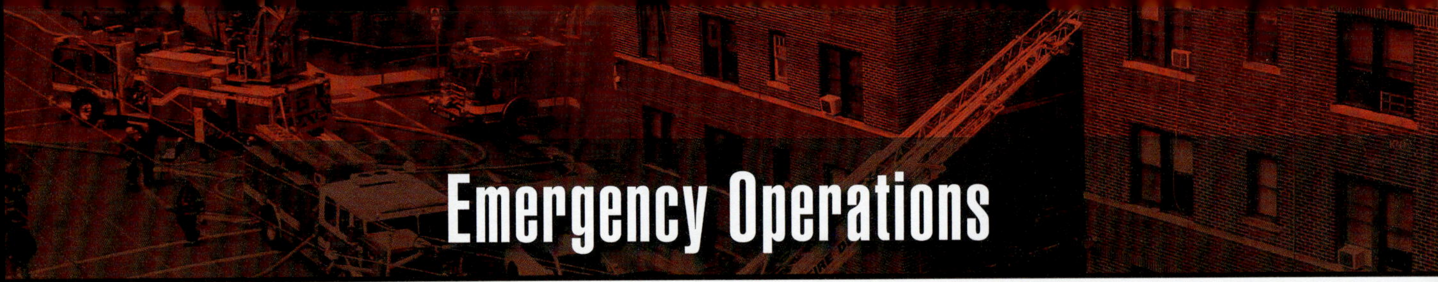

Emergency Operations

Learning Objectives

After reading this chapter, students will be able to:

1. Review trends of on-scene responder injuries, exposures, and fatalities.
2. Explain the on-scene responsibilities of administration, supervisors, and individual firefighters.
3. Identify the necessary components needed for a successful and safe incident response.
4. Explain the importance of all stages of the size-up process in creating incident action plans (IAPs).
5. Identify the components of the National Incident Management System-Incident Command System (NIMS-ICS).
6. Describe how to mitigate unsafe on-scene responder behaviors.
7. Differentiate between a post-incident analysis and a post-incident critique.

Chapter References

Hylton J.G. Haynes and Joseph L. Molis. 2017. "United States Firefighter Injuries – 2016". National Fire Protection Association. Accessed online at nfpa.org.

Rita F. Fahy, Paul R. LeBlanc, and Joseph L. Molis. 2017. "Firefighter Fatalities in the United States – 2016". National Fire Protection Association. Accessed online at nfpa.org

Chapter 9
Emergency Operations

Fire and emergency services organizations respond to a wide range of emergency situations and incidents. Each incident presents its own hazards, placing responders' health and safety at risk. Statistics collected and analyzed by the National Fire Protection Association (NFPA) and the National Institute for Occupational Safety and Health (NIOSH) indicate the following (Fahy, et al, 2017) (Haynes and Molis, 2017):

- Most fatalities at emergency incidents involved sudden cardiac death and internal trauma.
- 22 percent of fatalities occurred during fireground operations.
- The leading injury types in 2016 were strains and sprains.

Each type of emergency incident involves risk, and many incidents involve multiple risks. Emergency responders must be prepared for the unexpected and should protect themselves in the following ways.

- Use proper PPE
- Adhere to departmental standard operating procedures or guidelines (SOPs/SOGs), maintaining situational awareness
- Follow orders given by the supervisor or officer in charge
- Recognize a dangerous order and communicate concerns to the Incident Commander
- Speak about safety issues
- Maintain situational awareness

Figure 9.1 Fire fighters train in the proper use of PPE and equipment.

This chapter discusses trends in fatalities and injuries at emergency scenes and outlines the responsibilities that firefighters, supervisors, and the administration have to reduce them. It describes physical safety components for the incident scene as well as operational considerations that an Incident Command System provides. Other topics include behaviors that contribute to emergency scene casualties and the importance of a post-incident analysis.

The act of being safe in a hazardous situation takes training, awareness, preparedness, and use of proper PPE (**Figure 9.1**). An emergency responder, who becomes injured or incapacitated during an incident, increases the risk to fellow firefighters, drains resources from the incident, and reduces the efficiency and effectiveness of the responding units. Everyone on the scene is responsible for trying to prevent situations that will result in emergency scene casualties.

Emergency Scene Trends

Annual on-duty fatalities have declined since the late 70s and early 80s, when 100 firefighter deaths per year was the norm. Since 1981, when the NFPA began keeping injury statistics, there has also been an overall decrease in firefighter injuries. In 1981, there were over 100,000 injuries reported. In 2016, there were 62,085 injuries reported. Between 2015 and 2016, the number of injuries dropped 8.8 percent.

Injuries

Although the number of injuries has generally decreased for the 1981 to 2016 period, so has the number of fires. Therefore, the rate of injuries per 1,000 fire calls has remained consistent. Injuries for this period stand have averaged between 20 and 25 per 1,000 fires, but dropped to 18 in 2016 (Haynes and Molis, 2017).

For the same period, injuries at nonfire emergency scenes increased from 9,600 in 1981 to 12,780 in 2016, a 33.1 percent increase. However, nonfire emergencies have also increased by 340 percent, due to additional medical calls. Factored for actual injury rates, injuries at nonfire emergencies have declined from 1.24 in 1981 to 0.38 per 1,000 calls.

Firefighter injuries are classified under five types of duty:

- Responding to or returning from an incident (fire and non-fire)
- Fireground (at structure, vehicle, and brush fires)
- Nonfire emergency (rescue, hazardous, and natural disaster calls)
- Training
- Other on duty (inspections, maintenance)

In all five types of duty classifications, the injury categories of strain, sprain, or muscular pain continue to account for most injuries. The second most common injury category for all types of duty was wound, cut, bleeding, or bruise.

Exposures

In 2016, infectious diseases such as hepatitis, meningitis, and HIV resulted in 0.4 exposures per 1,000 emergency runs by fire departments. Hazardous conditions such as asbestos, radioactive materials, chemicals, and fumes accounted for 32.9 exposures per 1,000 emergency runs in 2016. The NFPA suggests that this 34 percent increase in exposures from the year before may be related to increased awareness about cancer and other chronic illnesses in the fire service and to the importance of documentation.

Fatalities

According to the 2016 NFPA report, 15 of 69 on-duty firefighter deaths occurred at fire scenes. This is the lowest number of fireground deaths since the NFPA began its study in 1977. The 26 sudden cardiac deaths in 2016, with onset while the victim was on-duty, is also the lowest number reported since 1977. However, according to the NFPA, sudden cardiac death continues to be a serious concern and the number one cause of on-duty firefighter fatalities in the U.S.

With the exception of 2001 (the World Trade Center) and 2013 (Yarnell Hill wildland fire and a fertilizer plant explosion), the number of fireground fatalities has been declining steadily since 1999. 2016 is the third consecutive year that the total has been below 25. The majority of the fireground deaths occurred in residential and wildland fires.

Emergency Incident Responsibilities

As in all other situations involving safety, department administrators, supervisors, and responders have varying safety responsibilities during an emergency incident. Each group should remain aware of others' responsibilities.

Administration

The administration is ultimately responsible for the safety of all fire department employees at all activities, at all times, and in all locations. However, every member of the department should appropriately communicate unsafe situations or behaviors to their supervisors to avoid injuring themselves and others. This responsibility is especially important during emergency incidents.

A healthy, proactive culture of organizational safety is encouraged when administrative personnel make time and offer support for staff and line personnel to research, plan, train, and execute the best known practices for safe operations. These practices include:

- Developing and continually reviewing standard operating procedures or guidelines (SOPs/SOGs)
- Developing adequate training budgets
- Acquiring quality PPE and tools
- Budgeting for staff to perform ongoing risk management programs
- Working toward an organizational framework of response, public education and prevention
- Consistently using the incident management system (IMS) during emergency response

These practices allow for healthy environments that can support safety culture development. A natural outcome of that development is a work force operating with more respect for each other at emergency incidents.

Supervisor

Supervisors are responsible for implementing the adopted rules, regulations, and SOPs/SOGs of the department. Safety of the responding crew, either in the supervisor's company or in other units, is the first consideration at any emergency incident.

At emergency incidents, supervisors must operate within the adopted policies, procedures, and guidelines. This includes the IMS, continuous risk management, care of personnel, and communication with other resources. When supervisors assume the role of **Incident Commander (IC)** at an emergency incident, they should:

- Assume Command in an effective location
- Establish Unified Command to work with other agencies as necessary

Incident Commander (IC) — Person in charge of the Incident Command System and responsible for the management of all incident operations during an emergency.

- Perform size-up and ongoing risk management to include assignment of Incident Safety Officer (ISO)
- Maintain control and develop a communications plan **(Figure 9.2)**
- Declare an overall strategy
- Develop an incident action plan (IAP)
- Employ an accountability system to track all personnel and assets at the scene of an incident
- Communicate incident objectives and assign companies to complete them
- Maintain span of control and request additional resources, as needed
- Review and revise the IAP
- Maintain, transfer, or terminate Command

Figure 9.2 An Incident Commander communicates with other officers on scene. *Courtesy of Chris Mickal, District Chief, New Orleans (LA) FD Photo Unit.*

Individual

Individuals can show a commitment to safety by participating in training and continuing education, modeling safe behaviors for peers, and maintaining a zero tolerance for unsafe acts. Those who disregard safe practices can draw resources away from an incident and create additional risk.

Individuals at emergency incidents should have a specific objective from Incident Command, use proper tools, work as a team, and see that the team is part of an effective accountability system. Individuals should also be competent at caring for themselves or others during **Mayday** scenarios **(Figure 9.3)**, be willing to rehab before becoming overly fatigued, and demobilize through the IMS designated channels.

Mayday — Internationally recognized distress signal.

Components

The fire department administration is responsible for the physical safety components that are used at emergency incidents. The administration must develop, fund, provide training, implement, and monitor the application of these components. SOPs/SOGs, resources, personnel positions or functions and training must take into consideration the following factors:

- Staffing
- Control zones
- Rehabilitation
- On-scene medical assessment
- On-scene lighting
- Visibility
- Exposure
- Environmental issues
- Building construction

Figure 9.3 Mayday emergencies include entanglement, SCBA failure, distress, and disorientation.

Staffing

Staffing levels have been a topic of conversation within the fire service for decades. The need for adequate staffing at emergency incidents is not questioned. The debate arises over what is adequate staffing or the minimum level of personnel to perform functions such as rescue, extinguishment, and spill containment, among others. As research findings prove that differing staffing levels for specific fires impact time to rescue and time to extinguishment, the debate moves from the importance of adequate staffing to the timing the community wants to financially support.

In career and combination departments, the administration is responsible for describing how budgets will translate into personnel utilization and deployment. The political leadership of the jurisdiction, based on local needs and mandates, in effect determines the following:

- Number of people hired by the department
- Number of individuals assigned to each apparatus during a normal work shift
- Minimum number of people needed to place the unit in service

Before requesting outside assistance, volunteer departments determine the minimum number of positions to be filled and the minimum number of firefighters required to respond effectively to an emergency. Staffing levels are driven by national standards, such as NFPA 1500, 1710, and 1720, OSHA two-in-two-out, National Institutes of Standards and Technology (NIST) Landmark High-Rise and Residential Studies, and insurance industry mandates. Staffing levels may also be mandated by local labor-management agreements.

Another consideration is when other organizational needs temporarily affect staffing levels. For instance, removing personnel from the front line for training, health assessments, public education, or other administrative needs can reduce staffing levels if they are not replaced with equally qualified personnel. This practice is common and may amount to several companies being out of service at the same time. The expectations of the community do not change because they are unaware that their local fire service has been reduced.

Control Zones

Control Zones — System of barriers surrounding designated areas at emergency scenes intended to limit the number of persons exposed to the hazard, and to facilitate its mitigation. At a major incident there will be three zones — restricted (hot), limited access (warm), and support (cold).

There are numerous reasons for establishing **control zones** at an emergency incident scene. First, controlling the perimeter facilitates the use of a personnel accountability system. Secondly, it helps in accounting for occupants or victims. Finally, it helps to keep the scene free of curious spectators.

Establishing three operating control zones is the most common and effective way to control the perimeter of an incident scene **(Figure 9.4)**. At a major incident there will be three zones — restricted (hot), limited access (warm), and support (cold). Control zones should be established as soon as resources are available for this assignment. Although not always practical, zones can be cordoned off or indicated with utility rope or fire line tape.

NOTE: Local terminology may differ. For example, warm zone is also known as *contamination reduction zone*.

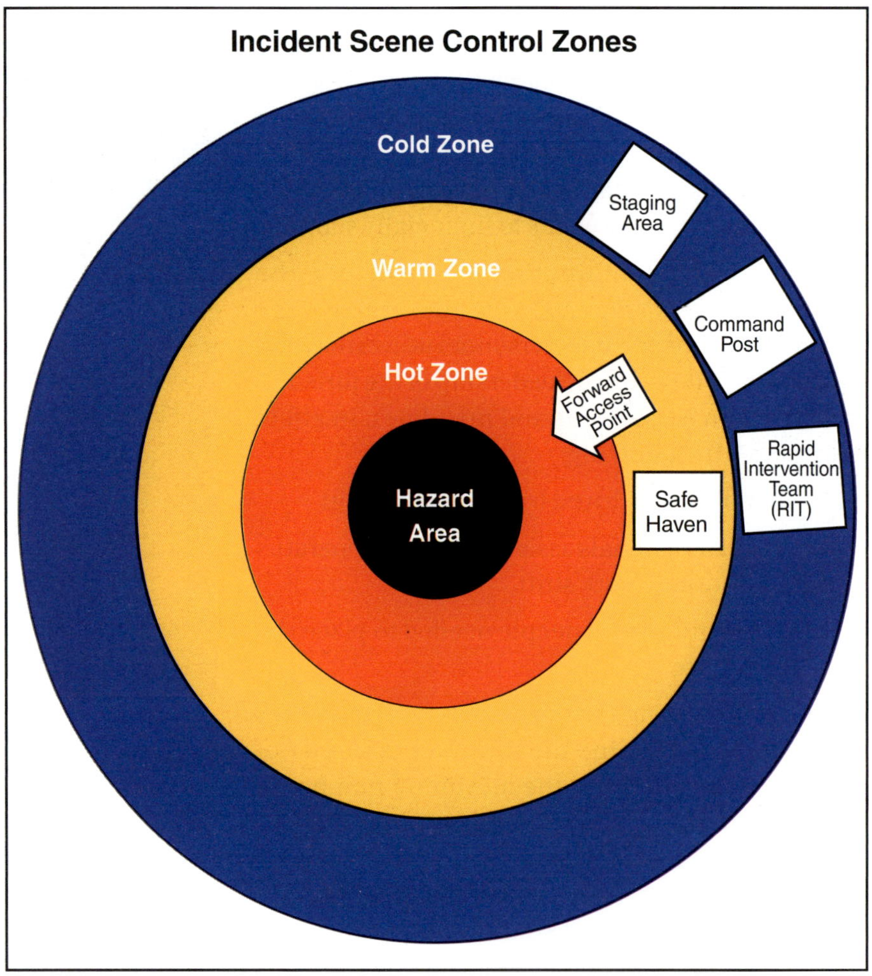

Figure 9.4 Hot, warm, and cold operative zones control the perimeter of an incident scene.

There is no specific distance or area that should be cordoned off for each zone or from the total incident scene. Emergency responders should establish zone boundaries by taking into consideration:

- Amount of area personnel need to work
- Degree of hazard presented by elements involved in the incident
- Wind and weather conditions
- General topography of the area

The three zones can be described as follows:

- ***Hot zone*** — Area where resolving the problem takes place — fires are suppressed **(Figure 9.5)**, hazardous materials (hazmat) releases are controlled and contained, and vehicle extrication is performed. To limit crowds and confusion at the most critical area of the scene, only personnel who are directly involved in mitigating the problem are allowed. The size and shape of the zone may vary greatly, depending upon the nature and extent of the problem. *Personnel requirements*:
  - Trained appropriately to manage the situation
  - Attired in complete PPE designed for the specific hazard
  - Participated in a personnel accountability system implemented by the IC or ISO

Figure 9.5 Fire suppression takes place in the hot zone, the most critical area of the emergency scene. *Courtesy of Bob Esposito, Pennsburg, PA.*

- ***Warm zone*** — Area immediately outside the hot zone for personnel who are directly supporting the work being performed by those in the hot zone — limited to personnel who are operating hydraulic tool power plants, providing emergency lighting, and providing fire protection. These personnel are in full PPE and ready to enter the hot zone. In hazardous materials incidents, a decontamination station is normally assembled here.

- ***Cold zone*** — Area immediately surrounding the hot and warm zones — may include the Incident Command post (ICP) with a rapid intervention crew/team (RIC/RIT) nearby, public information officer's (PIO) location, rehabilitation area, and staging areas for personnel and portable equipment. It may also include:
  - — Backup personnel available to enter warm or hot zones
  - — Witnesses and family members of victims
  - — News media accompanied by the PIO or organization representative

  The outer boundary of this area would be the control line for the general public (crowd-control line).

Collapse Zone — Area beneath a wall in which the wall is likely to land if it loses structural integrity.

In addition to the control zones, structural fires and structural collapse incidents may require the establishment of a **collapse zone**. Collapse zones get established when there is evidence that a structure or portion of it may collapse, and strategy shifts from offensive to defensive mode. Apparatus, personnel, and master streams must be relocated to a point of safety outside the designated collapse zone.

Collapse zones, sometimes referred to as *safety zones*, should also be established around structures that have large glass partition walls. These walls can be weakened or broken by internal fires or earthquake damage, presenting a hazard of flying glass around the structure or some distance from the structure if there is a wind current.

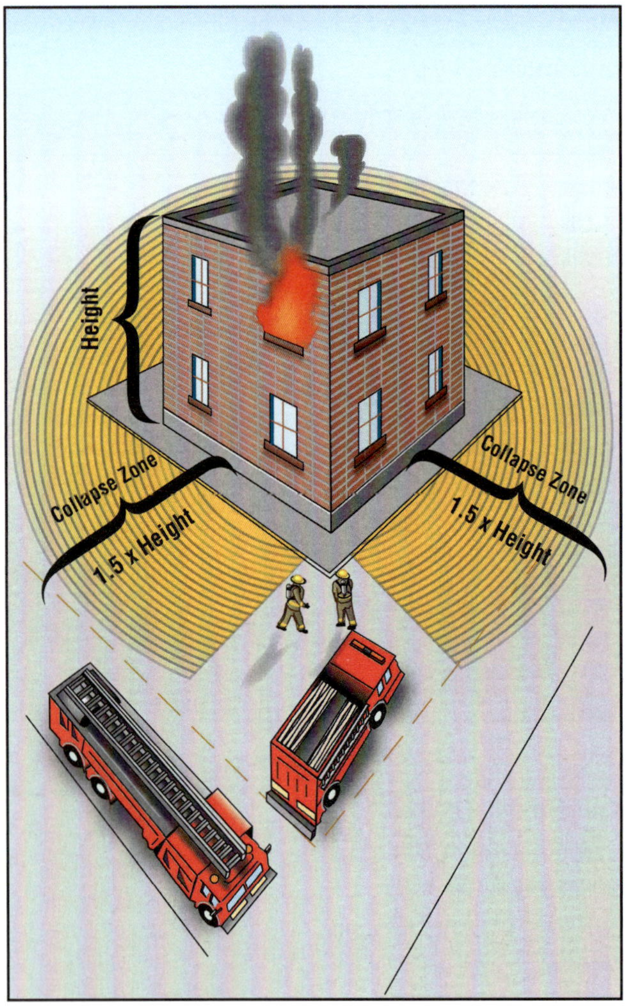

Figure 9.6 In addition to traditional calculations, the size of the collapse zone must take into consideration the type of building construction and other exposures. In most cases, the safest location for apparatus and personnel is at the corner of a building.

Calculating Collapse Zones

Traditionally, collapse or safety zones have been estimated by taking the height of the structure and multiplying it by a factor of 1½ **(Figure 9.6)**. For example, a three-story structure that is 30 feet tall (9.14 m) would require a collapse zone no less than 45 feet (13.72 m) from the base of the structure. As structures increase in height, this distance becomes impractical by creating a space that limits defensive operations. Therefore, consider the following guidelines when determining a collapse zone:

- Type I construction high-rise buildings are not as likely to collapse, making the primary concern the hazard of flying glass from windows or curtain walls. In Type I construction, the building contents are burning, not the structure itself. Collapse zones must be determined considering the direction and velocity of wind currents that can carry the glass shards. Structural collapse, if it does occur, will be localized and not structure wide.

- Type II construction consists of unprotected steel or noncombustible supports such as I-beams. When exposed to temperatures above 1,000°F (537.78°C), unprotected steel will expand and twist, pushing out walls and, when cooled, it will slightly constrict. These movements will cause floors and walls to collapse. Any type of construction that includes brick and block walls supporting unprotected steel bar joists and I-beams are involved in a large number of these collapses.
- Type III construction consists of two types: multi-story and ordinary. Multi-story buildings should have a collapse zone of 1½ times the height of the structure. For example, a building seven stories (70 feet [21.34 m]) will require a collapse zone of approximately 105 feet (32 m) during defensive operations. Ordinary construction, exterior load-bearing walls are made of concrete, brick, or masonry while interior loads are carried by wood, masonry, or unprotected steel. Masonry construction walls can collapse in one piece or crumble in many parts. When the debris strikes the ground, it can travel a distance and even cause the collapse of other structures or objects.
- Type IV heavy-timber or mill construction is one of the least likely to collapse. The weight-bearing capacity of the large dimension wood members will resist collapse unless they have been affected by a large volume of fire. A collapse zone should be established if the fire is intense or the structure has been weakened by repeated fires over time.
- Type V construction collapses are influenced by the style of construction. That is, a multi-story platform structure will generally burn through and collapse inward while a balloon structure can have the full walls fall outward in a single piece **(Figure 9.7, p. 294)**. Exterior masonry and veneer walls are not load-bearing and are placed over load-bearing wood walls. Brick veneer attached to the frame can curtain-fall collapse into a pile or fall as a unit straight out as the ties and supports fail. Although it is rare for a Type V structure to collapse outward, there is a great danger to firefighters due to interior collapses. Lightweight trusses will fail within five minutes when exposed to direct heat.

In North America, examples of structural collapse involving high-rise buildings or Type I construction buildings are few. Strict building codes ensure that structural members exposed to fire and high temperatures will remain sound until the fire is extinguished. Structural collapses due to earthquakes generally involve smaller buildings.

Collapse zones should be established when there is an indication that the structure has been weakened by prolonged exposure to fire or heat; when a defensive strategy has been adopted; or when interior operations cannot be justified. The size of the collapse zone must take into consideration the type of building construction, other exposures, and the safest location for apparatus and personnel. Church steeples, water tanks, chimneys, parapets and false facades that extend above the top of the structure must be viewed as a potential collapse hazard even if the structure is not. Most collapses usually involve brick or masonry block and may be structural components or veneer. Remember that structural collapses are not limited to the actual emergency and can occur well after the fire is extinguished. Building inspectors or structural engineers must ensure the structural stability of the site before entering it.

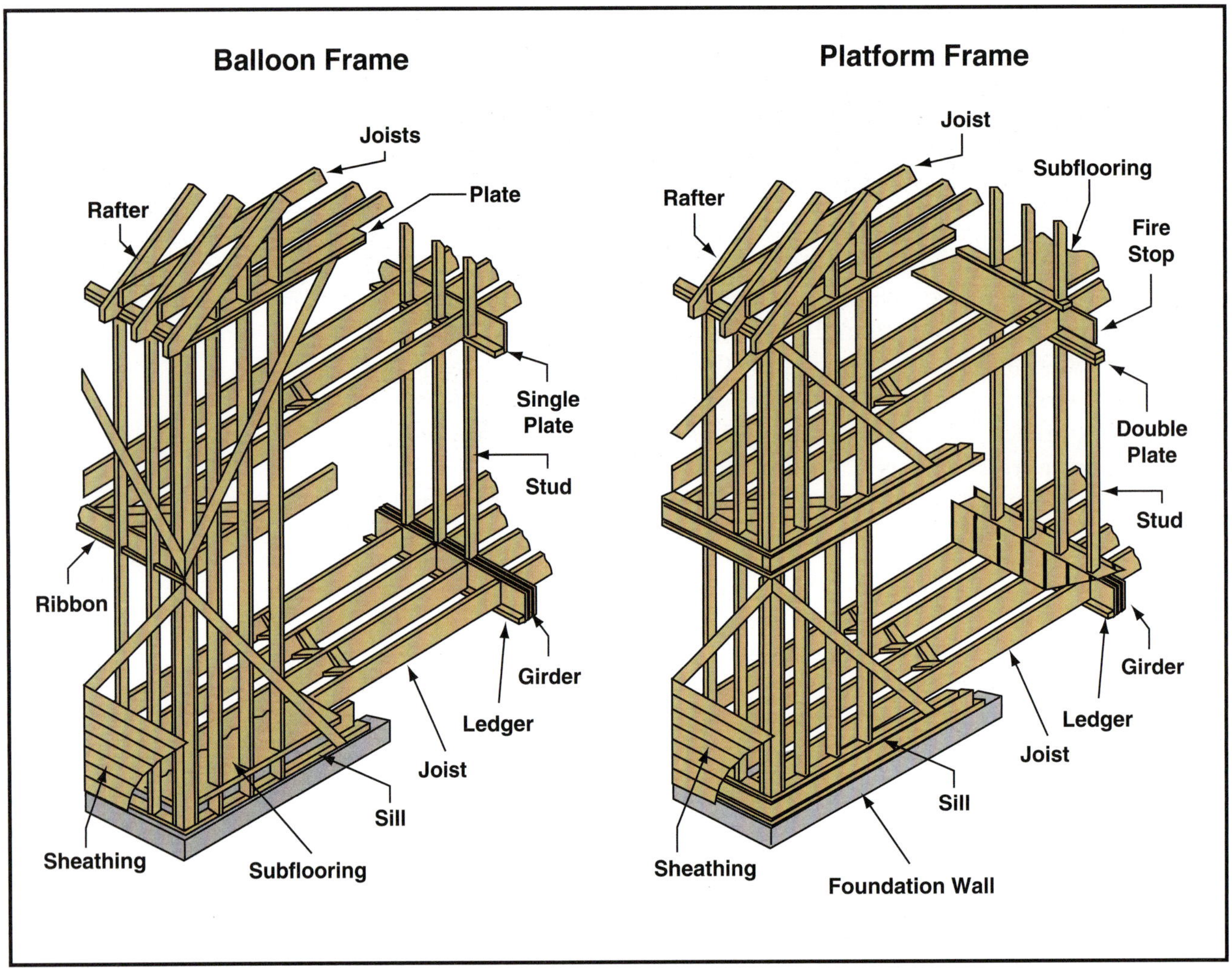

Figure 9.7 Balloon frame and platform frame construction are common in many jurisdictions.

Because the collapse zone is the full length of all of the affected walls, the safest location for defensive operations is at the corner of the building. Master streams and apparatus can be located in this 90-degree area, as long as they are far enough away that flying debris will not strike them.

Rehabilitation

Emergency scene operations place a great deal of physical stress on responders. Rehabilitation areas provide a place where personnel can rest, relax, rehydrate, take nourishment, and have their vital signs monitored by medical personnel. Rehabilitation areas, also known as rehab, are located in the cold zone and are generally incorporated with the on-scene medical assessment personnel. It is the responsibility of the administration to establish rehab procedures at emergency incidents and to fund the items needed. Water or fluid containers, folding cots, warming tents, and water-mist systems are just a few of the items generally associated with rehab areas.

The level of rehabilitation will be consistent with the size and scope of the incident **(Figure 9.8)**. Cots or warming tents may be needed for a 4-hour hazmat incident occ¬urring a January snowstorm in Massachusetts. A 2-hour fire in Phoenix, Arizona, may require medically monitored rehab for responders showing signs of overexertion.

Figure 9.8 The rehab section, located in the cold zone, provides water, rest, and monitoring for emergency personnel.

Supervisors must monitor the physical condition of their subordinates and send them to rehab before they become too fatigued. Responders must monitor their own condition and report the need for rehab to their supervisors. In practice, rehab may take two forms. The first form involves resting, removing wet or damp PPE, cooling down or warming up, and drinking fluids. The second form involves medical monitoring, treatment of injuries, rehydration, and/or transportation.

At structure fires or incidents that require the use of self-contained breathing apparatus (SCBA), personnel are normally sent to rehab when they have used two 30-minute air bottles or one 60-minute air bottle. Other factors, such as weather conditions, type of work, or fire behavior will affect this estimate. For a detailed look at rehabilitation, see *Emergency Incident Rehabilitation*, published by USFA/FEMA (February 2008) and NFPA 1584, *Standard on the Rehabilitation Process for Members during Emergency Operations and Training Exercises.*

On-Scene Medical Assessment

When the rehab area is near on-scene medical support, trained medical personnel, medical supplies, and transportation are located in one area. Medical personnel can monitor core body temperature, blood pressure and rate, hydration, and other vital signs **(Figure 9.9, p. 296)**. They can also provide immediate assistance and transportation if a firefighter shows symptoms of heat stress, cardiovascular problems, or other acute conditions. If the incident occurs too far from a medical facility for ground transport, an emergency helipad can be established at or near the on-scene medical assessment area.

On-Scene Lighting

Most fire apparatus have spotlights mounted on the four corners of the body and telescoping lights mounted near the pump panel. Newer types of apparatus lighting provide sufficient outdoor lighting for most emergency scenes. When power is shut off, portable generators and/or apparatus-powered lights can illuminate structure interiors.

Figure 9.9 Medical personnel monitor vital signs of firefighters and provide medical assistance or transportation to a medical facility as needed.

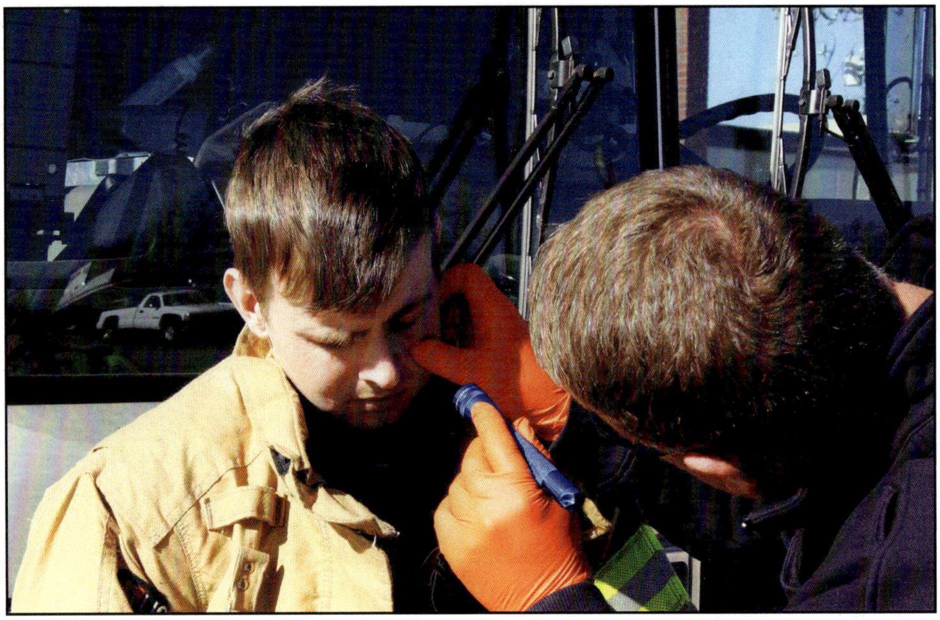

On-scene lighting can actually create a hazard if it is improperly set up. Lights should not be directed toward oncoming traffic. Especially at night, on-scene lighting can blind drivers, and prevent them from seeing apparatus and personnel along the roadway **(Figure 9.10)**. Obscured vision is more likely in inclement weather and should cause the IC, as well as each individual firefighter, to tighten possible breaches in the work zone defense perimeter.

Figure 9.10 On-scene lighting improves visibility for firefighters, but it can prevent drivers from seeing apparatus and personnel.

Visibility

Vehicle visibility is a crucial part of improving responder safely along the roadside. And, visibility studies have led to recommendations for marking apparatus with warning lights and reflective tape. NFPA 1500 has established the requirement for minimum reflective trim on PPE and the U.S. Department of Transportation (DOT) has established requirements for emergency scene warning signs, traffic cones, and the use of reflective vests during highway emergencies **(Figure 9.11)**.

Figure 9.11 Traffic cones, signs, reflective clothing, and flashing lights help reduce the risk of personnel being struck by a motorist.

According to NFPA, one or more of the following should be used by firefighters at emergency roadway incidents:

- Fluorescent and retroreflective warning devices such as traffic cones
- Retroreflective signs stating, "Emergency Scene Ahead"
- Illuminated warning devices such as highway flares
- Other warning devices appropriate to warn oncoming traffic

While the administration must supply the reflective trim, apparatus markings, and signage, it is the responder's responsibility to wear and use them. Supervisors must verify that personnel don their reflective vests, place signage as required, and keep the reflective trim on their PPE clean and repaired. All emergency responders must maintain situational awareness while taking all precautions with PPE.

Exposure

Exposure to hazardous conditions is basic to the fire and emergency services. Firefighters need to protect themselves from hazardous conditions through the use of PPE, including respiratory and eye protection.

Exposure is not limited to smoke and unburned particles of combustion. Other respiratory hazards include the vapors from fluorescent ballasts (older ones may contain polychlorinated biphenyls/PCBs); toxic chemical or gas leaks; communicable diseases and infection; and engine exhaust fumes. Situational awareness is the first step in limiting exposure to hazardous conditions. Until a responsible officer determines that the atmosphere is safe, firefighters must wear respiratory protection during the overhaul of fire scenes.

Protecting the Environment

Efforts to protect the environment should be part of emergency response. Most states/provinces have restrictions on flushing contaminants, including Class A and Class B foams, into storm drains. Gasoline, oil, radiator fluid and contaminated extinguishing agents must be contained and properly removed from motor vehicle accidents. PPE must also be decontaminated following any incident involving hazardous materials. The water used to clean PPE and equipment must also be properly disposed of, according to jurisdiction requirements. NFPA 1851, *Standard on Selection, Care, and Maintenance of Protective Ensembles for Structural Fire Fighting and Proximity Fire Fighting* and OSHA 29 CFR provide additional procedures for the inspection, care, cleaning, and decontamination of PPE.

Building Construction

Lightweight construction consists of vertical and horizontal structural members made of wood or light-gauge steel components that are lighter in weight than conventional solid wood. These lightweight components include:

- Trusses
- Glue-laminated beams
- I-joists
- Engineered wood
- Engineered I-joists
- Structural composite lumber
- Structural insulated panels
- Wood structural panels

Lightweight structural members have replaced dimensional lumber in many applications and are used to span wide areas without the need for vertical supports.

The three most common truss systems are:

- Lightweight or engineered wooden roof and floor
- Steel roof and floor
- Heavy-timber roof and floor

More than 60 percent of roof systems are constructed using a truss system. Lightweight and composite wood truss systems can contribute to increased structural fuel loads, create concealed spaces, and contribute to early collapse. Truss systems are designed to meet the minimum engineering requirements of the building code and are tested under controlled conditions in a laboratory setting.

Because actual fires are unpredictable, the failure time of truss systems cannot be predicted. When exposed to high temperatures in a compartment fire, lightweight steel trusses can fail rapidly. In addition, lightweight and composite wood truss systems can deteriorate from prolonged exposure to moisture in the environment, a condition that may not be readily apparent during visual inspections.

Firefighters must learn about trusses and their performance under fire conditions. According to NIOSH, UL, and the National Research Council of Canada (NRC), fire fighting tactics must be altered to meet the accelerated burn-through characteristics of lightweight construction. Learning to identify engineered roof and floor truss systems is key to firefighter safety. Preplanning and vigilant watchfulness of new or remodeled construction are the first steps in identifying lightweight construction in response districts. Secondly, have a general idea for when a building was built. Lightweight construction became prevalent in the 1980s. Next, use thermal imagers and small openings to look for concealed fires over or under fire fighting positions and withdraw immediately if it is determined that fire is in these spaces. Communicate all findings to the IC. Lastly, be aware of large spans above, as these are likely to be open and unprotected trusses. Expect early failure of lightweight construction.

Operational Size-Up

The emergency responder's job is to arrive safely at the incident scene, contain and control the hazard, and ensure that the incident is properly terminated. Fire officers and supervisors also establish command, assess the situation, develop an **Incident Action Plan (IAP)**, and allocate resources to mitigate the incident. To do this effectively and efficiently requires an accurate **size-up**: collecting information, evaluating it, and determining the correct strategy and tactics.

Incident Action Plan (IAP) — Written or unwritten plan for the disposition of an incident; contains the overall strategic goals, tactical objectives, and support requirements for a given operational period during an incident. All incidents require an action plan.

Size-Up

Size-up is used at all types of incidents, not just at structure fires **(Figure 9.12)**. The more firefighters practice sizing up incidents, the better they will become at reading clues and making informed decisions. An effective and thorough size-up includes information gathered before the incident occurs, during dispatch, upon arrival, and throughout the incident until all resources are placed back in service. This section describes a process for gathering information and provides important considerations for performing a size-up.

Size-Up — Ongoing evaluation of influential factors at the scene of an incident.

Figure 9.12 The on-scene size-up begins when responders arrive.

Pre-incident

The pre-incident portion of the size-up process combines knowledge gained from training, study, and experience with site-specific information learned from **pre-incident surveys** or planning. Firefighters must know about fire behavior, resource capabilities, building construction, and hazardous conditions, as well as the ICS, SOPs/SOGs, strategy/tactics, and fireground procedures. Pre-incident knowledge includes specialized training in:

- Vehicle extrication
- Rescue techniques

Pre-incident Survey — Assessment of a facility or location, made before an emergency occurs, in order to prepare for an appropriate emergency response.

- Confined-space entry
- Emergency medical care
- Disaster response
- Hazardous materials mitigation

Pre-incident surveys provide basic information about a site or structure than can be used to develop a pre-incident plan **(Figure 9.13)**. The tour is performed periodically, in accordance with local requirements. A quality pre-incident survey offers advance knowledge of important issues such as:

- Site access
- Structure
  - — Age and construction type
  - — Occupancy type
  - — Fire protection system(s)
  - — Content fuel load
  - — Unique characteristics or hazards
  - — Access
  - — Utility shutoff locations
- Available water supply
- Resource needs

Figure 9.13 A team of firefighters conducts a pre-incident survey of the exterior of a structure.

If time and resources allow, it is good practice to perform pre-incident surveys periodically on all types of structures with the exception of single-family dwellings. Although pre-incident surveys are rarely performed on single-family dwellings, these structures may have common floor plans and characteristics. Single-family dwellings may be informally surveyed during in-service home inspections and/or EMS responses. Crews should familiarize themselves with home styles, designs, and layouts common to their community or response area. Every department should have SOPs/SOGs for the type of strategy and tactics used in single-family dwellings.

Follow local SOP/SOGs to determine survey choices and frequency. The inspection criteria are usually set by the potential fire and life safety hazard of the site. Referred to as **target hazards**, these sites contain life safety concerns for firefighters as well as occupants; hazardous processes or storage that may have a high frequency or severity of fires; and high contents or structure value. Examples include:

Target Hazard — Facility in which there is a great likelihood of life or property loss in the event of a fire, terrorist attack, or natural disaster.

- Life safety concerns
  - Schools, colleges, and universities
  - Auditoriums, theaters, and restaurants
  - Places of worship
  - Hospitals, nursing homes, and daycare centers
  - Multifamily dwellings, hotels, motels, and dormitories
  - Institutions, jails, and prisons
- Hazardous processes or storage
  - Spray paint operations
  - Metal plating
  - Chemical plants
  - Automobile fueling stations
  - Compressed-gas storage areas
  - Paint storage warehouses
- High contents/structure value
  - Mercantile occupancies
  - High-rise structures
  - Office buildings
  - Warehouses

Generally, surveys are conducted annually, but they may be scheduled biannually due to a change of occupancy type or renovations or alterations to a structure. Surveys should also be made at construction sites to learn how the building is being built. Because construction sites are vulnerable to fires, surveys are a way of addressing that possibility.

The department may already have a survey check list. The information gathered will be used to develop a **pre-incident plan** or operational plan **(Figure 9.14, p. 302)**. The survey should concentrate on gathering the following information listed in NFPA 1620, *Standard for Pre-Incident Planning*:

Pre-incident Plan — Document, developed during pre-incident planning, that contains the operational plan or set procedures for the safe and efficient handling of emergency situations at a given location, such as a specific building or occupancy. Also known as Preplan.

- Construction
- Building management systems and utilities
- External site conditions or considerations
- Internal and external security features
- Fences, walls, elevator shafts, pits, and other physical barriers or hazards
- Occupant load and occupant accountability methods
- Assistance for the disabled
- Protect-in-place strategies and areas

City of LaFosse Fire Department
Preincident Site Survey

Date ______________

Company__________________ Officer-in-charge ____

Site Address____

Owner ____

Occupant ____

Emergency Contact Name ____

Phone Number ____

Number of site occupants: Day _______________ Night ___________ Weekend ________________

Occupancy Classification ____

Construction Type ____

Roof Construction Style _______________________ Material _____________________________

Number of stories: Above grade _________________ Below Grade __________________________

Stairways _________________ Hallways _________ Elevators _____________________________

Roof access _______________ Vertical openings ____

Site access ____

Portable fire extinguishers ____

Automatic fire suppression system ____

Fire pumps ____

Special agent suppression system ____

Water supply source ____

Fire department connection, location ____

Standpipe system classification ____

Fire detection and alarm system ____

Water flow alarm ____

Monitoring company ____

Fire control panel or room location ____

Location of main electrical shut-off switch ____

Location of main gas shut off ____

Location of water shut off ____

Fire doors, automatic ________________ Manual _______________________________________

Hydrant locations ____

Fire flow, estimated ____

Hydrant pressure ____

Structure Contents ____

Exposures type, distance ____

____

Additional comments or observations ____

____

Figure 9.14 A sample pre-incident survey checklist includes line items for pertinent information.

- Water supply
- Hazardous material amounts and locations
- Plans for managing vacant buildings that follow local ordinance requirements
- Responsible party to contact

Response

The size-up process continues when the alarm is sounded or other notification goes out. Pay careful attention to radio broadcasts or electronic messages. They describe the current condition as reported by witnesses at the scene or the automatic detection and alarm system at the site. Any updates are equally important because they may state that all occupants are reported to be out of a structure or someone is trapped in a vehicle.

Some information given at the beginning of the assignment may not be accurate. Prior to the arrival of fire personnel, the only immediately verifiable information may be the time of day, weather conditions, and the capabilities of a department's response. Even law enforcement officers, not trained as firefighters, have been known to report a building to be fully involved that later proved to be only a small kitchen fire. Local SOPs/SOGs generally establish the criteria for information given during a dispatch broadcast, including:

- Time — Exact time of dispatch
- Situation — Type of emergency reported to the dispatch center
- Location — Address of the emergency situation
- Resources dispatched — Units responding

Time is an important factor in the size-up process. Although people may not know the time the fire started and how long it has been burning, calculating response time and observing visual clues on arrival will help in determining the stage of the fire. Estimating timelines is also key to situational awareness.

On Arrival

The most intense part of the size-up process occurs upon arrival at an emergency incident scene. The scene may be utter chaos, or there may be no visible clues of an emergency. In addition to the emergency situation and those individuals directly involved in it, numerous spectators may have gathered at the scene, making it difficult to distinguish them from occupants or victims. These bystanders may be irrational or screaming for responders to do something **(Figure 9.15)**. Some may be attempting to extinguish the fire, assist victims, or perform a rescue — actions that can place them in direct danger.

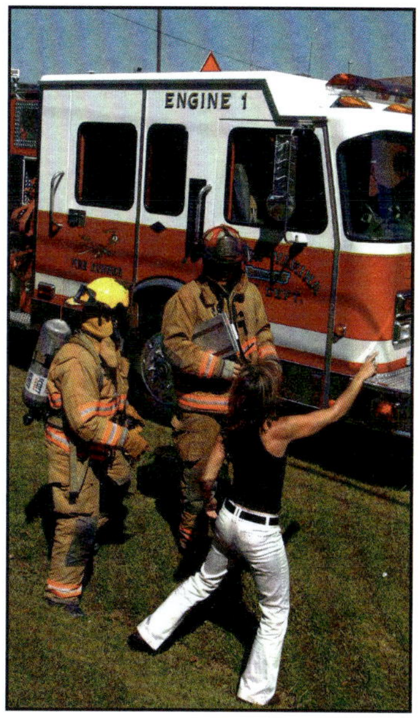

Figure 9.15 Bystanders can provide valuable information, but their expressed anxiety and excitement can harm the decision-making process.

Arriving units must assess and communicate critical information to other responding units and the dispatch center. Scene assessment begins before the apparatus comes to a stop and anyone steps from the apparatus. The incident may be categorized by the severity, extent, and dynamics of the incident. Size-up activities depend on the type of emergency incident, such as:

- Structure fire
- Nonstructure fire
- Medical response
- Hazardous materials incident
- Motor vehicle incident
- Rescue

Structure fire. At structure fires, there are two general situations: First, there may be nothing evident, which will prompt an investigation of the structure. Second, smoke or fire may be visible, prompting the deployment of hoselines and the initiation of a fire attack **(Figure 9.16)**.

At structure fires, these two situations are the core of the arrival or condition report (conditions on arrival or size-up report in some jurisdictions). Many fire and emergency services organizations mandate this type of report in their operations SOP/SOG. The arrival report is an essential communication that informs other responders and establishes a foundation upon which to build the incident. These reports also apply to all other types of nonfire emergency incidents. An arrival report states:

- What we have (incident description)
- What we need (additional resources)
- What we are going to do (deployment of resources and actions to be taken)

Figure 9.16 Firefighters attack a structure fire. *Courtesy of Bob Esposito, Pennsburg, PA.*

Following the arrival report, responders need to perform a physical size-up by looking at the scene from all sides. If a 360-degree walkaround is not possible, use other means to learn conditions on all sides of the incident. While checking the structure you are looking for:

- Any indication of occupants or the need for rescue
- Any indicators of fire location
  - Identify the flow path
  - Use a TI to look for heat signatures on the top of window frames
- Lowest floor of fire involvement
- Smoke conditions
- Forcible entry requirements
- Special hazards: propane tanks, chemicals, electrical lines, or elevation changes
- Building construction

When the structure is small, such as a single-family dwelling, it may be possible to perform a 360-degree walk-around size-up. That is, walk all the way around the structure to view it from four sides. In the case of large structures, strip malls, row houses, structures with limited access, or big-box retail stores, it may be necessary to order other arriving units to provide information from the other sides. In commercial buildings, the roof is often included in this request as aerial apparatus or ground ladders are used to view the roof area. Other incoming units may be assigned to various sides of the structure and report back to the IC.

Initial scene information is used to answer this question: Can the resources at the scene and en route handle this situation? If the answer is no, or even maybe, then additional resources must be requested immediately. Addressing this critical question is why the initial size-up sets the tone for the balance of the incident.

If there are not enough resources initially available, personnel are likely to start the incident behind the curve. Therefore, the initial size-up must be made with an attention to detail, a focus on the desired outcome, and the resources needed to accomplish that outcome. Considerations must be given to the level of acceptable personnel risk and which operational strategy to initiate. Properly interpreting condition indicators can help in decision-making and increase the safety of everyone on-scene.

Once the walk-around is performed, determine the appropriate course of action, develop an IAP, and assign duties to the crew and other personnel resources. Many departments have SOP/SOGs for routine calls that spell out duties for initial-arriving companies. For example, the second engine called is responsible for water supply. Keep in mind that additional units may not arrive for some time, making it necessary to prioritize activities. While the crew is setting up operations, the fire is increasing in size and extension. A risk assessment is necessary to ensure that tasks are carried out safely. Remember that while the crew may assist with information collection and have input, it is the IC who is responsible for making all decisions.

Before committing any resources to an interior fire attack, sufficient personnel must be on scene to adhere to the two-in-two-out rule. For the first two firefighters to enter the **immediately dangerous to life and health (IDLH)** area, two firefighters in full PPE and wearing SCBA must be outside prepared to respond to a firefighter rescue. These individuals may be the pump operator, the captain, or another firefighter. If it is not possible to implement two-in-two-out, personnel must remain outside the structure unless there is a known rescue, incipient stage fire, and/or the fire can be extinguished with a Class 1 fire extinguisher.

NOTE: Remember that the presence of four firefighters at a structure fire does not automatically ensure that the two-in-two-out rule has been met. If it is necessary for one or both of the outside members to leave the primary duty area, they may not be available for rescue. They may not leave their post if it will further endanger the incident scene. Decisions must be driven by local SOP/SOG and by industry best practices if national, state/provincial, or local laws/ ordinances do not apply.

Immediately Dangerous to Life and Health (IDLH) — Description of any atmosphere that poses an immediate hazard to life or produces immediate, irreversible, and debilitating effects on health; represents concentrations above which respiratory protection should be required. Expressed in parts per million (ppm) or milligrams per cubic meter (mg/m3); companion measurement to the permissible exposure limit (PEL).

OSHA Two-in-Two-Out Rule

The following is the applicable portion of 29 CFR 1910.134(g)(4) used to establish the two-in-two-out rule:

"***Procedures for interior structural fire fighting***. In addition to the requirements set forth under paragraph (g)(3), in interior structural fires, the employer shall ensure that:

1910.134(g)(4)(i)

At least two employees enter the IDLH atmosphere and remain in visual or voice contact with one another at all times;

1910.134(g)(4)(ii).

At least two employees are located outside the IDLH atmosphere; and

1910.134(g)(4)(iii).

All employees engaged in interior structural fire fighting use SCBAs.

Note 1 to paragraph (g): One of the two individuals located outside the IDLH atmosphere may be assigned to an additional role, such as Incident Commander in charge of the emergency or Safety Officer, so long as this individual is able to perform assistance or rescue activities without jeopardizing the safety or health of any firefighter working at the incident.

Note 2 to paragraph (g): Nothing in this section is meant to preclude firefighters from performing emergency rescue activities before an entire team has assembled."

Nonstructure fire. Nonstructure fires generally involve:

- Wildland or ground cover
- Materials stored outside
- Rubbish
- Sheds
- Dumpster fires

The size of the incident may determine whether it is necessary or possible to do a 360-degree walkaround. The size-up and condition report should consider:

- Type and quantity of burning materials
- Exposures
- Weather conditions
  - — wind direction and velocity
  - — humidity
- Topography

Medical response. Departments that normally respond to medical incidents have an established protocol of situation size-up and patient evaluation, which includes the overall situation and other potential hazards. Considerations include the location of the patient; the need for extrication or removal to a safe site; bystanders who might pose a threat to emergency responders; or physical hazards including ice, rising or rapid water, hazardous vapors, or structural collapse. Officers can use this information in company training.

Hazardous materials incident. The organization must have an established protocol for size-up and reporting of known hazardous materials (hazmat) incidents. However, not all hazmat incidents are reported as such **(Figure 9.17)**. Emergency responders must look for clues that the situation is more dangerous than the dispatch information indicates. Visible clues may include:

- Occupancy type
- Product containers
- Location
- Time of day
- Weather
  - Wind direction
  - Wind velocity
- Unusual signs
  - Vapor clouds
  - Spilled material
- Behavior of occupants or employees
- Topography

Figure 9.17 A protocol for size-up and reporting prepares firefighters for hazardous materials incidents that are more dangerous than initially reported. *Courtesy of Chris Mickal, District Chief, New Orleans (LA) FD Photo Unit.*

Motor vehicle incident. Incidents involving motor vehicles may range from a minor collision to a multivehicle incident. Size-up will determine the need for:

- Additional PPE
- Specialized extrication equipment
- Water supply or appropriate fire suppression agent
- Medical assistance
- Hazardous materials specialists

Rescue. Sizing up a rescue situation depends on numerous factors including:

- Type of incident reported
- Location and condition of victims -- rescue, search, or recovery
- Visibility
- Weather
- Topography
- Condition of structure
- Access to incident site
- Bystanders or witnesses
- Need for specialized rescue equipment
  - Watercraft
  - Earth movers
  - Structural braces
- Additional personnel

During the Incident

After the IAP is implemented, emergency responders will be busy performing assignments and progressing toward resolving the incident. During this phase, the situation changes, either improving or worsening. In either case, initial decisions that were based on the initial size-up may or may not remain valid. Continuous feedback from crews will enable supervisors to follow the progress of the IAP and continue to size-up the scene.

Plan of Operation

Depending upon the nature and scope of the incident, the plan of operation or IAP may be simple or complex. As mentioned earlier, the plan does not need to be in writing on relatively small, routine incidents involving only an initial assignment. However, the use of a tactical work sheet can be useful **(Figure 9.18)**. There should be a written IAP for long-duration or complex incidents. A written IAP is very useful when command changes or multiple jurisdictions are involved. How the plan is implemented is the operational phase. The plan starts by assigning tactical objectives to company officers in divisions or groups. It is implemented by selecting the highest priority assigned first and working down the list as more crews arrive or on-scene crews complete their original tasks.

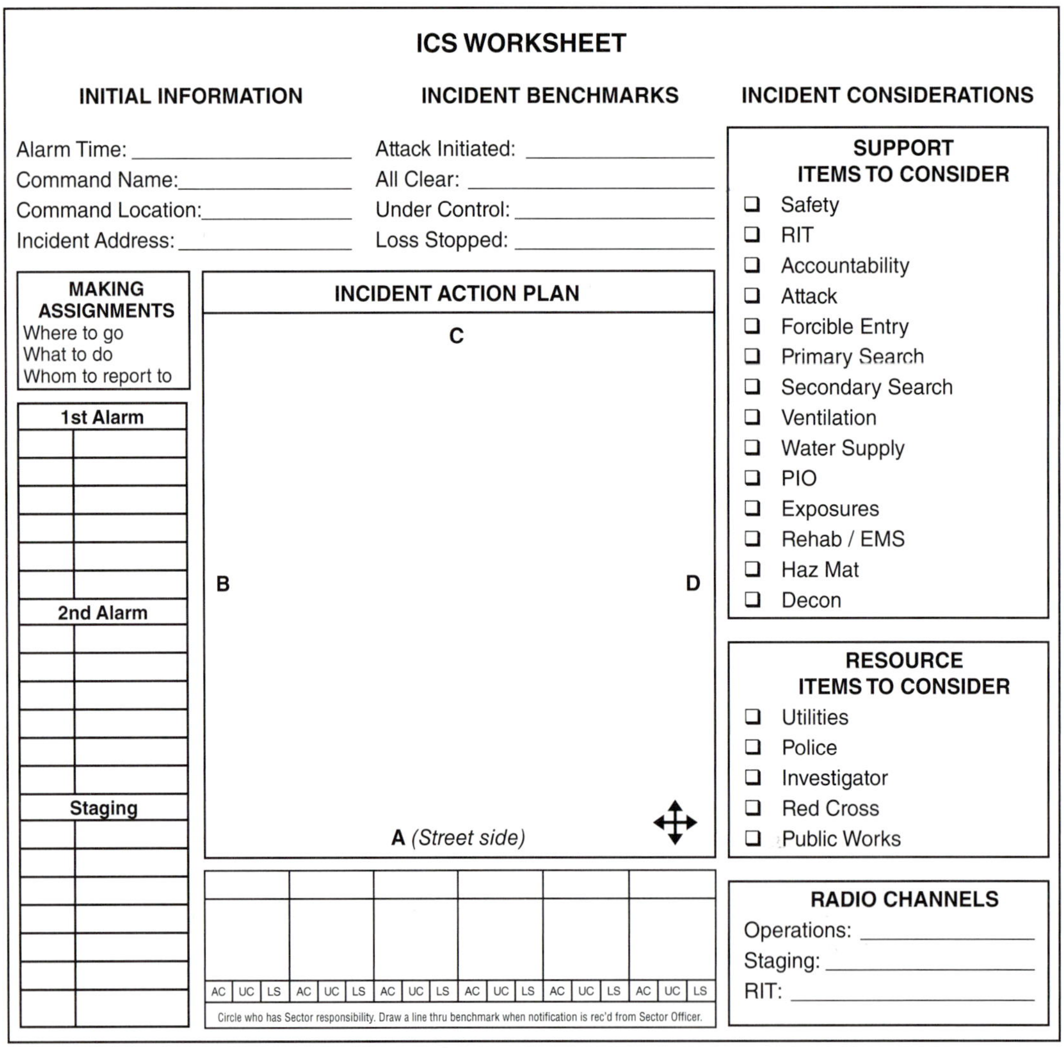
ICS WORKSHEET

INITIAL INFORMATION
Alarm Time: ____
Command Name: ____
Command Location: ____
Incident Address: ____

INCIDENT BENCHMARKS
Attack Initiated: ____
All Clear: ____
Under Control: ____
Loss Stopped: ____

INCIDENT CONSIDERATIONS

SUPPORT ITEMS TO CONSIDER
❑ Safety
❑ RIT
❑ Accountability
❑ Attack
❑ Forcible Entry
❑ Primary Search
❑ Secondary Search
❑ Ventilation
❑ Water Supply
❑ PIO
❑ Exposures
❑ Rehab / EMS
❑ Haz Mat
❑ Decon

MAKING ASSIGNMENTS
Where to go
What to do
Whom to report to

1st Alarm

2nd Alarm

Staging

INCIDENT ACTION PLAN
C
B
D
A (Street side)

AC UC LS | AC UC LS | AC UC LS | AC UC LS | AC UC LS | AC UC LS

Circle who has Sector responsibility. Draw a line thru benchmark when notification is rec'd from Sector Officer.

RESOURCE ITEMS TO CONSIDER
❑ Utilities
❑ Police
❑ Investigator
❑ Red Cross
❑ Public Works

RADIO CHANNELS
Operations: ____
Staging: ____
RIT: ____

Figure 9.18 A sample tactical worksheet is useful for smaller, routine incidents.

Incident Command System

Controlling all available resources and their activities is critical at an emergency incident. Responders must implement the locally adopted Incident Command System (ICS) based on the National Incident Management System-Incident Command System (NIMS-ICS). The ICS should be used, in varying forms, at all incidents **(Figure 9.19)**. This section will provide an overview of the NIMS-ICS in its current form.

The ICS model adopted by many jurisdictions in North America is based on NFPA 1561, *Standard on Emergency Services Incident Management System and Command Safety*. The ICS provides guidance and direction for the management and control of all types of emergency incidents. In March 2004, the U.S. government officially adopted ICS as part of the NIMS, and it must be used by all federal agencies or state and local agencies that receive federal funding.

Additional information on the NIMS-ICS model and its application may be found in the U.S. Department of Homeland Security (DHS) National Incident Management System document and the *Model Procedures Guide* series developed by the National Incident Management Consortium and published by Fire Protection Publications.

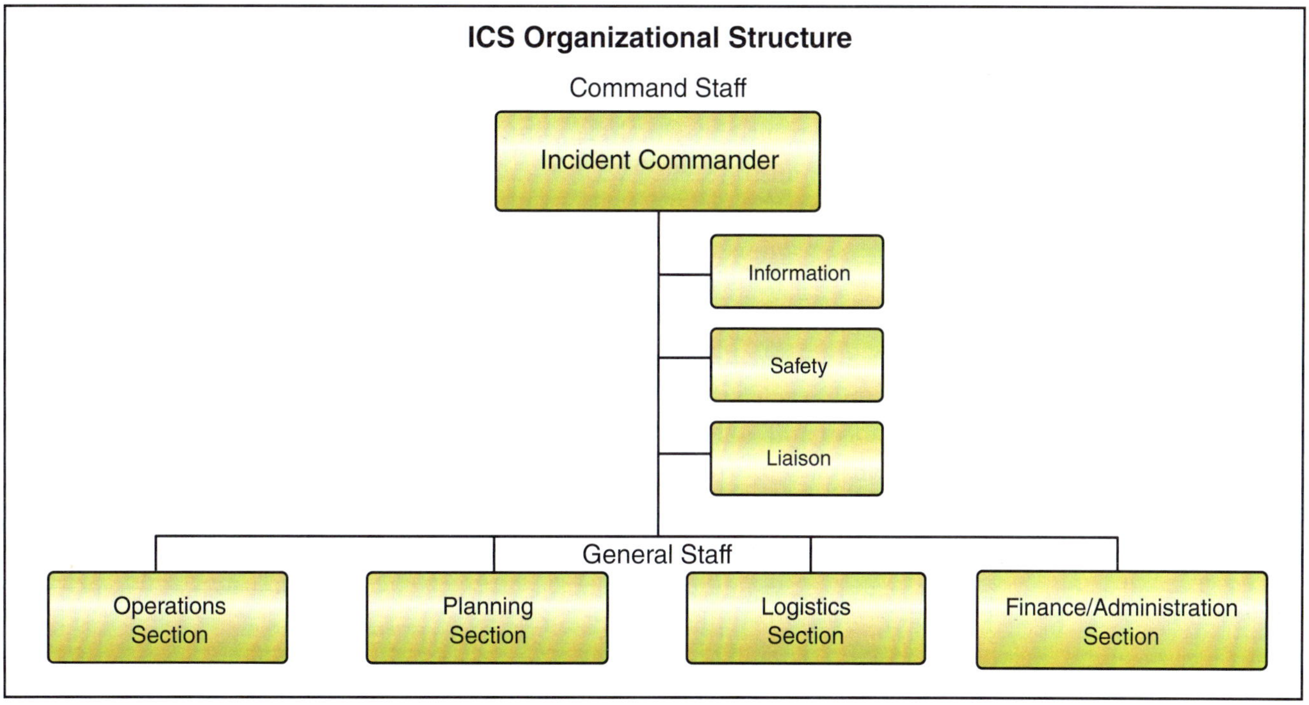

Figure 9.19 Major sections of the ICS structure include Command, Command Staff, and General Staff.

NIMS-ICS Training

All emergency response personnel should receive NIMS-ICS training as part of their entry-level training, recurring proficiency training, and professional development. NIMS-ICS courses are offered through the National Fire Academy (NFA), the Federal Emergency Management Agency (FEMA), many state/provincial, tribal, and local agencies, and online.

NIMS Compliance

An individual's responsibility in the organization determines the required level of ICS training, as shown in the following list:

Command and General Staff. Type I and II County and State Fire Coordinators Fire Chiefs, Deputy Chiefs	FEMA IS-700 and 800 ICS-100, 200, 300, 400
Chief Officers and others who have Command and General staff positions at the local level. Type III Special Team Leaders for USAR, TRT, IMAT	FEMA IS-700 ICS-100, 200, 300
Officer who have supervisory responsibility at the local level. Type IV and V Special team members	FEMA IS-700 ICS-100, 200
Entry level first responders, disaster workers Firefighters, Police Officers, Dispatchers	FEMA IS-700 ICS-100

Source: NIMS Training Guidelines for FFY 2014.

Company officers who have completed Fire Officer I certification training should be familiar with the basics of the ICS used in their jurisdiction. As a Fire Officer II, knowledge of the ICS evolves as the responsibility for larger and more complex incidents increases. This section includes a review of some of the basic information.

NIMS-ICS establishes an organizational structure for all types of emergency incidents, regardless of size. Every organization member, especially ICs, must be familiar with the system and trained in its application. All agencies with mutual or automatic aid agreements must know and use the same system. This system may require extensive cross-training at all organizational levels among units of the participating agencies. These levels may include independent EMS providers, law enforcement agencies, and public works.

Common characteristics of the NIMS-ICS are as follows:

- Common terminology for functional structure
- Modular organization
- Common communications
- Unified command structure
- Incident Action Plan (IAP)
- Manageable span of control
- Predesignated incident facilities
- Comprehensive resource management
- Personnel accountability

Operational Positions

To understand the application of ICS, fire officers should know the major operational position descriptions within the ICS structure. These include Command, Command Staff, and the General Staff sections. Officers may be assigned to any of these positions based on knowledge, skills, and ability.

Command

The IC is ultimately responsible for all incident activities, including the development and implementation of a strategic plan. This process may include making a number of critical decisions and being responsible for the results of those decisions. The IC has the authority to request resources to the incident and to release them from it. If the size and complexity of the incident requires it, the IC may delegate authority to others who, with the IC, form the Command and General Staffs **(Figure 9.20)**. At single-unit incidents, the IC will perform the functions normally assigned to a Command Staff.

Figure 9.20 Officers discuss on-scene incident activity. Command-staff functions may be performed by the IC at single-unit incidents. *Courtesy of Ron Jeffers, Union City, NJ.*

Command Staff

The Command Staff reports directly to the IC by providing management support in functions that are not directly involved with the operational tasks. Command Staff positions include the Safety Officer, Liaison Officer (LO), and Public Information Officer (PIO).

- **Safety Officer** — Assesses hazardous conditions and unsafe situations and develops an incident safety plan that ensures the safety of personnel at the incident. The Incident Safety Officer (ISO) has the authority to alter, suspend, or terminate activities that are imminently dangerous to the life of an occupant or responder.
- **Liaison Officer** — Acts as a point of contact for other agencies and jurisdictions and coordinates activities with them.
- **Public Information Officer** —Provides accurate and complete information about the incident to the media. During larger incidents, the PIO provides information to other governmental agencies as needed.

General Staff

As an incident becomes larger and more complex, additional tasks must be performed. This requires the addition of more personnel who will form the General Staff. The General Staff represent the major functional sections, including Planning, Operations, Logistics, and Finance/Administration. NIMS also provides for an Intelligence Officer, Section, or Unit, which may be part of the Command Staff, the Planning Section, or a completely separate section.

Planning — Planning is responsible for the collection, evaluation, dissemination, and use of information concerning the development of the incident. Planning is also responsible for tracking the status of all assigned resources and their situation status. Command uses the information compiled by Planning to develop strategic goals and contingency plans.

Operations — Operations may be divided into divisions or groups. Divisions have area responsibilities, while groups have functional responsibilities. NIMS-ICS recommends limiting subordinates to five. If an Operations Chief has more divisions/groups than are reasonable to oversee, branches may be organized with divisions/groups assigned to them.

Logistics — Logistics is responsible for providing the facilities, services, and materials necessary to support the incident. There are two branches within Logistics: Support Branch and Service Branch.

Finance/Administration — Finance/Administration has the responsibility for tracking and documenting all costs and financial aspects of the incident. Generally, Finance/Administration is activated only on large-scale, long-term incidents. Day-to-day mutual aid and automatic aid responses are usually considered to be reciprocal and do not require interagency reimbursement. Administration is also responsible for the legal aspects of the incident, such as monitoring contracts with vendors, agreements with other agencies, and compliance with state/provincial and national laws.

Developing a Risk Management Plan

Company officers are responsible for the safety of all personnel assigned to the unit or under their command at an emergency incident. This responsibility exists throughout the work shift and includes emergency and nonemergency situations. The officer uses a risk management model to determine the appropriate responses to life safety risks. The components of the risk management plan are described in the following section.

- **Risk Identification** — To identify the typical risks encountered at a structure fire, the fire officer compiles a list of all the operations and duties that occur at such an event. Ideally, the officer should take into consideration the worst possible conditions or potential events, including the need for rescue, rapid fire development, or structural collapse.
- **Risk Evaluation** — Once the risks are identified, they can be evaluated from both frequency and severity standpoints. Frequency, referred to by OSHA as incidence rate, addresses the likelihood of occurrence. Severity addresses the degree of seriousness of the incident. It can be measured in a variety of ways such as lost time away from work, cost of damage, cost of and time for repair or replacement of equipment, disruption of service, or legal costs. Incidents of high frequency and high severity are given the highest priority in the risk analysis, while those of low frequency and low severity receive the lowest priority. The method for calculating the risk may vary from one organization to another.
- **Risk Prioritization** — Taken in combination, the results of the frequency and severity assessments help to establish priorities for determining action. Any risk that has both a high probability of occurrence and serious

consequences deserves immediate action and is considered a high-priority item. Incidents with a low likelihood of occurrence are lower priorities and can be placed near the bottom of the incident action plan (IAP).

- **Risk-Control Techniques** — When the risks have been prioritized, it is time to apply risk-control measures. Several approaches can be taken in risk control, including the following:
  - **Risk avoidance**: The best risk-control choice is risk avoidance. Simply put, avoid the activity that creates the risk. At a structure fire, this may be the choice.
  - **Risk transfer**: This approach can be accomplished by physically transferring the risk to someone else or by purchasing insurance. Transfer of risk may be difficult at a structure fire.
- **Implementation** — Implementation of the risk management plan requires coordination with crew members and other units as well as communicating the plan to them.
- **Monitoring** — The effectiveness of the plan becomes evident through monitoring. Any problems that occur in the process have to be revised or modified. The intent of the risk management plan is to develop a strategy for reducing the inherent risks associated with fire and emergency services operations. For instance, if the plan begins with an offensive interior attack, changes in the location and intensity of the fire may require a rapid redeployment of resources and a transition to a defensive strategy.

NFPA 1500 8.3.2 lists the following four principles to guide on-scene risk management of personnel:

"(1) Activities that present a significant risk to the safety of members shall be limited to situations where there is a potential to save endangered lives.

(2) Activities that are routinely employed to protect property shall be recognized as inherent risks to the safety of members, and actions shall be taken to reduce or avoid these risks.

(3) No risk to the safety of members shall be acceptable when there is no possibility to save lives or property.

(4) In situations where the risk to fire department members is excessive, activities shall be limited to defensive operations."

Fire and emergency service responders are expected to take only calculated risks to provide for life safety, incident stabilization, and property conservation. Calculating risks means gathering information through size-up to determine what level of risk is acceptable for the given situation.

Calculation of risks versus benefit includes what has been considered under the organization's risk management plan. For instance, if the department rarely experiences swiftwater issues, and therefore does not train for this type of incident at the fire chief's direction, firefighters will have an increased risk in trying to manage an incident of this type. The IC at the incident may opt to wait for a qualified team after considering numerous other size-up considerations.

Unfortunately, firefighter risk at fire incidents has remained unchanged because the fire service continues to have the same rate of LODDs. In 2001, the International Association of Fire Chiefs (IAFC) developed a model policy called The 10 Rules of Engagement for Structural Fire Fighting to offer a risk-benefit model that might help decrease LODDs at fire incidents.

10 Rules of Engagement for Structural Fire Fighting

Risk Tolerance

1. No building or property is worth the life of a firefighter.
2. All interior fire fighting involves an inherent risk.
3. Some risk is acceptable, in a measured and controlled manner.
4. No level of risk is acceptable where there is no potential to save lives or property.
5. Firefighters shall not be committed to interior offensive fire fighting operations in abandoned or derelict buildings.

Risk Assessment

1. All feasible measures shall be taken to limit or avoid risks through risk assessment by a qualified officer.
2. It is the responsibility of the Incident Commander to evaluate the level of risk in every situation.
3. Risk assessment is a continuous process for the duration of each incident.
4. If conditions change, and risk increases, change strategy and tactics.
5. No building or property is worth the life of a firefighter.

In the Ten Rules of Engagement, the prohibition for committing an interior attack in an abandoned or derelict building can generate a great deal of discussion. For instance, because a structure is abandoned or derelict does not mean that there is no life safety hazard present. Someone who is homeless, demolition or construction workers, or security personnel may be inside. One approach is to consider all structures as occupied until proven otherwise. Conversely, although the "Primary All Clear" benchmark belongs to the fire department, Incident Commanders must make appropriate assumptions about each fire call and not apply blanket policies to every fire. Once personnel have been committed to an interior attack, the structure is occupied. These decisions should not be taken lightly. Any one of them could be life or career changing.

Firefighter safety should be everyone's primary concern. Decisions must be based on this model, which should be applied to initial size-up. When dealing with an abandoned or derelict structure, a quick attack on the seat of the fire may be the most efficient and safest way to control the fire. Doing this can decrease property damage, reduce the exposure hazard, and reduce the risk to firefighters.

Implementing the System

The first-arriving fire officer at the emergency incident establishes the ICS and remains in Command until relieved. This officer begins by establishing the incident priorities which include:

- **Life Safety**
  - What hazards are present and what risks do they pose to personnel?
  - What hazards are occupants, witnesses, and other people exposed to?
  - Are any occupants trapped or injured?
  - Who are the most severely threatened occupants?
- **Incident Stabilization**
  - What has occurred?
  - What is the current status of the incident?
  - What is likely to occur?
  - Can the incident be confined or controlled with the resources I have at my disposal?
  - What specialized resources or technical assistance do I need to mitigate the incident?
- **Property Conservation**
  - What actions must be taken to protect the property?
  - Is the property beyond saving or is the risk to personnel too great to attempt to save it?

Implementing the Incident Action Plan

If no life-threatening situation demands immediate action, the IC should begin to formulate an IAP. The plan should reflect the following priorities:

1. Ensuring personnel safety and survival
2. Rescuing or evacuating endangered occupants
3. Eliminating the hazard
4. Conducting loss control
5. Cleaning up and protecting the environment

The IC gathers enough resources, including personnel and materials, to handle the incident. ICs also organize information to ensure that orders can be carried out promptly, safely, and efficiently. Having sufficient resources on scene will help to ensure the safety of all personnel **(Figure 9.21, p. 316)**.

All incident personnel must function according to the IAP. Company officers, who may be designated as crew leaders, division or group supervisors, or branch directors, should follow SOPs. Personnel should direct every action toward achieving the goals and objectives specified in the plan. When all members (from the IC to the lowest-ranking member of the team) understand their positions, roles, and functions in the ICS, the system can serve to safely, effectively, and efficiently use resources to accomplish the plan.

Figure 9.21 The Incident Command System (ICS) controls and allocates multiple resources at large incidents. *Courtesy of Ron Jeffers, Union City, NJ.*

Incident Safety Plan

The ISO or the IC develops an incident safety plan that outlines the hazard control strategies required by the IAP and the specific incident. The plan is used during the incident and also as part of the post-incident analysis.

Figure 9.22 An initial IC relays information to the relieving officer. *Courtesy of Ron Jeffers, Union City, NJ.*

Chain of Command

Whenever the ICS is implemented, there can be only one IC, except in a multijurisdictional incident when a unified command structure is appropriate. Even when a unified structure is used, the chain of command must be clearly defined. One person should issue all orders through the chain of command to avoid confusion caused by conflicting orders. Based on the local ICS, the initial IC may retain command until the termination of the incident, or may transfer command to a higher-ranking officer or an officer with greater experience.

The initial IC provides the relieving officer with all of the information on the situation and the actions that have been taken **(Figure 9.22)**. This report can be accomplished by giving a situation status report, which is an updated version of the incident evaluation performed on arrival. The officer assuming command should acknowledge receipt of the information by repeating it back to the IC. If the reiteration is accurate, the recipient is ready to accept control of and responsibility for the management of the incident. The former IC can then be reassigned to an operating unit or retained at the command post (CP) as an aide or as a member of the Command or General Staff.

Situation Status Report

The situation status report should include the following information:

- Description of what happened
- Whether anyone was/is injured or trapped
- What has been done so far
- Whether the problem has stabilized or is getting worse
- What resources are on scene or en route
- Whether resources seem adequate for the situation or more resources need to be called
- What functions of the ICS have been implemented and what additional functions should be considered

It is common for a new IC to assess a situation and determine if the operation is reaching the objectives established in the IAP. Each group, division, or branch leader is contacted and a situation update requested. This will provide a newer IC with a current picture of the operational efforts from the perspective of those who are closest to the incident.

Tracking Resources

One of the most important functions of an ICS is providing tracking of all personnel and equipment assigned to an incident. In career departments, most units responding to an incident arrive fully staffed and ready to be assigned an operational objective. In the case of volunteer or combination departments, personnel may have to be formed into units as they arrive at the scene. In order to manage these resources, the IAP must contain a tracking and accountability system with the following procedures:

- Checking in at the scene
- Identifying the location of each unit and all personnel on scene
- Releasing units no longer needed

The IC must be able to locate, contact, deploy, and reassign the resources that are assigned to the emergency incident. At smaller incidents, the use of a tactical worksheet is recommended, while at larger incidents, this may be a function of the Planning Section or Resource Unit. As companies and crews arrive at the scene, the IC or Operations Chief assigns them to the area of the incident where they are most needed. The companies/crews may be held in a staging area until needed or released from the incident. If staging has not been implemented, company officers must check in with the IC and wait for an assignment.

Communications between resources will be through the jurisdiction's emergency radio communication system or face-to-face communication. As resources are assigned to the incident, the central communications center announces that the command frequency is in use or automatically places all radios on that frequency. Resources that have been assigned to the incident must contact the IC to ensure that they have complete communication with the CP. Face-to-face communication occurs as personnel without radios arrive at the incident and are formed into crews.

The IC can use a number of visual aids to help manage the arriving resources. The visual aid can be as simple as a preprinted form to sketch the incident scene and the location of companies/crews as they arrive, or it can be an elaborate tracking board with magnetic symbols identifying each of the units. Regardless of the tracking device, it should be simple to read and contain as much information as necessary about the activities of all the resources on scene, such as:

- Assigned radio frequencies
- Assigned companies/crews
- Activated ICS functions
- Site plan
- Staging areas
- Logistics location
- Control zones

CAUTION

Personnel who arrive individually at an incident must always check in at the command post or staging area, if one has been implemented. Personnel must not freelance or assign themselves to a task. Freelancing is unsafe and can result in fatalities or injuries.

Personnel Accountability System

Each organization must adopt an existing accountability system or develop a system of accountability that identifies and tracks all personnel at an incident. The organization should standardize the system so that it is used at every incident. All personnel must be familiar with the system and use it at an emergency incident. The system must also account for those individuals who respond to the scene in vehicles other than emergency response apparatus, including staff vehicles and personally owned vehicles (POVs).

Accountability is vital in the event of a change in the status of the emergency incident. At a structure fire, that change might be the extension of the fire through a concealed space, the rapid increase in the volume of fire due to the ignition of a flammable or combustible liquid, or a rapid fire development situation. The IC must know who is at the incident and where each person is located. An accountability system determines who and how many may be trapped inside or injured. SCBAs can malfunction or run out of air and firefighters can get lost in mazes of rooms and corridors. Firefighters have died because they were not discovered missing until it was too late to save them.

Wildland and urban interface fires have also claimed many lives when emergency responders were trapped by a change of direction of the fire. To effectively manage the various resources at the incident, the IC must know the status and location of all resources. Changes in the direction, magnitude, and behavior of the fire can be transmitted to the resources giving them time to relocate or take cover. Individual emergency responders have suffered fatal heart attacks because they could not be located in time to receive necessary medical treatment.

Company officers, or other ICS positions such as Division/ Group supervisors or strike team leaders, are responsible for keeping track of the members of their units. When operating in the hot zone, the unit members should stay within sight of each other. When the atmosphere is obscured, they should be in close proximity.

The IC is responsible for managing the personnel accountability system employed by the organization. The system may be assigned to an Accountability Officer through the Planning Section. Accountability systems should indicate:

- Individuals assigned to each apparatus crew or staff position
- Names of people responding individually, such as staff personnel and volunteers
- Time of arrival
- Assigned duty or unit
- Time of release from the scene.

Various systems, such as the tag system, SCBA tag system, and bar code readers, are available for tracking individuals at the emergency incident.

Figure 9.23 The tag accountability system is used to keep track of who is working at an incident. *Courtesy of Ron Jeffers, Union City, NJ.*

Tag system. A simple tag system can help account for personnel within the incident scene perimeter. Personnel can be equipped with a personal identification tag. Upon entering the outer perimeter, emergency responders leave their tags at a given location or with a designated person (CP, apparatus compartment, company officer, accountability officer, or ISO). Tags can be attached to a control board or personnel identification (ID) chart for quick reference **(Figure 9.23)**. Upon leaving the incident perimeter, the emergency responders collect their tags. This system enables officers to know exactly who is operating at the incident scene.

Figure 9.24 Each firefighter should activate the PASS device before entering a hazardous area.

SCBA tag system. An SCBA tag system provides closer accountability for personnel inside a structure. All personnel entering a hazardous atmosphere are required to wear full personal protective clothing with SCBA. Each SCBA is provided with a tag containing the name of the user and the air pressure in the SCBA air tank. Upon entering a building, personnel give their tags to a designated supervisor. The company officer records time of entry and expected time of exit. This officer also does a brief check to ensure that all protective equipment is properly worn, used, and activated, including the PASS device **(Figure 9.24)**. Responders retrieve their tags upon leaving the danger area so that the control officer knows who is safely outside and who is still inside the structure or danger area.

Bar code readers. Some organizations have adopted bar code readers for tracking personnel at emergency incident scenes. Similar to the inventory control devices used by retail outlets, the system assigns a unique bar code number to each member of the organization. When that person arrives at the incident, the bar code is scanned into a computer. When the person leaves the scene, the bar code is scanned a second time and the person's name is removed from the site. Depending on local policies, the bar code may be attached to the helmet, protective clothing, or other equipment that is permanently assigned to each member of the organization. It may also be on a personal identification card.

Air Management

The IC and the personnel entering an IDLH atmosphere must implement the organization's air management policy. At structure fires or hazardous vapor leaks, the IC must evaluate the size and complexity of the structure's interior to determine the time required to exit the structure from the **point of no return**. This calculation will determine the maximum amount of time that can be spent working in the IDLH atmosphere.

Point of No Return — Point at which air in the SCBA will last only long enough for an exit from a hazardous environment.

The ISO or HSO should monitor personnel entering the IDLH area by recording their starting air supply as a means of determining when they should be replaced and begin their withdrawal. A general rule of thumb for calculating air supply is to divide the capacity of the air cylinder (in cubic feet) by 2. This calculation equals the allowable working time, or a safety margin, for exiting the structure.

Individual firefighters who are assigned to work in the IDLH area are responsible for:

- Inspecting their SCBA when reporting to work
- Checking the air pressure on their cylinder prior to entering the IDLH area
- Reporting this pressure to their team leader or the ISO
- Monitoring air consumption and pressure
  - Continually while in the IDLH area
  - At regular intervals
  - At 10-minute intervals when notified by the communications office
  - At a change-of-work area
  - When passing major landmarks within the structure
  - At the completion of one assignment and before another assignment
  - As the situation dictates

Because there are a wide variety of SCBA types currently in use in the fire service, the local jurisdiction must establish an air management protocol in its operational SOP/SOG. Traditionally, the activation of the low-pressure alarm on the SCBA has been used as an indicator that it is time to exit the IDLH area. In small structures, this may be sufficient time to locate an exit or a source of fresh air. In high-rise or large-area structures, the low-pressure setting may not provide enough breathing air to leave the IDLH area. Other factors that can limit the breathing air supply are the individual firefighter's breathing rate (air consumption) and the intensity of the work being performed.

One suggested air management protocol requires that interior teams give an automatic air status report when the first team member's air supply reaches 50 percent. At that time, all members of the team must begin to leave the IDLH area BEFORE any low-air alarm is activated. Some SOPs/SOGs mandate that the activation of the low-air alarm requires an immediate Mayday response. A sample protocol when the low-air alarm activates is for the firefighter to transmit a radio signal to Command stating who, where, and status. Immediately identify Command upon exiting the structure.

NOTE: Many newer SCBAs alarm and/or illuminate at 33 percent, so departments may use that signal instead of 50 percent.

The IC may allow team members to continue working until the alarm cylinder indicator light activates under the following conditions:

- The fire is declared under control.
- A structural evaluation is made by the IC or ISO to determine that there is no structural collapse or entanglement hazard; the crew's visibility is not impaired; and a personnel accountability report (PAR) indicates the crew is intact and able to continue.

Emergency Broadcast Protocol

North American fire departments have many protocols for transmitting a radio signal for an emergency situation during an incident. Although the terminology may differ, the terms can be divided into two categories:

1. Mayday protocol – The broadcast of a Mayday indicates an emergency situation in which a firefighter is in trouble:
    - Firefighter in low-air-alarm situation and/or is disoriented or unsure of location
    - SCBA failure
    - Firefighter trapped, entangled, or unable to free themselves within approximately one minute
    - Finding a firefighter in distress
    - Command's discretion
2. Emergency Traffic protocol – The broadcast of emergency traffic is used when there is an increased hazard to all personnel such as a potential structural collapse, indication of an extreme fire condition (flashover), or a shift to defensive strategy and the need to evacuate the structure or IDLH area.

All nonessential radio traffic ceases or is shifted to another radio channel. In the case of a Mayday, RIT/RIC resources are committed to the incident while other personnel continue to perform their duties.

Releasing Resources

As control is gained over the emergency situation, personnel and units who have been actively involved in the incident should be sent to the rehabilitation unit prior to being released. This step helps to ensure that all personnel have an opportunity to rest, receive nourishment, and be examined by a medical professional. A monitored cool-down time, rehydration, and medical monitoring can help prevent heart attacks.

Before leaving the scene, the company officer must inspect units to ensure that they are ready to return to service. Firefighters must locate missing equipment, and damaged equipment must be replaced or repaired. If critical equipment is not available, the unit may have to remain out of service until it can be replaced.

Personnel or units may require decontamination before they can return or be placed back into service. This can take place at the incident site or when the unit returns to the station. Small tools, medical equipment, and PPE must be decontaminated in accordance with local policy and NFPA 1581, *Standard on Fire Department Infection Control Program.* Contaminated clothing and equipment must be cleaned in a designated decontamination area at the station and may not be cleaned in a kitchen, bathroom sink, or shower.

Assigning Multiunit Resources

The ICS is designed to permit the control and allocation of any number of units or agencies at an emergency incident. The IC may be responsible for an entire initial response consisting of multiple engines, ladders/trucks, rescue, and ambulance units **(Figure 9.25)**. If the initial assignment is insufficient to control the incident, more resources should be requested. The IC may be responsible for transferring command to a superior officer, assuming one of the ICS Command or General Staff positions, or commanding a single unit. The company officer must be familiar with the general configuration of a multiunit incident and terms used by the ICS to describe duties and functions.

Figure 9.25 A multiunit assignment may be managed by a designated IC, or command may be transferred to a superior officer. *Courtesy of Ron Jeffers, Union City, NJ.*

Command Organization

The command organization must develop at a pace that stays ahead of the tactical deployment of personnel and resources. The IC must be able to direct, control, and track the locations and functions of all operating units in order to efficiently manage the incident. Building a command organization is the best support mechanism the IC can use to achieve a balance between managing personnel and incident needs. The command organization has three levels: strategic, tactical, and task.

Strategic Level —The IC is responsible for the strategic level, which involves the overall command of the incident. The IAP should cover all strategic responsibilities, tactical objectives, and support activities needed during the operational period. The IAP may include the department's SOP for initial activities at specific incidents, such as High-Rise SOP, and HazMat SOP. The IAP is the basis for developing a command organization, assigning resources, and establishing tactical objectives.

Strategic level responsibilities include:

- Determining the appropriate strategy — offensive or defensive
- Establishing overall incident objectives
- Setting priorities and achieving benchmarks
- Developing an IAP
- Handling resource requests and assigning resources
- Planning
- Forecasting outcomes
- Assigning specific objectives to groups or divisions

Tactical Level — Tactical-level supervisors direct operational activities toward specific objectives. Tactical-level supervisors supervise grouped resources and are responsible for specific geographic areas or functions. A tactical-level assignment comes with the authority to make decisions and assignments within the boundaries of the overall plan and safety conditions. The accumulated achievements of tactical objectives should accomplish the strategy as outlined in the IAP.

Task Level —Individual companies, crews, or specific personnel normally accomplish task-level activities. Work gets done at the task level. Company officers routinely supervise task-level activities. The accumulated achievements of task-level activities should accomplish tactical objectives.

Basic Organization

The most basic organization combines all three levels of the command structure. For example, the company officer on a single-engine response to a dumpster fire determines the strategy and tactics and supervises the crew doing the task. An initial-response incident involving a small number of units requires only two levels of the command structure. The IC directly handles strategic and tactical levels. Emergency response units report directly to Command and operate at the task level.

Divisions and Groups are NIMS/ICS terms for tactical-level management components that assemble units and/or resources for a common purpose. *Divisions* represent geographic operations and *groups* represent functional operations. The following examples illustrate the use of these terms.

Divisions have responsibility for operations within a defined geographic area. The Division level is organizationally between Single Resources, Task Forces, or Strike Teams and the Branch. For situations where the incident has an odd geographical layout (no obvious North, South, East, or West), the front of the building is designated *Division A*, and the remaining sides are given a designation of B, C, or D in a clockwise manner **(Figure 9.26, p. 324)**.

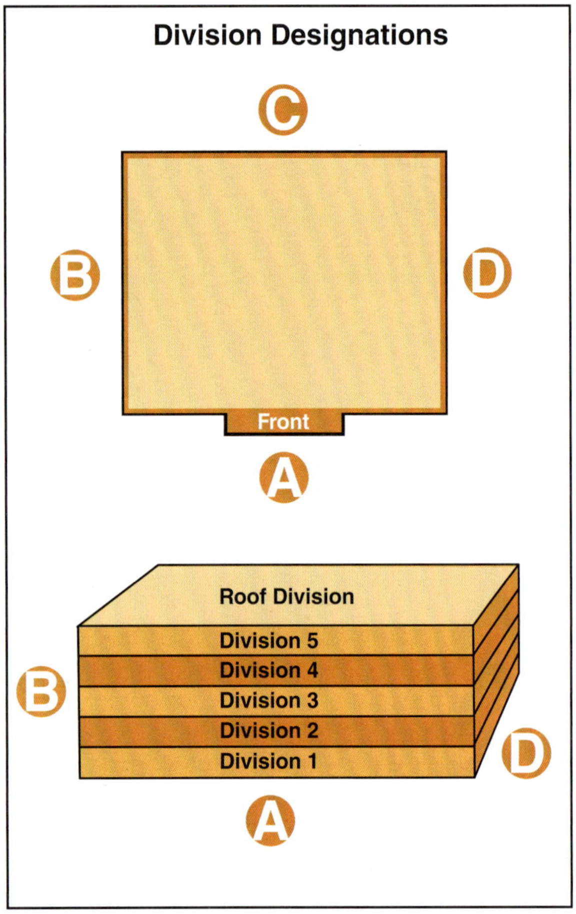

Figure 9.26 Sides and floors of structures are designated by Division as a means to assign companies and communicate with them.

For clarity of purpose during radio communications, the phonetic designations of *Alpha, Bravo, Charlie,* and *Delta* are suggested. For example, *"Command from Division Delta"* or *"Delta Division to Command."*

In multistory occupancies, Divisions are usually indicated by floor numbers (Division 6 indicates sixth floor). When operating in levels below grade, Divisions are usually indicated by a descriptive term, such as Basement Division 2 or Parking Division 4.

Groups are an organizational level responsible for a specific functional assignment. The Group level is also organizationally between Single Resources, Task Forces, or Strike Teams and the Branch. Examples are Salvage Group, Search Group, Rescue Group, HazMat Group, and Medical Group.

Establishing Divisions/Groups

A major incident will initially have more tasks than the available resources can accomplish. There is a tendency to start performing these tasks immediately upon arrival, thereby postponing the establishment of ICS. However, the lack of direction will result in confusion and lack of coordination. This increases the safety risks to emergency personnel and decreases the likelihood of a successful operation. The IC should begin to assign Divisions/Groups based on the following situations:

- Situations that involve a number of units or functions beyond the IC's span of control. As additional chief officers become available, they may relieve the company officer of responsibility for the area or function.
- When units are involved in complex operations, such as large interior or geographic areas, hazardous materials (hazmat) operations, high-rise fires, and technical rescues.
- When units are operating from tactical positions that Command has little or no direct control over.
- When the situation presents special hazards and close control is required over operating units, such as unstable structural conditions or heavy fire load

When establishing Divisons/Groups, the IC will assign/advise each unit of the following:

— Tactical objectives

— Radio designation (Division/Group)

— Identity of resources assigned to that Division/Group

Span of Control

Span of control is the number of direct subordinates one supervisor can effectively manage. An effective span of control runs from three to seven per supervisor; five is optimal. Supervisors can more easily keep track of subordinates and monitor safety if an effective span of control is maintained **(Figure 9.27)**. If the incident is developing quickly and risk is high to response personnel, the span of control should be smaller. Slower-developing incidents allow for a broader span of control.

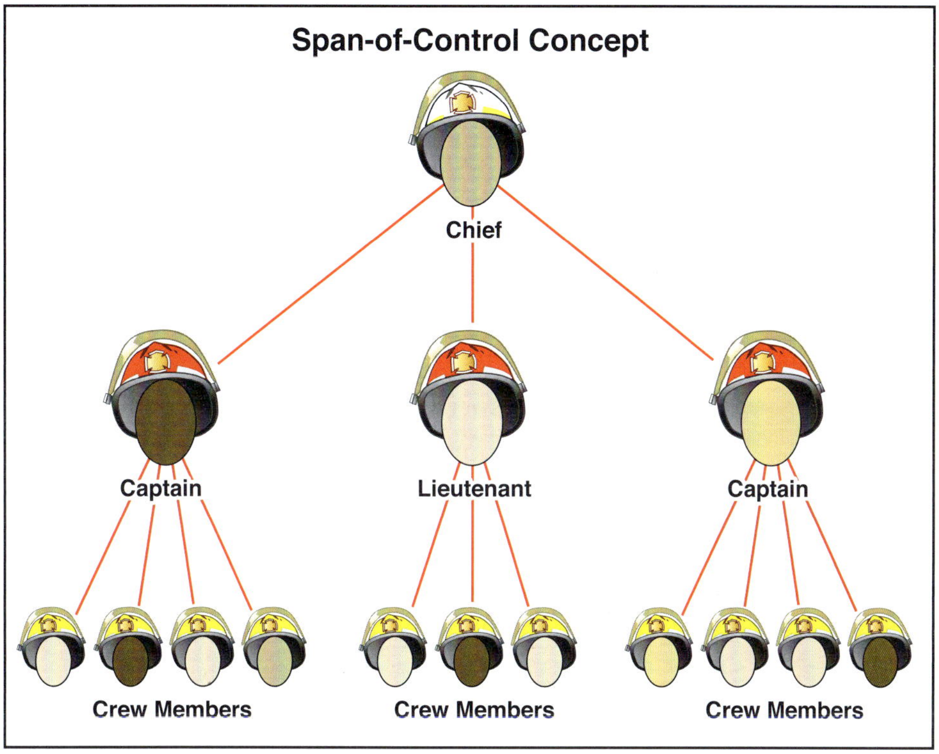

Figure 9.27 Three to seven subordinates per supervisor is an effective span of control.

Proper Use of Nonemergency Personnel

At many emergency incidents, especially in larger geographic areas, additional personnel will be available. These additional resources may come from:

- Law enforcement
  - Federal, including military
  - State/Provincial/Tribal
  - County/Parish
  - Municipal
  - Fire Police
- Emergency medical organizations
  - Public
  - Private
- Volunteer organizations/NGOs
  - Auxiliary
  - Fire Corps
  - Explorer Scouts
  - Red Cross
  - Salvation Army

The fire department's SOP/SOGs should define how these resources are to be used. For incidents where there is joint jurisdiction (motor vehicle accidents, crime scenes), the relationship with law enforcement agencies must be clarified beforehand.

Developing working relationships through unified training events, ride-alongs, and light social gatherings will have a positive effect on work at common incidents. These activities foster familiarity and trust, which makes working together more rewarding and efficient.

At emergency events, ICs must plan for resource assignments to keep ill-prepared or ill-equipped responders and volunteers out of harm's way. OSHA has fined law enforcement agencies that allowed personnel to get exposed to hazardous material spills, and fire departments for interfering with such law enforcement duties as arrests, DUI stops, and traffic routing.

Outside the hot zone, nonemergency resources may be used for a variety of functions including, but not limited to:

Figure 9.28 Emergency responders may be used for traffic control outside the hot zone.

- Traffic control **(Figure 9.28)**
- Deploying supply hose
- Loading hose and equipment
- Establishing rehabilitation sites
- Crowd control
- Evacuation of non-hot-zone areas
- Air management
- Logistics
- Record keeping

If the use of head, eye, hand, or hearing protection is required by the assigned function, that equipment must be provided by the responsible jurisdiction and used properly by personnel. Anyone performing a specialized task, such as refilling SCBA bottles, must be thoroughly trained in the function.

Installing an Operational Plan

Operational plans identify the specific resources needed to successfully deal with a variety of hypothetical incidents at a particular location or type of occupancy. The incident scenarios analyzed are those that are considered to be the most likely to occur at the location or occupancy in question. For example, if a building/occupant uses or stores large quantities of toxic or highly flammable materials, scenarios involving these materials are created and analyzed. Likewise, if a building houses many elderly or infirm residents, scenarios that require the evacuation of these residents or moving them to safety are created and analyzed. This process is applied to as many target hazards as time and staffing allow. Operational plans are created from these scenarios.

Operational Plans

Operational plans often include possible resource deployments. In the most likely scenarios, options for deploying the initial alarm resources are studied, as well as options for deploying additional resources if needed. Scenarios based on increasingly larger and more complex hypothetical incidents help planners to identify the resources that may be needed. Operational plans often include provisions for a number of contingencies. Some possible variations are:

- Operational plans might include responses based on unusually severe weather conditions that could greatly increase the potential for wildland fires to start and spread at a greater-than-normal rate. An operational plan might specify an increase in the initial alarm resources at certain points on the daily burning index.
- Operational plans might specify different initial alarm resource levels for a given occupancy if the typical number of people in the building varies significantly at different times of the day or night. For example, the operational plan for an elementary school might specify doubling the initial alarm assignment if an alarm is received from the school during normal operating hours. The plan might also specify a lesser response at other hours.
- The operational plans for occupancies that store large quantities of ignitable liquids might specify that crash/rescue vehicles from an airport be a part of the initial alarm assignment so their foam-making capabilities are available if needed.
- The operational plans for large industrial complexes might specify that the first-arriving unit is to respond directly to the fire alarm panel or the plant security station, but that other responding units are to stage at the facility gates and await directions from the first-in unit.

In essence, the data gathered at any particular target hazard is used to project the strategic and tactical possibilities and probabilities at that location. Based on these scenarios, the resources needed are compared to the resources available in the organization. If resources indicated are more than or different from what the organization has, planners can recommend purchases or develop mutual aid agreements with nearby organizations.

Operational Plan Components

The operational plan consists of information gathered from a number of sources. The initial component is the pre-incident survey combined with any fire and life safety inspection data. These two sources provide a picture of the actual conditions at the location or occupancy. This information must be current and accurate to be of value in the planning process.

The next component is the analysis of the fire and emergency services organization's assets and abilities. This is an inventory of such items as:

- Staffing levels
- Apparatus types and quantity
- Station locations
- Response times
- Training and certification of personnel

- Water supply
- Types and quantity of specialized extinguishing agents
- Specialized response units
- Automatic and mutual aid agreements
- Dispatch and communications systems

The final component is the jurisdiction's ICS. This is the organizational framework for all single and multiunit emergency incidents. It establishes functions and assigns duties based on the complexity of the incident. Emergency responders must be familiar with the system used locally and how it fulfills state/provincial and national requirements.

Operational Period Safety Concerns

Most emergency responses can be mitigated in a short time. Others require more time and resources. For this purpose, NIMS-ICS establishes operational periods as a parameter for setting and accomplishing goals and objectives. In general, operational periods may be as short as two hours but no longer than 24 hours. When an incident extends beyond one operational period, a planning meeting is held to define the goals and objectives for each operational period.

The following safety considerations are created by incidents that last longer than one operational period:

- Personnel rotation or availability of personnel
- Personnel rehabilitation
- Structural stability
- Environmental changes
- Equipment availability

Evacuation Commands

Firefighters know that an emergency incident can change without warning. Wind direction and velocity changes can force a wildland fire to change direction and trap crews. Introducing fresh air into a closed compartment with a smoldering fire can create a flashover or backdraft as the air and heated gases mix.

Because hazardous situations are unpredictable, fire departments must have SOP/SOGs that include protocols for emergency and nonemergency evacuation of personnel. 29 CFR 1910.120, *Hazardous Waste Operations and Emergency Response*, includes the requirement for evacuation plans that can apply to fire department operations.

Evacuation protocols, depending on the type of incident, must be written and communicated to all personnel. Then training is developed and presented, including the concept that evacuation should be done with a sense of urgency. Some evacuation pieces include:

- Specific radio commands, referred to as *emergency traffic* **(Figure 9.29)**
- Methods for withdrawal from the hot zone
- Accountability reports post-evacuation
- Air management

- Mandatory buddy system
- Rapid Intervention Crew/Rapid Intervention Team (RIC/RIT) activities during evacuation

NOTE: When withdrawing from a structure fire, the charged hoseline typically provides a path for withdrawal and protection from the fire.

Figure 9.29 An officer on-scene has the authority to issue specific radio commands, such as evacuation orders. *Courtesy of Ron Jeffers, Union City, NJ.*

NIMS-ICS Forms

NIMS-ICS provides approximately 26 forms to assist in the establishment of an Incident Command. The primary forms that are contained in most IAPs are the following:

- ***Incident Objectives (ICS 202)*** — Clearly stated and measurable objectives to be achieved in the specific time interval
- ***Organization Assignment List (ICS 203)*** — Description of the ICS table of organization, including the units and agencies that are involved
- ***Assignment List (ICS 204)*** — Specific unit tactical assignments divided by Branch, Division, and Group
- ***Incident Radio Communications Plan (ICS 205)*** — Lists the basic radio channel assignments for use during the incident
- ***Medical Plan (ICS 206)*** — Provides information on the location and staffing of the incident medical aid station, the types of ambulance resources available, the location of on-site ambulances, and the contact information for hospitals that are available to the IC.

Behavioral Factors

From the alarm sounding through the response, emergency responders receive an extra dose of adrenaline. The physiological effect results in a greater ability to focus on urgent, unforeseen situations. This condition is compounded by the emotional desire, supported by fire service culture, to save a person or structure from imminent danger.

Unfortunately, adrenaline can cause firefighters to lose their bearings and commit unsafe acts. Under stress, some individuals may feel the need to meet heroic expectations, or actually see themselves as a hero and take unreasonable risks. Others respond with a heightened sense of duty. These responses may end in tragic loss of life or loss of public trust. However, firefighters who get away with unreasonable risk may feel that mind-set reinforced for the next alarm. This is success-engendered optimism, and it can lead to increased risk-taking behaviors by individuals, teams, and organizations.

Other actions that can result from these conditions include an attitude referred to as Moth-to-Flame or freelancing. Additionally, unit or Incident Commanders may improperly assign tasks to unqualified personnel, thereby placing them at risk.

Moth-to-Flame Attitude

In nature, the moth is attracted to a bright light such as the flame of a candle. The moth is so focused on the light that it flies too close and burns its wings. For firefighters, it is the specific hazard they are responding to and the desire to control or overcome it. When this desire to act is strong enough, it can override situational awareness, causing responders to ignore indicators of danger. To avoid moth-to-flame attitude, it is important to stop long enough to properly size up an incident and confirm that safety measures are in place.

Freelancing

To prevent freelancing, all units and individuals must report to the IC, ISO, or accountability officer upon arrival at the incident. The IC will then assign them to function as a unit where they are most needed or hold them in reserve.

In most cases, freelancing occurs at the individual level. It is imperative that the integrity of assigned companies or crews be maintained at the incident. Crews or teams are composed of no less than two people. To ensure accountability, even simple tasks, like going to get a tool, should require two people.

Post-Incident Analysis

The post-incident analysis (PIA), evaluation (PIE), or review (PIR) is essential to the successful and safe operations of the fire and emergency services. Properly developed and written, the PIA determines the strengths and weaknesses of the organization's response to the emergency. It provides a training tool and a roadmap for organizational improvements. The PIA can also improve future operational plans for similar type calls and motivate change in policies and procedures that may be outdated or ineffectual.

Overview

The PIA is a written document compiled by the IC, a designated member of the Incident Command, or general staff such as the ISO. The PIA should be written for all incidents, small or large. If the results of the PIA indicate that a post-incident critique should be held, all participants should participate.

The objectives of the PIA are:

- To provide an opportunity for participants to review operations in a constructive manner
- To identify effective procedures (strengths) and areas needing improvement (weaknesses)
- To recommend changes that will improve effectiveness for emergency operations

The analysis is intended to focus on the activities of the participants, the elements of the emergency, and the decisions made that were intended to control the incident. The PIA is not intended to place blame or find fault with the participants, their decisions, or actions.

Post-incident analysis is sometimes confused with post-incident critique. For better understanding, this manual will use the term PIA to mean only the written report. The post-incident critique will be the meeting of participants that may be required, based on the results of the analysis.

Data Collection

The PIA is a critical and objective assessment of the emergency incident, based on all available information from the following sources:

- **Pre-incident survey or fire and life safety inspection report** — Background information on the structure, facility, and/or occupancy.
- **Size-up** — Information on the structure, facility, occupancy, or location based on the impressions of the first Company Officer and crew to arrive.
- **Incident Action Plan (IAP)** — Information on the actions planned to mitigate the incident.
- **Command and General Staff records** — Detailed view of the activities of each of the ICS functions during the incident.
- **Outside agency reports** — Information on the assistance provided by other agencies.
- **Interviews** — Narrative of the actions of each of the participants as well as the impressions of witnesses; some of this information may be provided by the fire investigator.
- **Site plan** — A visual aid in placing the responding units and personnel, as well as locating casualties and occupants.
- **Incident safety plan** — Information gathered by the ISO on all safety and health-related issues.
- **Personnel accountability system** — An accurate record of the participants and their times of arrival and release from the incident.
- **Weather reports** — Information on weather conditions that may have contributed to fire spread or hampered unit access, particularly important when analyzing wildland and urban interface incidents.
- **Communications records** — Accurate dispatch times and transcripts of radio communications at the incident.
- **Miscellaneous reports** — Additional information that may be of value, such as apparatus or equipment maintenance records of vehicles or other apparatus that may have broken down during the incident.

The use of the NIMS-ICS forms over a long-term or multi-response period event can assist in the collection of much of this data. These forms are designed for use in all types of emergency incidents and can be acquired from the FEMA website. The information from each of these sources is then compiled into a chronological report of the incident. Photographs and sketches can help to illustrate the situation both before and after the incident.

Analysis

When all the information gets compiled into one report, strengths and weaknesses become apparent. The following questions can help to focus on each aspect of the incident:

Dispatch

- Was the response time within the organization's minimum/maximum time criteria?
- How can the response time be reduced?

- Were there any extenuating circumstances that increased response time (weather, traffic, road construction)?
- Was all available information communicated to the responding units?

Initial attack

- Was the initial assignment adequate to mitigate the incident?
- What additional resources were required to mitigate the incident?
- Was the ICS properly implemented?
- Was the ICS adequate for the incident?
- Was all available information communicated from the incident scene to other responding units?
- What improvements could be made to the initial attack?
- What weaknesses should be corrected?
- What procedures need to be updated or changed?
- What strengths can be used as examples for similar situations?

Subsequent resource assignments

- Were additional resources requested in time to be effective?
- Was there a staging area established and a staging area manager assigned?
- Were IC duties delegated according to the ICS?
- Was adequate communication made between automatic mutual aid units and other support resources?
- Were there any difficulties in dealing with outside agencies or jurisdictions?
- What can be improved when dealing with outside agencies or jurisdictions?

Private fire protection systems

- How effective were the fire protection systems?
- How well did the responding units use these systems?
- Could the systems have been used to better advantage?

Health and Safety

- Did the activity assign an ISO?
- How was personnel accountability conducted?
- Did the activity necessitate the use of a rehab area? If so, was the location adequate?
- Did the operation identify operational periods and work/rest cycles?
- Was the two-in-two-out rule in place?
- Was there an assigned rapid intervention team (RIT) and/or a backup, if needed?

Post-fire activity

- Was the property properly secured or turned over to the Fire Investigator after suppression forces vacated the area?
- Did the Fire Investigations Unit respond in a timely manner?

Once these and other questions are answered, develop a set of recommendations based on the results of the analysis. The recommendations should be included in the PIA and in an executive summary that is provided to the organization's administration. The executive summary is a brief overview of the incident that includes the recommendations. If a critique is warranted, recommend that one be held.

CAUTION

Post-incident critiques must be positive. Personnel, units, agencies, or organizations must not be criticized for their actions, procedures, or policies. Facts should be simply stated, and recommendations must be focused on improving emergency responses and interagency relationships.

Post-Incident Critique

Post-incident critiques can be informal or formal. The informal critique can be held by the company officer and members of the unit following any single unit incident **(Figure 9.30)**. Weather permitting, this critique occurs immediately after termination of the incident and prior to leaving the incident scene. By discussing the incident while at the scene, the unit members can provide their impressions of the initial attack, suggest alternative approaches to the attack, and learn methods to improve their actions at similar incidents. In inclement weather, the critique can be held at the station and take place immediately after returning from the incident. Essentially, an informal critique serves as a training exercise for the unit.

Figure 9.30 A post-incident critique aims to improve operations and safety at future incidents by looking at what occurred at the last scene.

A formal critique is held if the PIA indicates that it is necessary, or if the incident involves an interagency or multi-jurisdictional response. If the incident involved only units from one jurisdiction, then all participants should participate in the critique. If the incident was multi-jurisdictional, representatives of each agency or jurisdiction should attend.

The analysis should be provided to participants prior to the critique to allow them time to read the report. The critique should be held in a classroom setting that provides comfortable seating, work space, audiovisual aids, and privacy. A set agenda and sufficient time are also important to a productive critique.

Following an overview of the incident, each unit commander or agency representative should present, in order of arrival, a brief report of their participation in the incident. The PIA recommendations can be reviewed, and procedures for implementing them can be developed. Any additional recommendations can be discussed and addressed. Whether the critique is informal or formal, the participants must be assured that necessary changes will be made and successes will be celebrated.

Chapter Review

1. What are the emergency response responsibilities for a fire department's administration, the department's supervisors, and individual crew members?
2. What activities are performed inside an incident's cold, warm, and hot zones?
3. How can supervisors ensure their crew's visibility during a night-time incident?
4. What are some common characteristics of NIMS-ICS that can be found across jurisdictions?
5. What types of accountability systems are availabe for tracking individuals at emergency incidents?
6. How are operational plans or incident action plans (IAPs) used in NIMS-ICS?
7. How can freelancing at emergency scenes be prevented?

Discussion Questions

1. What can a department's administration, HSO, and individual crew members do to reduce cardiac deaths in the fire service?
2. What specific types of pre-incident size-up surveys may be required in your jurisdiction?
3. How can collecting and reviewing incident data improve the outcome of future incidents?

Facility Safety

Chapter Contents

FESHE Learning Outcomes

Occupational Safety and Health for Emergency Services

5. Describe the relevance for safety in the workplace including the importance of PPE.
10. Describe the responsibilities of individual responders, supervisors, safety officers, and incident commanders, safety program managers, safety committees and fire department managers as they relate to health and safety programs.

Facility Safety

Learning Objectives

After reading this chapter, students will be able to:

1. Identify causes for injuries, fatalities, and illness that may occur in the facility environment.
2. Identify the design considerations for fire service facilities.
3. Describe the inspection and maintenance schedules and tasks for fire service facilities.
4. Describe individual firefighter behaviors that encourage on-site safety.

Chapter 10
Facility Safety

Fire and emergency services facilities include fire stations, administrative offices, dispatch centers, apparatus repair garages, equipment warehouses, training classrooms, burn buildings, aviation hangers, and driving pads. Throughout North America, these facilities vary widely in age, size, and design **(Figure 10.1)**. However, their goal should be the same: to provide a safe and healthy work and living environment.

Figure 10.1 Design and construction of fire department facilities must adhere to established building codes.

Design and usage are two key factors in facility safety. Safety features can be engineered into a facility during the initial construction or through updates and renovations. *Safety and Health Considerations for the Design of Fire and Emergency Medical Services Stations*, published by the USFA and FEMA, is a useful online resource for health and safety officers (HSOs) in the process of building or remodeling a fire department facility. In order to help reduce onsite accidents and injuries, this manual addresses design aspects of fire and EMS stations that pertain to the health and safety of their personnel.

However, facility safety also requires ongoing maintenance and a commitment to individual safety. The number of on-duty, non-fireground injuries has decreased over the past five years (NFPA, 2016) due, in part, to more fire departments incorporating safety and risk management into station planning, renovation, and design. Facilities in excellent condition benefit from cooperation on three levels:

- Governing agencies value the safety and health of their firefighters.
- Firefighters respect community property.
- Administration supports onsite safety and risk management.

This chapter discusses several safety design issues; addresses important components of station maintenance and inspection programs; and examines how the health and safety of any well-designed, maintained and inspected fire department facility also depends on the behavior of its occupants.

Facility Accidents

Safety and risk management are key elements in the successful planning, renovation, and design of safer environments for first responders. HSOs should conduct facility risk assessments, beginning with areas of highest perceived risk, to reduce or mitigate many onsite hazards. HSOs should also monitor accident, injury, and close call data for facility-related incidents and how to prevent them.

Injuries and Fatalities

According to the NFPA, there were over 62,000 firefighter injuries reported in the United States in 2016. Nearly 37,000 of those injuries were not sustained on the fireground and involve strain, sprain, or muscular pain. Although national data does not specify facility-related injuries and fatalities, individual departments continue to report onsite incidents and consider how best to prevent them. A reduction in the total number of on-duty, non-fireground injuries over the past five years suggests that safer design, maintenance, and usage of fire and emergency services facilities can make a positive difference.

Illnesses

Job-related illness may result from worksite exposure to patients, other emergency responders, or hazardous contaminants. If safety procedures are not followed, illnesses can spread from the fireground or medical response to the facility environment. Other illnesses are directly related to unsafe practices in the facility environment. For example:

- **Respiratory viruses from colds and the flu** — Germs pass through direct physical contact, by touching surfaces where they have accumulated, or by breathing the airborne particles emitted by sneezing.
- **Bacterial infections from biological contaminates** — Molds and mildew can cause bacterial infections when they are inhaled, due to poorly maintained HVAC systems.
- **Allergies** — Airborne dust, pollen, mold, mildew, and certain foods can cause unpleasant or dangerous immune system reactions.

- **Food poisoning** —Disease-causing microbes, pathogens, harmful toxins and chemicals can contaminate food that is spoiled or improperly cooked. Contaminated plates, silverware, cutting boards, countertops and other surfaces can also spread foodborne illnesses.
- **Cancer** — Poorly insulated stations, doors, and windows can allow carcinogens to enter station living spaces. Proper disposal of chemical-protective clothing and cleaning of personal protective equipment (PPE) is essential to keep the facility free of cancer-causing agents and other occupational health hazards.

Design Considerations

This section looks at the legal requirements for facility design and construction, including those addressed in NFPA 5000, *Building Construction and Safety Code*®. It also reviews the types of facilities normally operated by fire and emergency services organizations.

Legal Requirements

Building, electrical, mechanical, and plumbing codes are developed by professionals as industry standards. They give architects, engineers, contractors, and others in construction a way to adhere to criteria. The model codes and standards must be adopted by the authority having jurisdiction (AHJ) before they can be enforced. National codes evolve, so AHJs must regularly update them.

Amendments to the code are developed locally and put in place with national codes. Once adopted by the AHJ, the local adoptions and national building code become the design standard for all construction and renovations. All employers are subject to Occupational Safety and Health Administration (OSHA) state or federal regulations when planning, building, and inhabiting work sites.

Agencies contemplating construction or remodeling should also seek legal advice. The AHJ for building construction may not be the agency that will occupy the building. A municipality, county, or even state may have legal jurisdiction over the construction of the facility. In some cases, local or regional planning boards must authorize the location and even the appearance of fire facilities.

Local and state codes frequently refer to other standards, such as NFPA 70E, *Standard for Electrical Safety in the Workplace*®, NFPA 101, *Life Safety Code*®, or NFPA 13R, *Standard for the Installation of Sprinkler Systems in Low-Rise Residential Occupancies.* The AHJ (and/or occupying agency) may determine that additional safety requirements should be included in the building. For example, some agencies protect their facilities with automatic sprinkler systems, even if these systems are not required by local building codes.

Fire codes may also be adopted by the AHJ to ensure that structures built under the jurisdictional building code are used safely. Fire codes establish guidelines for safe behaviors and are intended to prevent or mitigate the effects of fires. These codes are generally enforced by the local fire department.

NFPA 1500, *Standard on Fire Department Occupational Safety, Health, and Wellness Program,* lists safety standards for fire facilities in Chapter 9. It requires all fire departments to comply with all legally applicable health, safety, building and fire code requirements. The federal government has regulations for the safe use of facilities where personnel are employed. These are contained in the U.S. Code of Federal Regulations and enforced by OSHA or similar state agencies. Most states, Puerto Rico, and the U.S. Virgin Islands have OSHA-compliant programs modeled on national regulations.

Another federal regulation that applies to the design of fire and emergency services facilities is the Americans with Disabilities Act of 1990 (ADA). The 2010 ADA Standards for Accessible Design, published by the U.S. Department of Justice, set minimum requirements for government, public, and commercial facilities to be readily accessible to and usable by individuals with disabilities **(Figure 10.2 a and b)**.

Figure 10.2 a and a The Americans with Disabilities Act requires that government facilities be easily accessible to people with disabilities.

Structures and nonpublic areas may be exempted if the local authority can demonstrate that access is not necessary.

OSHA regulations require the use of signs and symbols to define specific hazards that can harm workers or the public. In particular, these regulations apply to facilities that store or use hazardous materials. NFPA 704, *Standard System for the Identification of the Hazards of Materials for Emergency Response,* presents a system of symbols that are required to indicate the presence of hazardous materials and their flammability, health, stability, and reactivity hazards. Safety Data Sheets (SDS) must be maintained for each chemical that is used or stored on a property. Finally, all chemical containers must be properly labeled in accordance with OSHA regulations.

Other required signage includes the following:

- Warnings about wearing hearing and eye protection
- Warnings about electrical power equipment
- Prohibiting smoking or providing designated smoking areas
- Warnings about slick areas **(Figure 10.3)**

Facility Types

Safety requirements, whether or not they are mandated, exist for each type of emergency facility. Fire stations may include space designed and designated for the following health and safety functions:

- **Station facilities** —Structures intended to house emergency response apparatus, like land-based and marine (fireboat) stations, which may include personnel living quarters. Marine stations require special attention to mitigate water hazards, especially to station visitors. Newer station designs include integrated training props for firefighters and areas for community groups to meet. Community rooms must comply with ADA standards.
- **Apparatus and equipment maintenance facilities** — Structures designed for the repair and maintenance of apparatus, vehicles, tools, and equipment.
- **Training facilities** — Structures intended for training personnel. These facilities include classrooms, offices, smoke buildings, and live burn buildings or props **(Figure 10.4)**. If they are located adjacent to a station, care should be taken to assure that station operations do not introduce additional hazards to training operations or vice versa. Many new fire stations are designed to incorporate training facilities. This combined location increases efficiency by reducing staff downtime, cutting costs, and keeping firefighters in their first-due response areas.

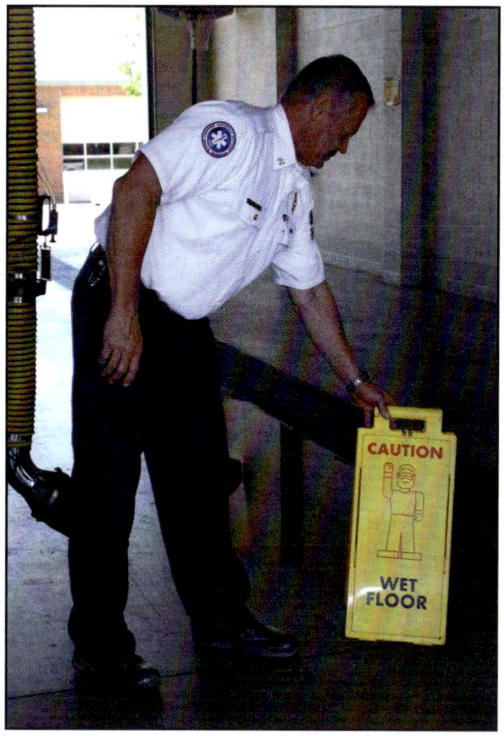

Figure 10.3 If there are slick areas in a station, a caution sign should be posted.

Figure 10.4 Some fire stations have on-site or nearby training facilities like this burn building.

- **Administrative facilities** — Structures that contain offices intended to house administrative and staff personnel.
- **Storage facilities** — Structures designed to provide logistical support to the organization.
- **Communications centers** — Structures that house telephone, radio, and other communications equipment, including emergency dispatch for the organization.

- **Aviation facilities** —Structures located on airport property that provide hangar, maintenance and storage space for helicopters and/or fixed-wing aircraft used in wildland fire fighting. If connected to a station or located adjacent to a station, aircraft operations must not expose station personnel to hazardous conditions, and safe aircraft operations must not be compromised by station operations.

Safety Requirements

Each of these facilities will have safety requirements specific to the activities that occur in them. They will also share many of the requirements listed in the following sections:

Indoor Air Quality

Figure 10.5 Apparatus exhaust capture systems reduce air pollution in apparatus bays.

The sick building syndrome describes a situation in which occupants of a building experience acute health or comfort-related effects that seem linked to the amount of time spent in that building. The HSO is responsible for establishing the types of inspections performed at each facility. An inspection should be performed at least annually to comply with NFPA 1500, 9.1. Monthly inspections help to promptly identify and correct more immediate safety and health hazards.

Air quality requirements may include:

- Separating office and living quarter HVAC systems from apparatus and turnout storage HVAC systems
- Designing living quarter HVAC systems to be at higher pressures than apparatus bays
- Using an exhaust capture and filtration system **(Figure 10.5)** for apparatus bays
- Prohibiting smoking in all facilities
- Providing ventilation hoods in areas that may produce toxic or flammable vapors
- Installing passive ventilation systems in apparatus bays
- Installing carbon monoxide detectors in sleeping and living areas of all existing and new fire department facilities
- Installing windows that open to allow ventilation
- Incorporating a climate controlled area for medical supplies, drugs, and other medications

Decontamination and Cleaning

NFPA 1500, Chapter 9, requires that fire departments provide facilities for disinfecting, cleaning, and storage in accordance with NFPA 1581, *Standard on Fire Department Infection Control Program.* OSHA, local health departments, and fire departments may require additional space and procedures for reducing or eliminating work-related illnesses. These safety measures include:

- Installing a stainless-steel lined decontamination area
- Installing a designated area and separate laundry equipment for cleaning and drying PPE

- Requiring and issuing individual sets of bed linens to each occupant
- Installing emergency decontamination showers and eyewash stands
- Storing PPE away from living quarters
- Storing PPE in a separate room off the apparatus bay with a motion detector on the light so it turns off automatically
- Cleaning PPE or medical equipment away from kitchen and bathroom sinks
- Providing labeled medical waste containers
- Providing stainless steel sinks and countertops in kitchens **(Figure 10.6)**
- Installing antibacterial hand pump stations throughout the facility
- Enforcing proper hygiene in kitchen and bath areas
- Installing an apparatus bay drain system to capture and filter contaminated water and fluids

Figure 10.6 Stainless steel appliances and countertops are designed for easy cleaning and to prevent the growth of bacteria.

Fire Protection

Depending on the local building code, fire protection systems may be required in fire and emergency services facilities. NFPA 101, *Life Safety Code®,* addresses life safety from fire in both new and existing structures. It can be used in conjunction with a building code or alone in jurisdictions that have no building code in place. Fire protection systems include:

- Automatic and manual detection and alarm systems
- Kitchen ventilation hood fire suppression systems
- Kitchen stoves and ovens with automatic shutoffs
- Automatic sprinkler systems for the entire facility
- Paint booth fire suppression systems
- Fire extinguishers **(Figure 10.7)**
- Fire separation barrier between apparatus bay and office or living quarters
- Automatic fuel shutoff valve on cooking appliances (engaged whenever the station is dispatched)

If older fire stations are not protected by an approved, supervised automatic sprinkler system, they must have approved smoke detectors in every sleeping room. When a smoke detector is activated, a general evacuation alarm sounds throughout the building.

Figure 10.7 Fire extinguishers may be required in emergency services facilities.

Hazardous Materials

Hazardous materials found at fire and emergency services facilities often include flammable/combustible liquids, compressed/liquefied gases, and corrosives. Most are stored in small quantities and used for cleaning and maintenance. OSHA and local fire codes regulate the use and storage of these materials. Safety requirements for hazardous materials include:

- Maintaining a SDS book on-site
- Labeling all containers
- Constructing a flammable/combustible storage room or using an approved storage cabinet
- Storing hazardous materials in approved containers
- Providing the proper PPE for use with hazardous materials
- Providing containment systems for hazardous materials

Ergonomic Concerns

Ergonomics is the scientific study of people and their working conditions. It is concerned with designing and arranging things that people use in the workplace in order to improve efficiency, effectiveness, comfort, and safety. Ergonomics can be applied to fire and emergency services facilities by:

- Designing office workstations that relieve stress to backs, arms, and wrists
- Locating tools and equipment within easy reach
- Providing mechanical assistance for lifting heavy objects, such as cylinders in ambulances and ARFF apparatus

Figure 10.8 Emergency exit lighting serves as one component of an overall lighting strategy.

Lighting

Proper lighting reduces eyestrain and associated headaches. It is critical to prevent injuries and to complete work safely. General room illumination is established by building codes and OSHA, depending on the type of occupancy and work area. There are a number of lighting needs to consider in the design of a fire and emergency services facility:

- Installation and location of exit and emergency lighting **(Figure 10.8)**
- Installation of auxiliary lighting powered by batteries or generator
- Installation of suitable lighting in each room or work area
- Connection of lights to emergency alert systems in fire stations
- Installation of traffic warning and control lights
- Installation of accessible work (drop) lights in apparatus and shop areas
- Installation of exterior security lighting around the building perimeter and parking areas

Floor Surfaces

Floors in fire and emergency services facilities should provide a nonslip walking surface, reduce noise levels, and reduce physical strain on legs, back, and lower body. Options include vinyl, carpet, and ceramic tile.

A popular choice for floors of apparatus bays is smooth-troweled concrete with a sealant. In apparatus bays, walkways and backing guideline stripes should be designated by striping or a contrasting color. Floor drains should be located where apparatus are parked, and the surface under a truck should slope directly toward the drain to prevent liquids from pooling on the floor and creating a slipping hazard. Maintenance pits should be avoided when possible. If necessary, they need to be gated and secured.

Floors in physical fitness areas should be covered with rubber or vinyl tiles that absorb impact and moisture. The floor area should be large enough for exercise areas that are free from equipment. Sound-suppressing carpet or tiles reduce noise in office, living, and sleeping areas. Kitchen and bathroom areas should be covered with ceramic tile or similar easy-to-clean surfaces. Tile and grout must be sealed to prevent staining.

Stress Reduction

Heart attacks continue to be the leading cause of death among firefighters on duty. Repeated exposure to loud alarm systems contribute to on-duty stress. Startling alarms and bright lights, especially those that awake sleeping firefighters, can cause heart rates to soar. Alert tone technology and illumination methods can counteract and reduce stress. Additional ways to minimize stressors include:

- Testing for noise levels in all facilities
- Installing noise-reduction materials on floors, walls, and ceilings
- Installing an alert system that uses colored lights to illuminate rooms and an audio alert that gradually increases in volume from soft to loud

General Requirements

Many safety requirements do not fit a specific category; they include the following:

- **Roadway access** — Station driveways should be long enough to accommodate the longest apparatus housed there **(Figure 10.9)**. This distance should provide ample time for the driver/operator to assess the traffic, activate warning devices, and prepare to enter the roadway. It should also provide time for oncoming traffic to see the apparatus and stop. Traffic control lights and audio warning devices linked to the alert system should be installed to warn motorists. Station driveways should be visible in both directions of traffic flow.

Figure 10.9 A station driveway should be long enough to accommodate the longest apparatus in a fire station. *Courtesy of Chris Mickal/District Chief, New Orleans (LA) FD Photo Unit*

- **Medical equipment** — Emergency equipment for use by facility occupants must be located where it is easily accessible. Medical equipment includes first aid kits, defibrillators, and oxygen inhalators **(Figure 10.10, p. 348)**. Equipment and a direct source of communication should be available to citizens who meet in community rooms that are connected to fire stations.

Figure 10.10 Defibrillators and other emergency medical equipment must be accessible in all facilities.

- **Evacuation plans** — All facilities, especially those open to the public, must develop evacuation plans and post them along travel routes. Structures designed for shelter-in-place protection during a weather emergency should designate the safe area on their plan.
- **Slide poles** —Slide poles have been used for more than a century to move firefighters from upper stories to the apparatus bay of multistory stations. Unfortunately, use of these poles has resulted in foot, ankle, and leg injuries. Firefighters have also died from falls through the opening on the upper floor. Slide poles in older facilities must be equipped with required safety accessories, including a cushioned mat at the base of the pole, a removable covering for the hole, and a guardrail with retractable gate around the hole and pole. In some states, no new installations of slide poles are permitted.
- **Security** — An ever-increasing safety concern is security for public facilities. Vandalism of unattended facilities, theft, and violence toward fire and emergency services personnel have led to increased security measures that include:
  - — Locking stations when they are unattended
  - — Providing security card swipe doors to all exterior and apparatus bay entrances
  - — Requiring visitors to use a single point of access and requiring they be escorted while on the property
  - — Requiring vendor preapproval through internal governmental security departments
  - — Installing bulletproof glass in exterior doors and windows
  - — Installing security fences and gates around parking areas
  - — Installing burglar alarms and panic alarms for staff working in offices
  - — Installing external security bars or grills on windows and doors
  - — Hiring personnel to provide security
  - — Providing a secure, locked cabinet or closet for controlled drugs or medical supplies
  - — Installing video surveillance camera systems
- **Storage** — Injuries from falling objects are common. Items stored on mezzanines, lofts, or high-storage shelving can be dropped or fall on their own. The following modifications can prevent injuries related to storage locations and practices:
  - — Require that all vertical file cabinets with three or more drawers have a file interlock mechanism that prevents opening more than one drawer at a time.
  - — Place a lip on the open edge of storage shelves, and a retaining wire or bar at six-inch intervals along the edge of elevated storage areas, to keep items from falling.

— Secure all compressed gas cylinders to a back wall to prevent them from toppling over.

— Require a wall or guard rail at least 42 inches (1.1 m) high on all mezzanine storage areas.

Usability

When designing a facility or renovating one, usability is another important consideration. Usability, or user-centered design, means that the structure meets the goals, mental models, tasks, and requirements of its occupants. Usage of fire and emergency services facilities has changed drastically over the past 50 years. New services have been added and personnel requirements have changed. Usability issues include the following:

- Life expectancy of structure
- Station height
- Drive-through versus back-in station
- Gender privacy
- Auxiliary power
- Physical fitness equipment
- Public accommodations

Life Expectancy of Structure

Fire and emergency services facilities are usually built to last at least 50 years **(Figure 10.11)**. However, because of the high cost of replacing stations, a building may stay in service for more than a century.

Figure 10.11 In many parts of the country, fire stations are more than 50 years old.

Extending the life of an old building can pose serious problems. It can be difficult to predict the demographics of the service area, and to anticipate changes in services and personnel. As land use surrounding a station changes, apparatus and personnel needs must be reanalyzed along with the utilization of stations. Fire agencies that once assumed responsibility for ambulance service no longer meet the requirements for that role. Facilities in other agencies may lack proper accommodations for women. It is also possible that station designs of today will not be able to accommodate larger apparatus or crew sizes in the future. Advances in technology may make current station designs obsolete.

In areas where rapid growth is expected, some departments set up temporary stations consisting of manufactured homes and large metal sheds. Care must be taken that these temporary facilities do not compromise personnel safety.

Station Height

Most fire stations constructed in recent years have been single-story facilities. Departments look at predicted population growth and purchase parcels of land sufficient to accommodate single-story facilities.

Drive-Through vs. Back-In Station

There is less opportunity for an accident in a drive-through station, but it takes a considerably larger lot and budget **(Figure 10.12, p. 350)**. Without a corner lot and side street, a driveway must be built to connect the street to the rear

Figure 10.12 Fire stations with drive-through bays all but eliminate the possibility of accidents that occur when apparatus are backed into stations.

apron. This driveway permits a driver to reach the rear apron by making only right turns and provides the best view of potential obstructions. A driveway is also needed on the left to serve as an alarm bypass or a staging area.

The lot size requirement is smaller for back-in stations, and experience is essential for safe backing procedures. Rear- and side-view cameras, floor paint, and lighting can assist drivers in safely backing fire apparatus. Driver experience, communication, and situational awareness are also essential to preventing backing accidents.

Gender Privacy

One of the most significant changes in station design is the growing need for gender-specific, gender-neutral, or gender-friendly facilities. As more women enter the fire service, departments either retrofit facilities or design facilities to meet the increased need for privacy in living quarters. Areas that have the most need for privacy are sleeping areas, locker and changing areas, restrooms, and showers. In addition to increasing privacy and improving sleep quality, the change from dormitory sleeping arrangements to private cubicles or private rooms may also reduce exposure to illnesses, such as colds and flu.

Auxiliary Power

Because modern fire and emergency services facilities depend on electricity to power more equipment than in the past, high-capacity auxiliary generators are essential. Today's generators operate with natural gas, propane gas, or diesel fuel, and they can provide power for an entire facility. This increased capacity ensures that firefighters will have uninterrupted power in case of an outage. It may also convert the station into a place of refuge for displaced citizens. Periodic testing evaluates how well a station reacts to a power disruption and which internal equipment or systems continue to function.

Solar power is also becoming a viable source of energy for daily facility use. It can reduce the size of backup power sources, thus reducing overall costs.

Physical Fitness Equipment

Because cardiac arrest is the leading cause of firefighter fatalities on the job, fire department administrations and labor organizations are increasing their requirements for participation in physical fitness programs. To accommodate these programs, designs for new facilities include space for physical fitness equipment and floor areas for core workouts.

To address safety needs, physical fitness areas should be large enough to allow personnel to perform exercises; to access each piece of equipment; and to provide a clear passage through the room. Floor coverings should be easy to clean, soundproof, and cushioned. The space should be well-illuminated and offer exterior air ventilation when weather permits. Large windows to interior high traffic areas enable coworkers to see who is exercising.

Public Accommodations

Unless they are privately owned, fire and emergency services facilities are public property. As governments attempt to cut net expenses, non-emergency-related activities have started to take place in facilities. Administrative facilities, training centers, and, in particular, fire stations are increasingly being opened for public use. Nongovernmental organizations have also collaborated with fire and emergency services organizations to provide services through these facilities. Uses include the following:

- **Community use** — Community meeting rooms or areas are added to a facility for voting, public meetings, health examinations or vaccinations, CPR classes, and other civic events. If possible, this area should have its own access and public restrooms.
- **Place of refuge** — During emergencies or natural disasters, fire stations and other facilities are used as places of refuge. They offer accessibility and space; auxiliary power and sanitary facilities; and sufficient storage space to hold emergency supplies. When fire stations are used in this manner, firefighters should have secure storage for their own belongings and separate living quarters to shower and dress.
- **Safe Place** — The Safe Place program provides access to help for youth in need. Its yellow and black sign outside a fire station or other facility designates the site as a place where victims of abuse, lost children, and those in danger can seek assistance (**Figure 10.13**). Station personnel are trained to assist victims and refer them to the proper authorities. Those authorities go to the Safe Haven site to talk with the youth and determine if he or she needs professional help.

Figure 10.13 A Safe Place sign at a fire station means the site provides immediate help for youth in need.

Facility Protection

Facilities in regions that experience tornadoes, hurricanes, earthquakes, or other catastrophic natural disasters must be designed to protect their occupants from these hazards. Some areas may also need structural protection from potential terrorist attacks. Safe rooms and storm shelters may be added to structures. Video surveillance of all exterior entrances ensures that firefighters know who is on the other side of a door before they open it. Bollards can be installed to protect fixed objects around the facility, such as fuel pumps.

Inspections and Maintenance

Facilities must be inspected and maintained regularly to prevent deterioration. The responsibility for both of these duties may be spread through the fire and emergency services organization or shared with another department of local/state government. Administrators, supervisors, and individual firefighters all participate in inspection and maintenance procedures.

Figure 10.14 Station inspections are generally performed on a weekly or monthly basis by personnel assigned to the facility.

Inspections

Performed properly, inspections ensure a safe and healthy work environment. According to NFPA 1500, 9.3., each fire department "shall have an established system to maintain all facilities and to provide prompt correction of any safety or health hazard or code violation."

The least formal and most frequent inspection is one that every member of the organization can perform. In a career or combination organization, a shift change is the first opportunity for facility inspection. Immediately following the formal change of command or during station cleaning, personnel informally inspect the condition of the facility and note any maintenance needs. Through the chain of command, repairs are made or reported to the proper authority.

During their shift, personnel repair or report any needs they have noted. As a form of preventive maintenance, they make small repairs before urgent problems develop. In a volunteer organization, this type of inspection may occur whenever the unit holds mandatory training or work sessions.

A more formal inspection is usually performed weekly by the company or chief officer **(Figure 10.14)**. These inspections may coincide with a thorough cleaning of the facility. This level of cleaning and inspection allows an officer to note more significant maintenance needs, such as painting, caulking, and replacing filters. Monthly inspections are generally performed by a district/battalion chief. The thoroughness of these inspections is similar to the weekly inspection. Monthly inspections may result in more detailed repairs or alterations and require greater expenditures or budget requests.

Finally, the HSO and/or a fire prevention officer performs an annual inspection. It ensures compliance with health and safety policies and with fire and life safety codes. The focus of the annual inspections is on the condition of the facility and on the safe behavior of its occupants. The HSO may inspect air quality, noise levels, tool guards, chemical storage, oil buildup on floors, and door seals throughout the facility.

The HSO should test the air quality in the building annually, or when the need arises, to reduce the potential for illness from germs or excessive levels of carbon monoxide. Testing equipment samples the air and determines the parts per million of contaminants, such as dust or pollen. Inspection results may recommend cleaning air ducts or replacing filters. Apparatus room contamination should also be tested to determine if apparatus exhaust systems need to be installed.

The HSO is also responsible for testing the noise level in the facility. Noise-testing instruments determine the level in decibels (dB). Facility noise originates from the HVAC system, kitchen range ventilation hood, auxiliary power generator, power tools, and equipment. The greatest source of noise is the apparatus in the bay. Although a maximum allowable noise level in the cab of the apparatus is limited to 90 dB by federal law in the United States, the level outside the vehicle may be higher. If the noise level cannot be reduced, the department may require the use of hearing protection while the apparatus is running.

Inspection checklists or reports provide a record of tests, maintenance, and repairs. These records are generally maintained for the life of the facility, or for the minimum retention time prescribed by local policy. Inspections that result in maintenance requests must be forwarded to the appropriate authority. Maintenance requests should be tracked to ensure that work is completed in a timely fashion. Repairs related to health and safety should take priority. A sample facility inspection checklist form is located in **Appendix D**.

Maintenance

Like inspections, facility maintenance is the responsibility of many people and departments. In volunteer organizations, volunteers are responsible for most of the work. In large departments, maintenance work may be assigned to another department or to a private contractor.

Basic maintenance begins with continual cleaning and care of the facility. Cleaning is usually performed at the start of each work shift. Some facilities, such as administrative offices and training sites, are cleaned by private contractors during off-duty hours. Cleaning reduces germs in the air and on surfaces that can cause illnesses and allergic reactions. A thorough cleaning of all cooking and food-preparation areas prevents the growth of microbial germs. These surfaces may need to be tested periodically to ensure that they are properly disinfected.

While local policies may vary, minor maintenance is generally performed by the facility personnel. Replacing light bulbs or HVAC filters are part of this responsibility. Major maintenance and repairs are normally performed by the organization's supply or support division, the jurisdiction's public works department, or a private contractor. Painting, roof repairs, glass replacement, plumbing and electrical repairs, and training prop repairs fall into the category of major maintenance work. In many volunteer departments, trained members of the department also perform these tasks.

Emergency personnel normally perform ground maintenance, although a local parks department or a private contractor may also provide this service. Mowing the lawn, trimming shrubs and trees, and maintaining planting beds improve the appearance of the facility and reduce the potential for vermin to nest in weeds and tall grass. However, ground maintenance can also result in injury when personnel operate power tools and equipment.

Personal Behavior

Strategic design, regular inspection, and timely maintenance all contribute to safety and health in a fire and emergency services facility. However, the behavior of its occupants determines whether or not those conditions will last.

Figure 10.15 Personal protective equipment is kept in ventilated storage lockers in the apparatus bay.

Safe practices on the part of individual firefighters is initiated on common sense, good training, and adherence to safety policies and procedures. Common sense actions and best practices for on-site firefighters include:

- Avoid horseplay.
- Keep PPE in the apparatus bay, designated storage areas, and cleaning areas; away from living quarters **(Figure 10.15)**.
- Keep structural turnout gear out of living quarters and personal vehicles.
- Use approved ladders for reaching high areas.
- Use mechanical aids or additional personnel to lift heavy objects.
- Wear approved PPE when operating power tools and equipment.
- Practice personal hygiene.
- Thoroughly clean all cooking and food-preparation surfaces.
- Report all facility maintenance needs immediately.
- Report any illnesses or injuries that may be caused or aggravated by the facility's condition or design.
- Use only approved cleaning and maintenance materials.
- Update and read the SDS for all cleaning and maintenance materials.
- Adhere to the safety policy and procedures manual.
- Use physical fitness equipment according to instructions.
- Adhere to the No Smoking policy.
- Practice good housekeeping in all facilities and work areas.

The HSO and the supervisor are primarily responsible for providing all safety-related training to firefighters. Firefighters are responsible for practicing the safety-related skills that are taught. Each firefighter is ultimately responsible for their safety in and around department facilities.

Chapter Review

1. How can unsafe practices in fire service facilities lead to illness?
2. What are the safety considerations for the different types of fire service facilities?
3 What are some reasons why fire service facilities are designed with facility or structural protection in mind?
4. How can the HSO, supervisors, and individual firefighters enhance the safety of fire service facilities?

Discussion Question

1. Why might regular and thorough inspections and maintenance of fire service facilities be as, if not more, important than inspections and maintenance of apparatus and PPE?

Safety and Health Program Administration

Chapter Contents

chapter 11

Key Terms

FESHE Learning Outcomes

Occupational Safety and Health for Emergency Services

3. Compare the difference between standards and regulations.
4. List and describe the components of risk identification, risk evaluation, and incident management.
6. Apply the knowledge of an effective safety plan to pre-incident planning, response, and training activities.

Safety and Health Program Administration

Learning Objectives

After reading this chapter, students will be able to:

1. Explain the legal requirements, standards, and regulations that oversee fire service safety and health programs.
2. Identify the information sources used to build safety and health programs.
3. Review the roles and regulations governing safety and health programs.
4. Describe how to utilize a risk management plan to guide department operations.
5. Describe how common fire service hazards are mitigated by safety and health plans.
6. Compare control plans for infection and hazardous materials incidents.
7. Explain the basics of the investigation/review process.

Chapter 11
Safety and Health Program Administration

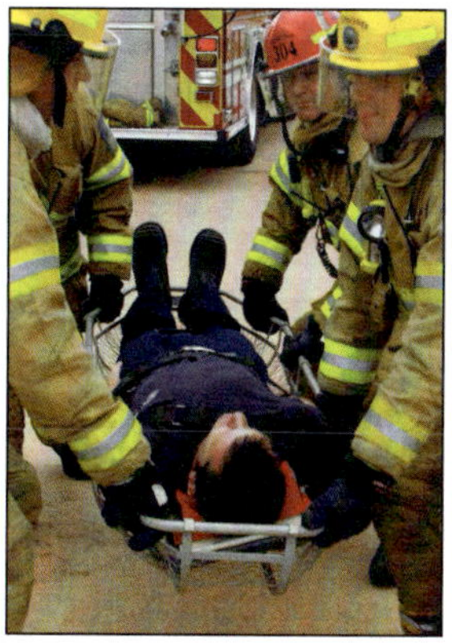

The goals of safety and health program management in the fire service are threefold. First, due to changing regulations and standards, public agencies that protect workers need to know what fire departments are doing to help workers meet response challenges. Secondly, departments have the moral and ethical responsibilities to ensure everything possible is done to protect firefighters. Third, good program management provides the Health and Safety Officer (HSO) with a way to understand which risks have been managed and which need work. With this information, the HSO can set goals to change behaviors to benefit fire service culture.

This chapter addresses legal requirements, data collection, the role of a Safety and Health Committee, risk management planning, prevention of injuries, exposure management, quality control processes, and the accident/injury/exposure/near-miss investigation/review process.

Legal Requirements and Industry Standards

The occupational safety and health policy, as required by Chapter 4 of NFPA 1500, *Standard on Fire Department Occupational Safety, Health, and Wellness Program,* establishes an infrastructure to develop and implement the organization's occupational safety and health program. The policy aims to clarify the responsibilities of each member and of the organization as a whole in developing and maintaining safety and health programs.

Based upon the fire and emergency service organization's risk assessment, members are exposed to hazards that create a complex set of occupational safety and health issues. The objective of an occupational safety and health policy is to provide the necessary control measures to help members perform assigned tasks safely, efficiently, and effectively. For a policy to be effective, it must be communicated to all members of the organization, monitored for desired results, reviewed periodically, revised as necessary, and applied fairly.

According to NFPA 1500 and 1521, *Standard for Fire Department Safety Officer Professional Qualifications,* the organization's designated HSO is responsible for the safety and health program based on the policy. In this chapter, the term *HSO* is used regardless of the rank of the person with that function. This position may be held by a chief officer or other officer, or a nonuniformed health and safety professional who meets the educational and skills requirements of NFPA 1521. Smaller agencies may meet this requirement by using training

Figure 11.1 The Health Safety Officer develops plans to help the department meet its policies and goals.

officers as the HSO with assistance from staff members, or by contracting private or outside agencies to provide the HSO function. Small departments with financial constraints may share an HSO with neighboring organizations. However, it may not be an option in all jurisdictions.

The HSO develops a plan to address the goals, exposures, and risk of the organization's occupational safety and health policy **(Figure 11.1)**. A comprehensive program includes the following:

- Risk management plan
- Accident, injury, and illness prevention and rehabilitation program
- Medical exposure management program
- Employee physical fitness and wellness program
- Scene rehabilitation
- Employee assistance program (EAP)

In addition to the NFPA standard requirement for a safety and health program, there are other legal mandates. By citing NFPA standards, other laws or ordinances may incorporate them. For instance, a local ordinance that establishes the fire department may adopt NFPA 1500 by reference or through an enabling ordinance. Adoption by reference means that the local jurisdiction will follow the written standard. Meeting a standard through an enabling act allows a jurisdiction to select portions of the standard that fit local needs and delete others. Check to see which edition is referenced, as NFPA is frequently the catalyst for Occupational Safety and Health Administration law.

Regulations — Rules or directives issued by authorized administrative agencies.

Standards — Criterion documents developed to serve as models or examples of desired performance or behaviors that contain requirements and specifications outlining minimum levels of performance, protection, or construction. No one is required to meet the requirements set forth in standards unless they are legally adopted by the authority having jurisdiction, in which case they become law.

Simply stated, a **regulation** is a rule or law that may be based on or incorporate a **standard**. Safety and health requirements may be the result of federal or state/provincial acts, laws, or regulations. Federal laws will usually take precedence over state/provincial or local laws unless the state/province enacts similar legislation. In the case of safety and health requirements, the federal regulations generally apply because they must be consistently applied throughout the country.

The OSHA enforces worker-protection requirements for federal employees and in states that have approved OSHA regulations. A General Duty Clause may be used to enforce OSHA regulations nationwide. As part of OSHA's Right-to-Know Regulations, safety data sheets (SDSs) must be provided for all potentially hazardous materials **(Figure 11.2)**. SDSs must be available in the workplace, and employees must be trained in the mitigation of hazards caused by the materials.

The labor/management agreement with the fire and emergency services organization (or jurisdiction) will contain safety and health clauses. These clauses may include requirements for personal protective equipment (PPE), a wellness program, an employee assistance program (EAP), or annual medical examinations. The labor/management agreement may directly cite language in a NFPA standard, essentially adopting that standard for the duration of the agreement.

The Value of Documentation and Research

Development of an effective safety, health, and wellness program depends on current, accurate information. Based on raw data, information must be gathered from a variety of sources. Gathering information and using large databases, also called data mining, requires significant time and effort. Once gathered, the data must be categorized by type and analyzed. To effectively research a topic, learn what to look for, where to look for it, and how to validate collected data.

Figure 11.2 Employers must provide employees access to safety data sheets for all the hazardous materials in their workplace.

Data Collection

The project topic helps determine the data to be collected. For example, if it involves developing a safety program, the data collected will include the following:

- Local accident, injury, exposure, near-miss, and line-of-duty death (LODD) reports
- Similar reports for other fire departments in the region or of similar size
- National statistics on accident, injuries, exposures, near misses, and LODDs
- Statistics on injuries and deaths in similar high-risk occupations
- Insurance industry statistics
- Fire and emergency responder pension program statistics
- Worker's compensation data
- Data compiled by governmental agencies

It can be difficult to determine what to include in a search. If the focus of the project is injuries and fatalities, as in this example, then data on illnesses will not be needed.

The first step in the process is to write down the research topic. List all topics that resemble the main topic. Next, list the types of data that may support the topic, such as internal reports, regional or national reports, legislation, product reviews, and cost estimates. Then list the possible sources for information. Index cards can be used to record information and sources. It can also be placed in a database for easy retrieval and sorting. The information types may include the following:

- **Expert opinions** — Statements by experts or someone who has analyzed and/or experienced a similar situation. Prior knowledge is an acceptable starting point when looking for expert opinions.
- **Trends** — Patterns that can be traced through time and used for forecasts. Trends may be developed from raw data, such as incident scene strains and sprains. However, raw data must be the same type for each entry on the time line.

- **Models** — Frameworks composed of accepted practices that an organization can modify to fit needs and resources, such as a fitness program.
- **Similar situations** — Information from interviews about how situations were addressed. Sources for this type of research include industry journals, newspaper and online articles, or networking with peers through professional organizations.
- **Data** — Raw numbers, such as lost-time injury leave or cost of overtime.
- **Examples** — Similar situations, processes, items, or other material a researcher can use to illustrate concepts.
- **Analyses** — Detailed examination of a substance or structure that provides information on its nature, essential features or constituents. This information often serves as a basis for discussion, interpretation and/or additional research.
- **Demographics** — Collected information on the population in a specific area or of a specific group, such as the members of the fire and emergency services organization.
- **Recommended/suggested practices** — Suggestions for how to do something; usually based on trial-and-error experiences or scientific research.
- **Industry standards** — Generally accepted methods for accomplishing a task or function (similar to recommended practices).
- **Legal requirements** —Laws, codes, ordinances, and decrees that are legally binding requirements created at most levels of government to ensure the safety and welfare of a society.

Once the information type has been identified, locating sources becomes the next step. If necessary, be prepared to seek assistance from a professional research firm to assist in finding relevant sources of information. The cost of hiring researchers is often worth the time, effort, and frustration it can save department members.

Information Sources

Finding information sources depends on the research topic. Data may come from the Internet, government agencies, libraries, educational institutions, testing and standards organizations, nonprofit organizations, and vendors/manufacturers. The greatest challenge in locating reliable data can be validating sources. Data provided by the government may be written to justify an agency's actions or existence. Information from vendors tends to emphasize the positive results from using their products, but third-party testing agencies provide unbiased results.

Internet

Finding credible information on the Internet has always been a challenge. Some is proprietary, such as the formula for Class A foam concentrate. Other information can be considered vital to national security, and its access is restricted.

Government Agencies

Canadian and U.S. government agencies are the best sources of raw data on fire and emergency services topics. Laws, ordinances, and statutes affecting the fire service and emergency services exist at national, state/provincial, and local levels. In addition, the U.S. government maintains data of fire and emergency responses through the National Fire Incident Reporting System (NFIRS) and on incidents involving firefighter fatalities through the National Institute for Occupational Safety and Health (NIOSH). Agencies responsible for transportation, manufacturing, and natural resources all maintain information on incidents on highways, airways, railroads, and ship channels, as well as in factories, national forests, and grasslands. The Centers for Disease Control and Prevention (CDC) and the World Health Organization can provide information on health and wellness topics. The U.S. federal government funds research into fire and life safety hazards.

The U.S. Department of Labor and state agencies can provide data on workers' compensation claims, including number and type of injuries, cost of treatment, and rehabilitation time required. Insurance companies can provide injury compensation data based on an experience modifier. The experience-modification factor compares an organization's actual loss experience with an expected loss experience over the past three years. If the organization's losses are lower than expected, it should reduce its insurance premium. However, if losses are higher than expected, this usually results in higher insurance premiums for the organization.

As is the case with all information collected by the government, it is subject to manipulation, biased analysis, and out-of-date and incorrect collection and recording processes. In some cases, data that is supposed to be collected never gets to the agency responsible for it, which makes it impossible to accurately forecast or analyze a trend.

NOTE: For reasons stated above, it is vital that departments enter all information and data into NFIRS reports completely and accurately.

Libraries

Some library collections, like the U.S. National Fire Academy Learning Resource Center, are specific to the fire service. As most libraries are networked and have sharing agreements, materials unavailable locally can be acquired from other libraries through interlibrary loan. The Online Computer Library Center's WorldCat can be a valuable resource for hard-to-find materials. Materials listed on the WorldCat website can be obtained through a participating library system, which includes many municipal libraries.

Figure 11.3 University libraries are excellent resources for data. Many university libraries may be able to provide departments with Internet access to their materials.

Educational Institutions

Vocational schools, community colleges, and universities can be sources of information **(Figure 11.3)**. Their libraries may be linked to public libraries through interlibrary loan agreements. In addition, many of these institutions have research projects underway.

Professors may be involved in research on topics from fire science to social/political issues. Educational institutions can provide assistance in applying for and obtaining grants.

Professional Organizations

Fire and emergency services professional organizations may have safety-related data, information, and model programs. These organizations include the following:

- International Association of Fire Chiefs
- Canadian Association of Fire Chiefs
- International Association of Fire Fighters
- Canadian Volunteer Fire Services Association
- National Volunteer Fire Council
- Fire Department Safety Officers Association
- American Society of Safety Engineers
- Canadian Society of Safety Engineers
- National Fire Protection Association
- National Fallen Firefighters Foundation
- National Council on Compensation Insurance
- Occupational Safety and Health Administration

At the state/provincial level, organizations affiliated with these groups may be able to help gather data, but a membership may be required to access their services. The Volunteer Firemen's Insurance Services provides educational material through its subsidiary, the Emergency Services Education and Consulting Group. The Congressional Fire Services Institute and the Home Fire Sprinkler Coalition are additional sources of information. Although not directly associated with the fire service, the International City/County Management Association publications can provide information on local government, public management, human resources, and trends in laws affecting employment practices.

Testing and Standards Organizations

Third-party testing and certification organizations provide statistics on the quality of equipment and products, and on the ability of that equipment or product to perform a specific task. Most fire and emergency services personnel are familiar with the testing that is performed by Underwriters Laboratories, Underwriters Laboratories of Canada, Factory Mutual, and similar organizations. Test results, specifications, and performance criteria available from these organizations may assist in writing specifications or in establishing testing criteria for internal equipment evaluations. The NFPA maintains data on the cause of fires, as well as fire-related losses, injuries, and deaths. They also create consensus standards.

Nonprofit Organizations

Nonprofit organizations with safety-related missions are excellent sources of information for chief officers. Most of these groups work to reduce accident-related injuries and death by educating the public in proper safety activities.

These nonprofits include the Home Safety Council, the Association of State and Territorial Health Officials, Safe Kids, and other public safety councils at federal, state/provincial, and local levels.

Safety and Health Program

The organization's HSO is usually assigned the task of developing and implementing a safety and health program that meets the requirements of an organization. Developing a safety and health program becomes easier if the HSO uses current safety and health laws, codes, and standards as foundations. These standards have requirements for employers, and the regulations contain important requirements for employees. Each employee must comply with the occupational safety and health standards. Foundational documents are required to outline the program for the organization and state its purpose.

Examples of SOP/Gs, preincident plans, risk management plans, and safety and health plans are often located in the annex section of NFPA standards. NFPA 1500, *Standard on Fire Department Occupational Safety, Health, and Wellness Program*, stipulates the following criteria should be part of all fire department safety and health programs:

- Written standard operational procedures/guidelines that show the organizational structure, membership, roles and responsibilities, expected functions, and training requirements.
- Preincident plans.
- Written risk management plan.
- A written safety and health policy that identifies goals for the prevention and elimination of all types of occupational injuries, accidents, exposures and LODDs. It should also contain policy language for audits to the program once every three years and audits of the departments' progress to the fire chief annually.
- The roles and responsibilities of the HSO and any incident safety officer **(Figure 11.4)**.

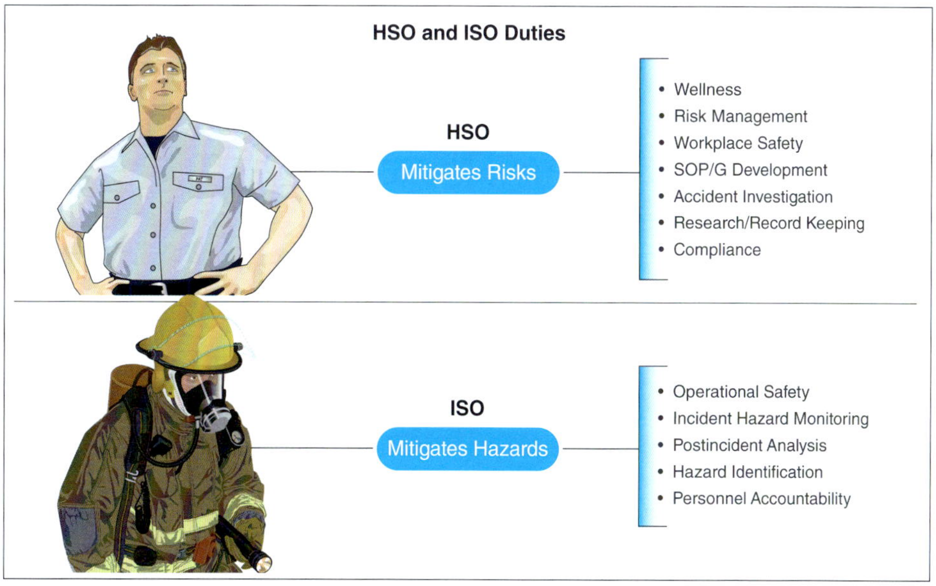

Figure 11.4 Both health safety officers (HSO) and incident safety officers (ISO) have safety functions in a fire and emergency services organization. HSOs more frequently deal with risk factors, while ISOs mitigate hazards at emergency scenes.

- Membership and function of the departments' Occupational Safety and Health Committee.
- Safety and health record management.

Occupational Safety and Health Committee

One of the instrumental and valuable components of a fire and emergency services organization's safety and health program is the development, implementation, and operation of an occupational safety and health committee. The organization's HSO runs this committee and facilitates its work. This person is the ultimate manager of the safety and health program and reports to the chief/manager of the organization.

Figure 11.5 The occupational safety and health committee membership should include people from all levels of the organization.

The occupational safety and health committee is a centralized group within the organization that functions as a clearinghouse for activities, problems, and issues relating to employee safety and health **(Figure 11.5)**. The issues that confront this committee are unlimited. Ensuring employee safety and health is a constantly changing process based on the needs of the organization and the administration's commitment to safety. Committee success depends on how actively it examines and manages safety and health issues within the organization.

One of the goals of the occupational safety and health committee is to resolve conflicts. Disciplinary action should not be a committee objective. The activities and issues addressed must stay within the scope of the committee. For example, occupational safety and health committee members may be required to perform the following duties:

- Identify situations that may be a source of danger to members.
- Investigate/review complaints of violations of the organization's safety policy.
- Make recommendations to the chief or manager of the organization on matters reported to the committee, and on rules and regulations promulgated by outside regulatory agencies that relate to safety issues.
- Evaluate safety rules and regulations established by the organization to ensure compliance with state, territorial, and/or provincial mandates.
- Review the annual injury and illness report for trends.
- Review the safety and health inspection reports to assist in correcting identified unsafe conditions or practices.
- Analyze incident investigations/reviews conducted since the last meeting to determine if the cause of an unsafe act or unsafe condition was properly identified and corrected.
- Evaluate the accident and injury prevention program and make recommendations for improvement where indicated.
- Make recommendations to the administration and to members for the improvement of safety and health throughout the organization.

- Recommend, maintain, and monitor safety and health programs and procedures.
- Consider forwarded reports from outside agencies that address safety and health in order to make recommendations to management regarding these issues.
- Take and post meeting minutes that include attendance and the topics of discussion.

For the foundation and function of an occupational safety and health committee NFPA 1500 stipulates the following:

- The fire and emergency services organization needs to establish an occupational safety and health committee that will serve in an advisory capacity to the organization's chief or manager.
- The mission of the occupational safety and health committee is to conduct research, develop recommendations, and study and review matters pertaining to occupational safety and health within the fire and emergency services organization.
- The occupational safety and health committee must hold regularly scheduled meetings and special meetings as needs arise. Regular meetings must be held at least once every six months. Written minutes of each meeting must be maintained and distributed to all members of the organization.

Regardless of size and type, every fire and emergency services organization should establish an occupational safety and health committee. The total number of members will vary depending on the size and type of organization. A career organization should select members from management and labor, as well as from each shift, each district/battalion, and each division. Volunteer and combination organizations should recruit a representative cross section of the organization that will provide sufficient diversity of views, opinions, and skills. Additional members may be appointed. Departments with sufficient resources should include the organization's physician.

As with any organization, it is important to effectively manage the time of the committee members. With training mandates, the key to participation is efficiency and effectiveness. An organization must emphasize the importance of good time management to make the committee process work. A checklist helps establish a safety and health committee. A sample checklist is located in **Appendix E**.

It can help to work with other local fire departments and hold regional safety committee meetings that can supplement or replace one agency's safety committee. This may lead to information sharing and database management sharing. Ultimately, some fire departments make intergovernmental agreements and share resources, such as one HSO for an area of fire departments.

Health and Safety Officer Roles and Responsibilities

The HSO, under the direction of the chief of the organization or another chief officer, must be able to apply analytical skills to the development and evaluation of a safety and health program. These skills apply to the program elements including the risk management plan; the accident, injury, and illness prevention program; the medical exposure program; and the employee physical fitness and wellness program. NFPA 1521, *Standard for Fire Department Safety Officer Professional Qualifications,* is specific in defining the roles, responsibilities, and job performance requirements (JPRs) of the HSO in the safety and health program. As a minimum standard, JPRs include (but are not limited to) the following:

Figure 11.6 Inspections should be conducted before all live-fire evaluations and should be part of the organization's safety procedures for live burns.

- Develop, manage, and implement an organizational Risk Management Plan
- Develop an operational risk management plan including treatment and transportation of an injured or ill member
- Establish SOP/Gs for an occupational safety and health program and provide written assessments of the adequacy of the program
- Implement safety and health program training and education
- Develop safety procedures for live burns **(Figure 11.6)**
- Assume overall management of the accident prevention program and an emergency vehicle safety program
- Submit annual audits of the safety and health program to the Fire Chief
- Manage the review or investigations program
- Record management and data analysis
- Recommend safety-related specifications for apparatus and equipment
- Develop and conduct a facility safety inspection program
- Coordinate and analyze health and wellness maintenance programs
- Serve as liaison to numerous department committees as the safety and health representative
- Assess and coordinate the department's infection control program

Other Safety and Wellness Issues

Safety and wellness involve many concepts and issues, including some that may not seem to belong in the program. The following issues are related to safety and wellness:

Staffing

Staffing directly affects accidents, injuries, and fatalities at emergency incidents. When a unit or group of units is forced to operate at a minimum staffing level, greater physical stress is placed upon the personnel. Units cannot function at their maximum efficiency and an operation takes longer to complete. When members are placed in "moved up" or "acting" positions, staffing can also impact the safety of the operation. Personnel in the acting position must be trained to assume the assigned duties.

Drug-Free Workplace

The use or abuse of drugs and alcohol before or during a work shift impairs judgment. It increases the likelihood of mistakes, and it slows reaction times. The use of some drugs can lead to cardiac arrest. Administrative policies of career and volunteer departments prohibit personnel from reporting to duty while under the influence of drugs or alcohol. Supervisors and members of the organization must learn to monitor the behavior of their coworkers and prevent them from attempting to work while impaired. The organization must also support the provision of medical and psychological assistance to any member who uses or abuses drugs or alcohol.

Safety Stand-Down

The National Safety Stand-Down to Prevent Falls in Construction is an annual event that coincides with the North American Occupational Safety and Health Week. According to OSHA, any number of people can hold a Safety Stand-Down. Participation can be as simple as a 15-minute toolbox talk or can involve several hours of training. The goal is to provide opportunities for participants to discuss hazards, protective methods, safety policies, goals and expectations. Certificates of participation can be accessed on OSHA and NSC websites.

Taking Care of Our Own

Through a Department of Justice grant, the National Fallen Firefighters Foundation offers Taking Care of Our Own, a free training program to help fire departments prepare for LODD or serious injury. This one-day course, also available as a self-paced, independent online module, targets senior career or volunteer fire officers, chief officers and senior deputies of federal and state government, honor guard commanders, chaplains, and officials of national and state fire organizations.

Developed by fire service personnel and families who have lost a firefighter in the line-of-duty, the course covers pre-incident planning, survivor notification, family and co-worker support, and resources available to the families. More information on hosting a training session, the online module, and other customizable resources are available from the National Fallen Firefighters Foundation and the Fire Hero Learning Network.

Overcoming Unsafe Cultural Practices

Culture – a group's shared attitudes, believes, and practices – challenges an HSO who must distinguish between positive and negative effects of cultural practices. Some practices support a safe and healthy work environment, while others can destroy it.

HSOs must advocate, defend, and change cultural practices to ensure the safety and health of their department members, but they can expect resistance. As an example, when one department HSO opposed a damaged equipment/PPE purchase practice, members were angered by the change. They had made a cultural practice of purchasing fire-melted or damaged helmets from the Resource Division once they had received new ones. Members grew to appreciate the change when cancer research revealed risks posed by residual toxins on used equipment.

Public Safety Officers' Benefits Programs (PSOB)

The Public Safety Officers' Benefits (PSOB) Programs are a combined effort of the U.S. Department of Justice; local, state, tribal, and federal public safety agencies; and national organizations. They provide death and education benefits to survivors of fallen law enforcement officers, firefighters, and other first responders, and disability benefits to officers seriously injured in the line of duty. Visit the PSOB website for more information and to file a claim.

Risk Management Plan

Regardless of the size of the community or service area, every fire and emergency services organization needs a risk management plan to guide its operations.

The department chief usually develops a risk management plan but may delegate this responsibility to the HSO. Regardless of whether it is the chief officer or an assigned officer, all members of the organization should become familiar with the plan. Those responsible for the risk management plan must be capable of performing skills such as training, investigating, evaluating, analyzing, implementing, and communicating. Each of these skills applies to tasks required of the HSO.

The risk management plan, as described in NFPA 1500, encompasses many aspects of the fire and emergency services organization. It says risk identification is an ongoing process, and a reasonable control plan has been followed. An effective risk management plan aids many areas of the department, including its operational, safety, financial, and legal operations. This plan must be reviewed and revised annually. The fire chief assigns responsibility for the plan's review and revision to the HSO or an occupational safety and health committee. A risk management plan may form part of the overall community/jurisdiction plan, becoming the responsibility of that organization's risk manager (the community health and safety manager or loss control manager).

Chapter 4 of NFPA 1500 states the requirements:

> *The fire department shall adopt an official written risk management plan that covers administration, facilities, training, vehicle operations, protective clothing and equipment, operations at emergency incidents, operations at nonemergency incidents, and other related activities. At a minimum, the plan shall include risk identification, risk prioritization, risk evaluation, risk control techniques, and risk monitoring.*

Risk management and system safety concepts must be understood by all fire officers and organizational HSOs. These concepts form the basis for most of the roles and responsibilities of officers who plan, develop, and manage the safety and health program.

Personnel Risk Analysis

Risk analysis is applied to the safety and health of the internal customers of an organization, its employees and members. Since injuries and fatalities occur during training and/or physical fitness sessions, risk analysis applies to all nonemergency activities of the organization in order to prevent and/or reduce occurrences. NFPA 1500, 4.2.3 contains the five elements of a risk analysis: identification, evaluation, prioritization, control, and monitoring **(Figure 11.7)**.

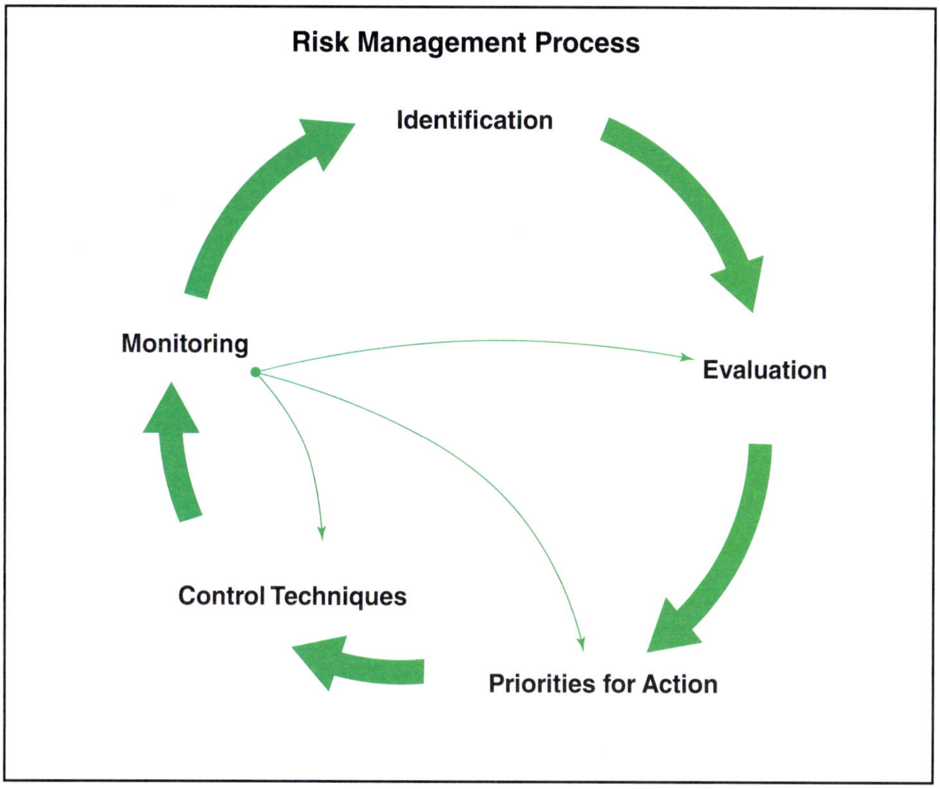

Figure 11.7 The risk management plan cycles through five steps outlined in NFPA 1500.

Risk Identification

To identify risks, an HSO compiles a list of all emergency and nonemergency operations and duties of the organization. Ideally, an HSO should take into consideration the worst possible conditions or events, including disasters **(Figure 11.8)**. There are many sources to assist with the risk identification process.

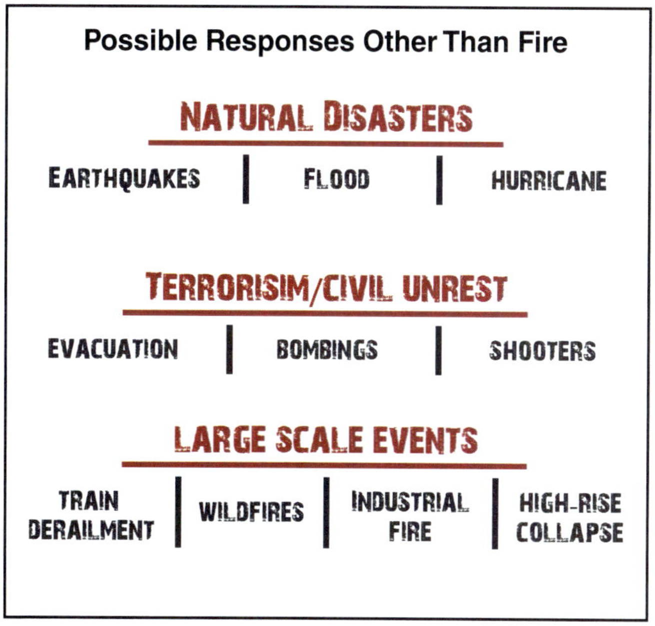

Figure 11.8 In an all-hazards response environment, risks other than fire need to be included in the risk management plan.

The first, and possibly the most effective, is the department's loss prevention data, a collection of annual fire-loss reports by categories that include occupancy type, loss value, and frequency. National averages and trends are available from NFPA and the National Fire Academy (NFA). However, national data is not always complete, accurate or current.

The HSO should seek input and ideas from department personnel, trade journals, professional associations, and other service providers to identify potential risks. When using information provided by other fire departments or organizations, the HSO should consider the effect of local circumstances on the set of emergency and/or nonemergency operations and duties. Other sources for risk identification include risk management plans developed by local industry and hazardous substance sites, vulnerability analyses, and the U.S. Environmental Protection Agency.

Risk Evaluation

Once the HSO identifies the risks, they can be evaluated for frequency and severity. Typically, if a particular type of incident, such as injuries related to lifting, has occurred repeatedly, it will continue to occur until a job hazard or task analysis has been performed to identify the root cause and effective control measures have been implemented. In this example, the HSO or occupational safety and health committee must develop and implement guidelines that outline proper lifting techniques and physical fitness requirements or provide mechanical aids for lifting.

Severity can be measured several ways, such as lost time from work, cost of damage, cost of and time for repair or replacement of equipment, disruption of service, or legal costs. Incidents of high frequency and high severity must have the highest priority in the risk analysis, while those of low frequency and low severity receive the lowest priority. The method for calculating the risk may vary from one department to another.

Risk Prioritization

Studied in combination, the results of frequency and severity assessments help establish priorities for determining action. Any risk that presents a high probability of occurrence and serious consequences deserves immediate action and is considered a high-priority item.

Risk Control Techniques

Once the HSO prioritizes the risks, it is time to apply risk control measures. Several approaches can be taken in risk control, including the following:

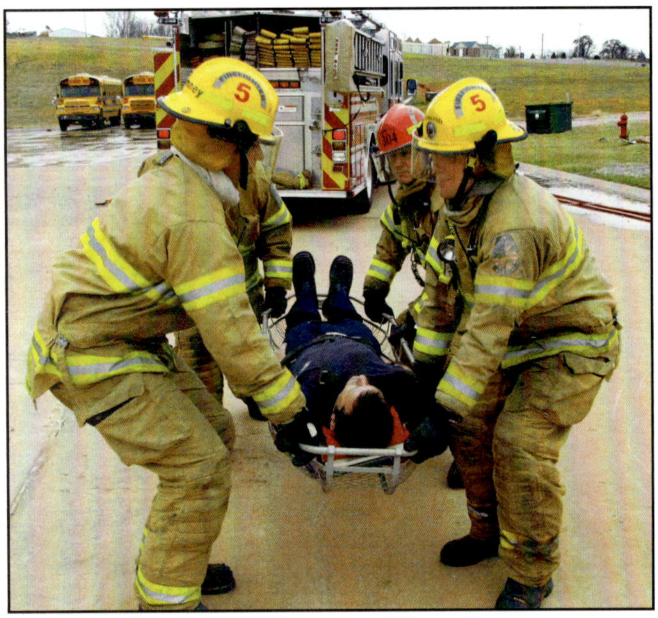

Figure 11.9 Risk avoidance includes training personnel in the use of proper lifting techniques.

- **Risk avoidance:** The best way to control risk is to avoid activities that create it. However, in a fire and emergency services organization, this approach can be impractical. For example:
  - — Lifting a stretcher presents the risk of serious back-injury, but personnel cannot avoid this risk and provide effective service. Training in the use of safe lifting techniques and/or using safer equipment would be a more acceptable solution **(Figure 11.9)**.
  - — Smoking presents the risk of lung cancer and other illnesses. Hiring only nonsmokers could eliminate the risk by ending any exposure to tobacco smoke in the workplace.
- **Risk transfer:** Activities associated with risk can also be outsourced or insured, although these options may be more difficult, if not impossible, for a fire or emergency medical services (EMS) organization. For example:
  - — The cleanup and disposal of hazardous waste could be outsourced, thereby transferring the risk of contamination/exposure to a private contractor who accepts liability for any activity-related damages.

- Purchasing insurance transfers the financial risk of an incident but does nothing to change the likelihood that the incident will occur. For example, buying fire insurance for the station, while highly recommended to protect the assets of the department, does nothing to prevent the station from burning. Insurance is no substitute for effective control measures, such as an automatic sprinkler system.

- **Control measures:** Control measures can reduce the risks and/or mitigate their severity. Effective control measures include safety, health, and wellness programs; ongoing training and education programs; well-vetted equipment and PPE acquisition programs, and well-defined SOPs/Gs). For example:
  - Risks associated with backing apparatus into station bays are well known. However, practical changes in station apparatus, bay design, and backing apparatus have helped to control risks. The simplest solutions include improved driver/operator training, painted guidelines on the apparatus bay floors, and a policy that requires a second person to guide the backing operation from the rear of the vehicle. A more expensive solution would be the replacement of older single-door stations with drive-throughs.
  - Control measures that help minimize injuries at an incident scene include adequate lighting at night operations, accountability systems, protective clothing and gear, mandatory respiratory protection plans, professional training and development sessions, and health and wellness SOP/Gs.

Monitoring Program Effectiveness

Once control measures get implemented, they must be evaluated to measure their effectiveness. Some common forms of monitoring include:

- Periodic review of accident or injury data
- Surveys of members after training or during post-incident critiques
- Inspection of maintenance records or resource management logs for damage and repair requests

Plan Implementation Procedures

Once the analysis process determines appropriate control measures, the plan gets implemented. The HSO or another assigned officer must implement, monitor, evaluate, report, and revise it. For the plan to meet its goals, administrative support in this process is essential.

Implementation

Implementation of the risk management plan requires communication, training, and application. The written plan is distributed to the organization membership as part of the organization's SOPs/Gs.

Distribution includes the public acknowledgement by the administration and the leadership of the jurisdiction of the importance of the plan. Administration support of the risk management plan is essential to membership acceptance and support. Communication of this support helps to ensure a positive response from not only the internal customers, but also the external customers and stakeholders.

Thorough training sessions on the plan involve communicating its importance, use, and intended results. Training occurs throughout the organization and begins with entry-level training for new employees. Sessions for teaching changes in the plan, as well as refreshers, are provided periodically to members of the organization. The effectiveness of the program depends on proper training.

Application of the plan takes place daily, as officers and members follow the prescribed policies and procedures. The risk control techniques must become second nature to all personnel. Whether these techniques involve applying proper lifting techniques when picking up heavy objects or putting on respiratory protection when entering a contaminated atmosphere, they must be performed naturally.

Monitoring

Monitoring shows the plan's effectiveness and ensures that the system is dynamic. It also facilitates periodic reviews of the entire program. Problems that occur during the process have to be addressed. The intent of the risk management plan is to develop a strategy to reduce the inherent risks associated with fire and emergency service operations. Regardless of the size or type of organization, every emergency services organization should operate within the parameters of a risk management plan. The implementation of this plan is a dynamic and aggressive process the HSO must monitor and revise as needed.

Evaluation

Applying the evaluation techniques discussed earlier, the chief or HSO compares the desired results with data from injury and fatality reports, safety training attendance records, fitness testing results, pre-employment physicals, and changes made to address preplan risks. Sources for this information include the target risks or hazards; policies and procedures intended to eliminate the risks; emergency incident reports; daily attendance rosters; medical leave requests; training records; and physical fitness reports.

The results of the comparisons - increase, decrease, or no change in the risks - determine the effectiveness of the risk management plan. For example, if there is a decrease in the risk, indicated by a reduction in medical leave taken or lost-time injuries, then the plan may be considered effective. If, however, the evaluation indicates there is no change or an increase in the number of injuries, then the implemented risk control techniques must be reviewed and alternate solutions applied.

Determine if the cost/benefit is appropriate. The cost of a risk control technique should be less than the cost of potential damages from the risk. For example, if the cost of altering apparatus storage compartments is greater than the cost of lost time due to back strain among employees working in that area, it may not be worth the control cost to alter the compartments.

Revision

Revision procedures get included in the risk management plan. Although the revision process follows the same steps as the initial risk management plan, it focuses only on those risks that need revision and not the entire plan. Fire officers should be aware of the reasons why revisions may be necessary. They should also be able to recognize these reasons when they appear in the plan evaluation. Reasons for revision may include:

- Increase in injuries, fatalities, or loss due to the target risks
- Increase in medical leave requests
- Increase in risk-related costs
- No apparent change in the risk results
- Ineffective cost/benefit
- Changes in the target risks
- Ineffective training

At the beginning of a mandatory physical fitness training program, there may be a temporary increase in on-the-job injuries. That may be due to improper warm-up techniques, peer pressure to keep up with others, or attempts to make the program fail.

Accident, Injury, and Illness Prevention Program

One of the elements of the comprehensive safety and health program is the accident, injury, and illness prevention program. Its purpose is to ensure the protection of members of the fire and emergency services organization by providing information and training regarding the hazards that exist in their work environments **(Figure 11.10)**. These environments include all facilities, apparatus, and emergency incident sites and training exercises. Hazards include those chemicals normally used and stored by the organization, and conditions that present a risk of physical injury, damage to the respiratory system, hearing loss, or other occupational illness.

Although the chief of the fire and emergency services organization is ultimately responsible for developing this program, the responsibility is usually delegated to the HSO and occupational safety and health committee. The HSO and committee members identify the hazards, develop the program, assist in its implementation, monitor its progress, evaluate its effectiveness, and recommend revisions as indicated. The result of their work is a written accident prevention program tailored to the particular operation and hazards that are present in their work place.

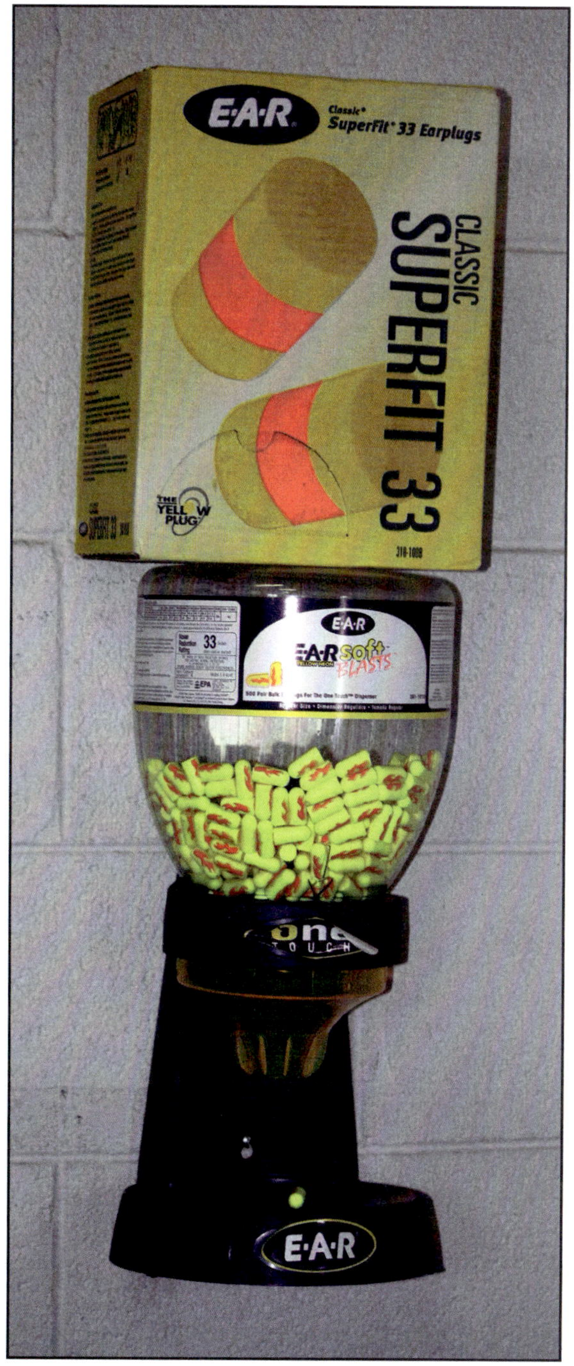

Figure 11.10 A comprehensive safety and health program strives to protect personnel from risks of physical injury, including hearing loss.

Identify Hazards/Corrective Measures

The first step in developing an accident, injury, and illness prevention program is to determine hazards present in the workplace. While hazards associated with fire suppression may be all too familiar to members of the fire service, dangers that exist within facilities may not. The most common hazards can be divided into the following categories:

- Physical injury
- Occupational illness
- Motor vehicle related accidents
- Hearing loss
- Respiratory injury or illness
- Hazardous materials exposure

These hazards may occur at emergency incident scenes, during training, or while working in the organization's facilities. Some hazards may be present in one form or another in each of these situations. Once the hazards have been identified and prioritized, the appropriate corrective measure — policy, procedure, or equipment required to protect personnel — is selected.

Physical Injury

Fire and emergency services personnel are responsible for following safety procedures, adhering to all safety policies, and using safety equipment and clothing. Information is provided to members during the initial entry-level training course and in annual refresher courses. The organization helps to ensure that members are aware of their responsibilities by providing training in the following areas:

- How and when to report injuries, including instruction as to the location of first-aid kits, oxygen inhalators, defibrillators, and decontamination facilities
- How to report unsafe conditions and practices
- Usage, care, selection and maintenance of required PPE, including respiratory, hearing, and eye protection
- Proper actions to take in the event of emergencies, including exit routes from areas during training events
- Description of the organization's comprehensive safety and health program along with the accident, injury, and illness prevention element
- On-the-job review of the practices necessary to perform initial job assignments in a safe manner

Injuries that may occur to fire and emergency service personnel include muscle strains and sprains, traumas, burns, electric shock, and abrasions. Some causes include improper lifting techniques, improper use or lack of protective equipment, and the lack of training. Written policies that may protect personnel from injuries or fatalities include the following:

- Lockout/tagout procedures to prevent unexpected energization, startup, or release of stored energy **(Figure 11.11)**

Figure 11.11 Lockout/tagout procedures help prevent injuries and fatalities from electrical hazards.

- Techniques (for two people or for safe operation of a mechanical devices) to safely lift objects over a predetermined weight
- Appropriate protective clothing when performing certain tasks; may include eye protection, hand protection, hearing protection, respiratory protection, or ice cleats worn on boots or shoes
- Heavy items mounted in lower compartments on apparatus

Based on an analysis of firefighter fatalities data collected over a period of 30 years provided by the U.S. Fire Administration, fatalities could be significantly reduced by emphasizing the following:

- Proactive wellness/fitness programs
- Annual medical fitness evaluations
- Safe driving practices
- Enforcement of seat belt requirements
- Ongoing incident strategy training for fire officers
- Training designed to prevent firefighters from getting lost inside burning buildings

Training reduces fatalities and all types of physical injuries among fire and emergency services personnel. Proper procedures should be established and all personnel trained in their use. Training topics may include:

- Proper lifting techniques
- Lockout/tagout programs
- Use of all types of personal protective clothing and equipment
- Use of first-aid kits mounted in the facilities and on apparatus

Occupational Illnesses

Occupational illnesses are directly related to the duties and tasks of fire and emergency services personnel, including heart attacks, cancer, strokes, respiratory ailments and stress ailments. While the risks for some of these can be limited with training and physical fitness, exposures must be controlled by the use of PPE that is specifically designed to protect wearers from airborne and bloodborne pathogens. Emergency medical responders are particularly susceptible to these types of exposures. Other exposures may occur at hazardous materials incidents or while using certain chemicals for maintenance and cleaning. Furthermore, occupational illnesses may not become apparent for years following exposure. Some personnel do not show symptoms until well into retirement.

Develop Policies. Policies may also be included in the respiratory protection and hazardous materials units of the accident, injury, and illness prevention program and the employee physical fitness and wellness program. Policies that may be initiated to reduce occupational illnesses include the following:

- Give medical examinations and evaluations at the time of employment and annually during service career.
- Report medical exposures within a set period following the suspected exposure.

- Include required information in the occupational illness report, such as **(Figure 11.12)**:
  — Location of the incident
  — Time of the incident
  — Number of fatalities or hospitalized members/employees
  — Contact person
  — Phone number
  — Brief description of incident
  — Type of symptoms or physician's report
- Investigate each occupational illness and report it to the appropriate authorities.
- Keep all records relating to the suspected occupational illness or medical exposure with requirements for privacy and duration of retention.
- Maintain safety data sheets (SDSs).
- Define the types of reportable cases of occupational illnesses.

Provide Training. Training in the established policies must be provided for all personnel. It should include recognition and immediate care of persons with symptoms of occupational illnesses. Additionally, training in the selection and use of personal protective clothing and equipment during emergency medical responses is essential for all personnel who respond to such incidents. All emergency and nonemergency personnel should be trained and certified in cardiopulmonary resuscitation (CPR) techniques in the event of a heart attack in the workplace. Oxygen inhalators should be available in all facilities and training provided in their use.

Motor Vehicle Related Accidents

While the number of firefighters killed in road vehicle crashes has declined in recent years, automobile accidents still pose a safety threat. Volunteer firefighters, who respond to calls in their vehicles, are especially at risk. HSOs should review national and local statistics to determine which type of apparatus may be most susceptible to accidents resulting in injury or death. With this data, they can develop effective driver operator training for their members. By evaluating responses in personal vehicles, HSOs can establish and/or modify driving policies to lower the risk of accidents, injuries and liability.

Develop Policies. NFPA standards, along with state, territorial, provincial, and local laws and ordinances, require the installation and use of seat belts in fire and emergency services apparatus. Organizations can improve safety by adopting the following policies and procedures:

- All persons riding in/on fire and emergency apparatus must be seated and secured to the vehicle by seat belts or safety harnesses while the vehicle is in motion.
- Personnel must be seated while riding in/on an apparatus in motion. Even when traveling at low speeds and/or during public events, riding on tailsteps, tailboards, or any other exposed position (sidesteps or running boards) is strictly prohibited.

City of LaFosse Fire Department
Medical Exposure Report Form

Name:________________________________ Work Phone:________________________________
Age:________ Sex: ❑ Male ❑ Female
Address:________________________________ Home Phone:________________________________
City, State, Zip:________________________________
Company or Assignment: ________________________________
Supervisor: ________________________________

SOURCE OF EXPOSURE

Date of Incident:______________ Time of Incident: ___________am / pm

Incident Location:________________________________

Reference Number to Incident (Dispatch Number):

Type of Incident ________________________________

Name of Source Patient or Individual: ________________________________
Age:________ Sex: ❑ Male ❑ Female
Address:________________________________ Home Phone: ________________________________
City, State, Zip:________________________________
Other Phone:________________________________

Other identification (e.g. operators permit number, vehicle license plates, etc.) ________________
____

Receiving Facility of Source Patient or Individual (e.g., hospital, mortuary, etc.,) ________________
____
Address:________________________________
City, State, Zip:________________________________
Phone: ________________________________

Patient's Attending Physician: ________________________________

Work Phone: ________________________________

Address: ________________________________

City, State, Zip:________________________________

Known Infectious Disease: ________________________________

Describe the Significant Exposure: ________________________________
____

Describe any action taken in response to the exposure to remove the contamination: ______________
____

What Personal Protective Equipment and Procedures were you using at the time of the exposure: ___
____
____

Any other information related to the incident: ________________________________
____

List witnesses to the exposure: ________________________________
____

Signature________________________________ Date________________

Figure 11.12 An Exposure Report Form can provide details of an incident and document any suspected exposure/ contamination.

- Before any apparatus or vehicle operator places a unit into reverse, a spotter must be correctly positioned and identified. In addition, the driver and the spotter must make eye contact. Back-up cameras and radios should also be used to assist in backing, but they are not a replacement for a spotter.
- All fire and emergency apparatus must be brought to a full stop and the parking/air brake set before personnel are allowed to unbuckle and dismount the apparatus.
- Apparatus parked at incidents must be positioned to protect personnel from traffic.
- Wheel chocks must be set whenever an apparatus is parked with the engine running.
- Driver/operator training and certification is required in the operation of specific types of apparatus (includes recertification criteria) **(Figure 11.13)**.

Figure 11.13 Proper driver/operator training should include tests along routes that include or simulate urban and rural driving conditions.

- Daily inspection, maintenance, and repair are required of all vehicles operated by the organization.
- Preventive maintenance programs (including record keeping for each vehicle) are required to record and track potential and/or ongoing problems.

Provide Training. Training for apparatus driver/operators should meet the requirements provided in NFPA 1002, *Standard for Fire Apparatus Driver/Operator Professional Qualifications*. Ambulance driver/operator training programs may be based on the requirements provided by the Commission on Accreditation of Medical Transport Systems (CAMTS) and the Commission on Accreditation of Ambulance Services (CAAS). All personnel should have defensive driving and driver safety training. Model training programs are available from organizations such as the National Safety Council (NSC), the Transportation Safety Institute (TSI), Canada Safety Council (CSC), and OSHA. After reviewing local accident records, training should adjust to address the agency's most common types of accidents.

Hearing Loss

Hearing loss is an ongoing health problem for fire and emergency services personnel. The causes of hearing loss include the following:

- Exposure to apparatus audible warning devices (sirens and air horns)
- Exposure to powered equipment noises
- Continued exposure to constant low-level noises, such as those from air-handling units of engines or apparatus pumps
- Lack of hearing protection
- Viral diseases
- Aging

Hearing conservation plan. As part of the accident, injury, and illness prevention component of the comprehensive safety and health program, a hearing conservation plan should be developed. Plan development begins by identifying the types of noises employees hear and determining what types of noise reduction methods may be used to reduce the effect on employee hearing. The accepted approach is a written hearing conservation plan designed to protect members from exposure to noises that top 85 dB, roughly the volume of an operating food blender.

Employees receive a hearing evaluation when they are hired. Periodic hearing examinations become part of the organization's medical examination requirements. They may be administered upon the member's request or following any exposure to noises in excess of the established minimum level. A hearing examination should also be performed when an employee leaves the organization. A licensed or certified audiologist, otolaryngologist, or other qualified physician or technician certified by the Council for Accreditation for Occupational Hearing Conservation, performs audiometric tests. A technician who performs audiometric tests must be under the direction of an audiologist, otolaryngologist, or other qualified health care professional.

Develop Policies. Policies should be established that require the use of hearing protection when operating equipment or riding on apparatus **(Figure 11.14)**. These usually include earplugs, earmuffs, or two-way communication systems that include earphones. Protective hoods or helmet earflaps can provide some protection at fire-related incidents.

Figure 11.14 Required hearing protection can reduce or eliminate occupation noise-induced hearing loss.

The HSO is required to perform sound-level tests in all facilities and for all types of apparatus and equipment to determine the level of noise present. Hearing protection is then provided, and it is used by all members when exposed to an eight-hour time-weighted average of 85 dB or greater or when exposed to noise in excess of 115 dB from power tools, vehicle engine warm-ups, drafting/ pumping, or other activities (except in situations, such as fire suppression where the use of protective equipment would create an additional hazard to the users).

Provide Training. A final component of the hearing conservation plan is training members on the effects of noise on hearing; the purpose of hearing protection; the selection, use, and care of all hearing protectors; and the purpose of audiometric testing. The training session(s) should be offered or repeated annually for each member of the organization. Information provided in the training session should remain consistent with changes in protective equipment, the work environment, and policies and procedures.

Respiratory Injury or Illness

Members of the fire and emergency services are regularly exposed to hazards that can affect the respiratory system. Sources of these hazards include the following:

- Toxic or hazardous atmospheres found at incidents
- Concentrations of dust, paint, and other particulates produced during cleaning and maintenance
- Airborne microorganisms that may be present during medical emergencies or in the air-handling systems of the organization's facilities

Symptoms may develop rapidly (as in the case of smoke inhalation or oxygen deficiency) or slowly. Each type of respiratory risk requires a different level of personal protection and a different type of equipment. Self-contained breathing apparatus (SCBA) can provide protection against all of these hazards. However, because of their weight and bulk, they are generally used to protect against only the toxic and hazardous atmospheres found at emergency incidents. Dust or particle masks provide the minimum level of protection and can be used in the workshop environment. High-efficiency particulate air (HEPA) filter-equipped masks can be used in the shop environment and in a medical emergency. All masks or respiratory protection must meet the minimum requirements for the employee's environment. NFPA provides standards that define the selection, use, and care of all types of respiratory protection, including those associated with chemical, biological, radiological, nuclear, and/or explosive events.

Respiratory protection plan. The HSO takes responsibility for the development of an organization's respiratory protection plan as outlined by OSHA in Title 29 CFR 1910. A respiratory protection plan requires the following elements:

- Selection criteria for equipment
- Proper-use policy
- Maintenance schedule
- Air-quality testing schedule
- Inspection criteria
- Individual fit testing for employees
- Training session
- Medical certification for those wearing the equipment

Equipment selection criteria can be established by the occupational safety and health committee or another designated body. The HSO should be part of the selection team to ensure safety requirements get met.

Considerations in the selection process are as follows:

- Types of hazards members may be exposed to
- Cost of equipment purchases
- Periodic testing and maintenance requirements
- Training requirements
- Design requirements established by NFPA

Respiratory protection equipment used in fire fighting activities must be third-party certified for structural fire fighting. Particle masks and HEPA-filter masks must meet the minimum requirements set by the NIOSH.

Develop Policies. Construct a written policy covering the proper use of all respiratory protection equipment as required by NFPA 1500 and NFPA 1404, *Standard for Fire Service Respiratory Protection Training*. This policy can be developed by the HSO or occupational safety and health committee. When adopted, it must be communicated to all members of an organization. Like all operating procedures, this policy must be monitored, reviewed periodically, and revised as necessary. Policies must also be established that define the use of all types of respiratory protection. Examples are as follows:

- Respirators will be provided for (and used by) all personnel working in areas where the atmosphere is hazardous; the atmosphere is suspected of being hazardous; or the atmosphere may quickly become hazardous.
- SCBA or supplied-air respirators (SARs) must be worn in all oxygen-deficient atmospheres (less than 19 percent oxygen content).

Fit testing is required of all employees who will use respiratory protection equipment. This testing helps to ensure a full seal of the mask to the face. Testing must be done annually and before any employee is allowed to use the mask in a hazardous atmosphere. The HSO may be assigned this testing activity.

Provide Training. All employees must be trained in the use of respiratory protection equipment. Training includes when to wear the equipment, how to put it on and take it off **(Figure 11.15)**; operation of the equipment; daily inspection procedures; and how to clean; maintain; and dispose of the equipment. This training begins during entry-level training sessions and is provided annually to all members of the organization. The HSO may find it necessary to work with the training officer or staff when providing this training.

Air-quality testing. Air-quality testing responsibility also falls to the HSO. This testing takes two forms: atmospheric testing in all facilities owned or operated by the organization and air-quality testing of the breathing air used in SCBA and SAR systems. Atmospheric testing in all facilities owned or operated by the organization aims to determine the presence of mold, mildew, or other airborne contaminants in the air-handling system of the facility. Apparatus exhaust emission must be tested and appropriate ventilation systems or policies established. Air-quality testing of the breathing air used in SCBA and SAR systems should occur quarterly to determine moisture content

Figure 11.15 All firefighters must be properly trained in donning and doffing SCBA.

and contaminates, regardless of whether the breathing air is generated by air-generating systems operated by the organization or purchased under contract from certified suppliers.

Hazardous Materials Exposure

Fire and emergency service personnel may be exposed to hazardous materials four ways: absorption, inhalation, ingestion and/or penetration. Exposure risks may occur at emergency incidents or when using common materials for cleaning, lubrication, pest control, and/or weed control around facilities. To better understand risks at emergency incidents, the HSO should survey chemical usage within the department and the likelihood of chemical encounters in the agency's jurisdiction. Additionally, chemical usage within the department should be evaluated for the types of PPE required. Often, less toxic agents may be used to reduce risk.

Hazardous communication plan. Protection against common workplace hazards is in the OSHA-required hazardous communication plan. Its aims to ensure that information on the hazards of chemicals used by fire and emergency services personnel is communicated to all members when the products are placed into service. This plan makes members aware of the location and use of hazardous materials in work areas. It also informs them of the requirements of the Hazard Communication Standard (29 CFR 1910.1200), including availability and location of the written hazard communication program, lists of hazardous chemicals, and safety data sheets (SDSs) provided with each product. The SDS relays chemical hazard information from the manufacturer to the user and is maintained at the workplace where the chemical is used.

Develop Policies. Protection against exposures at emergency incidents includes hazard recognition, policies and procedures governing hazardous materials responses, and the use of proper personal protective clothing and equipment. Since structural fire fighting clothing does not protect the wearer from all hazardous materials, it is necessary to establish a procedure that will contain the incident until trained, protected personnel arrive. Exposure protection around the station, offices and maintenance facilities may require the placement of wash stations, warning signs, eye and ear protection supplies, and laundry procedures for clothing and shoes.

Training. Training topics include the following:

- Protective equipment to wear when using specific products
- Product-labeling system
- Obtaining and using SDSs
- Physical and health hazards of chemicals
- Recognizing, avoiding, and preventing accidental exposure to hazardous chemicals in the work environment

Work place protection. Protection in the workplace includes the following items:

- Ventilation systems
- NFPA 704, *Standard System for the Identification of the Hazards of Materials for Emergency Response*, placards on storage rooms
- Spill protection and controlled drainage

- Fire protection systems
- Alarm notification systems
- First aid kits with appropriate ointments or medications for treatment of injuries

Rapid Intervention Teams (RIT)

Rapid intervention teams (RITs), also known as *rapid intervention crews (RICs)*, or *firefighter assistance and search teams (FASTs)*, add a layer of personnel safety at emergency operations for firefighters who get lost or injured An RIT is mandated by OSHA in the U.S. and the Workers' Compensation Boards in Canada. RITs are required as an element of any command structure for an incident where the requirement applies. The incident commander is responsible for assigning this duty and managing the assets.

To have an effective rescue team, it is necessary to equip and train RITs **(Figure 11.16)**. Command staff and firefighters need to practice RIT operations under a variety of conditions. Complex interiors and collapsed structures are some of the most difficult environments. RIT deployments at scenes can tax even the best-staffed departments, and it takes time and coordination to implement an organized RIT response.

Figure 11.16 RITs/RICs must be equipped and trained to operate under a variety of conditions.

Implement Program

Reducing accidents and occupational illnesses, injuries, and fatalities is the responsibility of all fire and emergency service members. Personal safety must be adhered to at every moment. Implementation of the accident, injury, and illness prevention program includes communication and training. Employees learn the requirements of the program and policies that mandate it. They also receive training in the proper use of equipment and safe procedures. New employees receive a safety orientation before beginning work. An introduction to the accident, injury, and illness prevention program begins on the first day on the job for all employees, including rehires and transfers.

An introduction to the organization's policies and rules includes a thorough safety briefing. A tour of appropriate facilities serves to acquaint employees with their areas of responsibility and the locations of safety-related equipment. A session to describe the fire and emergency services accident, injury, and illness prevention program should address these and other topics:

- How and when to report injuries, including where to find first aid facilities
- How to report unsafe conditions and practices
- Use and care of required PPE
- Proper actions to take in the event of emergencies, including emergency exit routes
- How to identify hazardous gases, chemicals, or materials, along with instructions for safe use and/or emergency action following accidental exposure
- Description of the organization's safety and health program
- On-the-job review of the practices necessary to safely perform initial job assignments
- Procedures to follow in the case of exposure to bloodborne pathogens or any contagious respiratory illnesses
- Proper procedures for lifting objects and back care
- Proper handling and use of power tools
- Rules and regulations regarding the use of vehicles in all situations
- Approved safety procedures involving trenching/shoring and confined-space rescue incidents
- Safety procedures designed to provide fall protection
- Use of respiratory equipment and the type of systems used
- Proper care, use, selection, and maintenance of PPE along with its limitations

The HSO should also begin noise-level testing of all apparatus, equipment, and facilities, as well as quality testing of air and water in all facilities. Breathing/air quality testing should be contracted to a certified third-party testing organization. Facility inspections for safety hazards should be implemented monthly and should include (but not limited to) tools, apparatus, extinguishers, protective equipment, and life-safety equipment.

The HSO (in coordination with the appropriate chief officer) should establish hose, apparatus, and equipment testing criteria based on NFPA requirements. Third-party testing agencies such as Underwriters Laboratories, Underwriters Laboratories of Canada, and FM Global usually provide annual certification tests. The department's insurance carrier may perform safety inspections and should coordinate them with the HSO. In addition, the worker's compensation carrier may perform an audit of recent injuries and recommend programs, reporting, and/or training options.

Monitor, Evaluate, and Revise Program

The accident, injury, and illness prevention program is continually monitored, evaluated, and revised. Like other programs implemented by the fire and emergency services organization, the results of the program are compared with the goals, and deviations are corrected. Revisions are dictated by:

- The identification of new safety hazards, such as the use of biological agents by terrorists
- The need for new or updated equipment to combat these hazards, such as the need for greater respiratory protection
- The need for updated safety procedures to ensure safety before, during, and after an incident, such as entering a structure containing smoke without proper respiratory protection

Fire officers and the HSO must monitor the respiratory protection plan to ensure it meets the changing challenges of the workplace. Organizations like NFPA, NIOSH, the International Association of Fire Fighters, the International Association of Fire Chiefs, and the CDC provide information on new respiratory protection equipment, requirements, and procedures, as well as updates on the rapidly expanding list of hazards that can damage the respiratory system. Fire officers must actively pursue this information and provide it to the organization administration and membership.

Medical Exposure Management Program

Some occupational illnesses may result from medical exposures to individuals with communicable diseases or to situations involving hazardous materials. To control and manage this threat to the safety and health of organization members, a medical exposure management program must be established. The HSO, together with the occupational safety and health committee and the organization's physician, is charged with developing and managing this program.

Infection Control

A written infection exposure control plan clearly explains the intent, benefits, and purposes of the program. The plan must cover these standards of exposure control:

- Education and training requirements
- Vaccination requirements for potential threats
- Reporting, documentation and record-keeping requirements
- Cleaning, decontamination, and disinfection of personnel and equipment **(Figure 11.17)**
- Exposure control protocols

Figure 11.17 Fire stations should have designated disinfecting areas that have appropriate cleaning agents, rinsing sinks, and drainage systems.

An infection exposure control officer ensures that an adequate infection control plan is developed and that all members are trained and supervised in the plan. The HSO or another qualified officer may fill this position. Departments may ask local hospital ERs to test patients who may have exposed firefighters to infection. Coordination and legal approvals may be necessary prior to obtaining samples from patients.

Exposure reports require lengthy retention of employee medical records. Some symptoms of disease may not appear until years after initial exposure. Consequently, OSHA 1910.1020 mandates these records be "preserved and maintained at least for the duration of employment plus thirty years." These records may only be released to the member or a designated representative. They may also be used for training in the proper use of PPE, exposure protection, post-exposure protocols, and transmission of infectious diseases. The HSO and safety and health committee members should annually review the infectious disease plan, updates, protocols, and equipment. Policies should be developed to specify the types of protection that prevent exposure.

Hazardous Materials Exposure Control

As the number of medical calls increases, incidents involving exposure to bloodborne pathogens and other infectious diseases also rise. Fire and emergency services personnel are regularly exposed to hazardous materials during emergency operations. The administrative tasks involved with hazardous materials safety are assigned to the HSO. The operational duties at emergency scenes are usually assigned to a hazardous materials technician. In either case, the skills necessary to perform these tasks are outlined in NFPA 472, *Standard for Competence of Responders to Hazardous Materials/Weapons of Mass Destruction Incidents*.

The HSO develops exposure protocols and implements them through the organization's SOPs/Gs. As with all other protocols, they must be reviewed and updated periodically and made a part of the organization's training sessions. During a known hazardous materials incident, the hazmat branch safety officer monitors potential hazards to emergency personnel and advises the incident commander. According to OSHA's CFR 1910.120(c)(7), the following risks must be considered during a hazardous materials incident:

- Exposures exceeding the permissible limits and published exposure levels
- Immediately dangerous to life and health (IDLH) concentrations
- Potential skin absorption and irritation sources
- Potential eye irritation sources
- Explosion sensitivity and flammability ranges
- Oxygen deficiency

Should the presence of hazardous materials become known following an incident, Title 29 *CFR* 1910.1200 requires a fire and emergency services organization provide affected employees with all data on the materials to which they were exposed. Record keeping for hazardous materials exposure incidents matches the guidelines for tracking infectious diseases.

Investigation/Review Process

NFPA 1500, 4.4 outlines the responsibilities of the fire department for developing an accident investigation procedure. It includes investigating and reviewing "all accidents, near misses, injuries, fatalities, occupational illnesses and exposures involving members." The HSO ensures a policy or procedure is developed and implemented in the organization. Investigative and review processes differ from Quality Control processes in focus. Still, both types of programs strive to increase the organization's safety and reliability.

Quality Control

The objective of a quality control (or quality assurance) program is to review types of work production and determine if procedures or protocols have been followed correctly. This program strives to maintain or improve the quality of service offered by the agency. Employee performance is measured by reviewing several critical factors to overall performance. To avoid supervisor bias against any employee, specific types of production are reviewed or random samplings are made. Although the quality of work must be reviewed after an incident has occurred, a quality control program is usually considered a proactive approach to organizational performance and reliability.

Investigations/Reviews (I/Rs)

The objectives of the I/R are to investigate an incident after it has occurred; to learn why those who were involved thought their choices were appropriate under the circumstances; to better understand the vulnerabilities of the organization; and to implement improvements. Accidents, injuries, occupational illnesses, exposures, and near misses must be investigated or reviewed according to NFPA 1500. Policies should describe who is responsible for conducting the I/R at each level of each event.

As I/Rs are necessarily reactive, participants often do not trust the person(s) conducting the I/R. Members may worry that these reports will result in punitive measures or a loss of credibility among peers. To avoid this perception, organizations can choose whether to call the discovery phase an investigation or a review. Both terms are used interchangeably in this chapter.

Errors

It can be difficult to define an error. For some, an error represents any deviation from operational protocols. Others believe errors are defined by an outcome. Consequently, the annual or quarterly measurement of error in quality control programs and I/Rs can be misleading. For fire departments, the result may create a false sense of security or alarm. A low error rate does not always mean that chances of experiencing a disastrous event have diminished by a similar percentage. Accidents, injuries or near misses may have low rates of error for a variety of reasons, including poor supervisor support for documentation, low reporting rates due to fear, harsh policies for employee reporting, and the failure to identify what should have been reported.

The HSO needs to recognize that the identification and reporting of errors is one small component of the overall organizational safety. A department should not rely on error rates to summarize a department's overall safety culture or climate. Each accident, injury, or near miss should be evaluated individually and scrutinized for the type or level of vulnerability it may represent.

Accidents

Accidents may result in injury, death, or property damage. The organization must develop and implement an accident investigation/review policy and procedure.

When an accident occurs, an I/R is conducted to determine the root cause, source, or origin. Accident I/Rs should be objective, impartial, and directed toward fact-finding (not fault-finding). These investigations aim to identify and document the following:

- Thought process behind the decision(s) that were made
- Behavior or condition that resulted in the accident (root cause)
- Previously unrecognized hazards
- Defects or design flaws in the apparatus or equipment **(Figure 11.18)**
- Additional training needs
- Improvements needed in safety policies and procedures
- Facts that could have a legal impact on the accident or case
- Historical trends

Figure 11.18 Accidents involving fire apparatus may be the result of equipment defects, bad behavior, or environmental conditions.

Accident Scene Control

When the incident involves serious personal injuries, the victim must first be stabilized and transported to a health care facility for medical treatment. Minor injuries may be treated at the scene. Next, the officer in charge of the incident ensures that the accident scene is secured and that all equipment, clothing, or vehicles are left in place.

If conditions require that vehicles be moved, an attempt should be made to photograph or sketch the location of each vehicle. The officer in charge must remember that anything involved in the incident is evidence for:

1. Reconstructing the accident and preventing future accidents
2. Supporting potential legal cases that may result

Accurate Data Collection

In conducting an I/R, officers collect basic information about the participants, event, or incident. The accident statistics used to analyze accidents and determine trends are only as accurate as the original information gathered by investigating/reviewing officers. All personnel must follow the prescribed procedures and properly complete accident I/R forms provided by the jurisdiction.

In a simplified data collection form, I/R officers fill in the blanks to complete a narrative account of the incident. Other forms are based on the NFIRS and require the use of predesignated codes to describe an incident. In either case, the form should be designed to provide the following:

- **General information**
  - Date and time of incident
  - Type of incident, illness, injury, or fatality
  - Location and emergency response type
  - Names of witnesses and their accounts of the situation
- **Employee characteristics (participant)**
  - Name and unit assignment (company/shift)
  - Age and gender
  - Rank/function
  - Personal protective clothing or equipment in use
- **Environmental information**
  - Weather and temperature
  - Day or night conditions
  - Noise and visibility
  - Terrain
- **Apparatus/equipment information**
  - Type of equipment involved
  - Age and condition
  - Location
  - Maintenance history
  - Distinguishing characteristics

A narrative description of the incident is the final portion of an investigation report. This narrative includes observations on the part of the officer; eyewitness reports; participant interviews; and information from additional sources, such as law enforcement reports and dispatch information.

Data gathered by the I/R officer is transmitted through the system to the HSO. The HSO should solicit the help of the Safety and Health Committee to analyze the reported event. Once the committee has analyzed the report, the division(s) most involved with the accident should use this analysis to help determine what the organization will do to inform all members and make necessary changes to prevent the incident from happening again. Once the implementation of this last phase is complete, the report should be placed in a database that will help give direction to the organization in the future.

Injuries, Illnesses, and Exposures

When a member of the unit experiences a job-related injury, illness, or exposure, the first duty of a fire officer is to ensure prompt medical treatment for the individual. When the injury, illness, or exposure is serious, the individual must be transported to the closest available medical facility. Once the officer has provided for the treatment of the member, the I/R process begins.

Information for I/Rs of job-related injuries, illnesses, or exposures is gathered in much the same way as information for accident I/Rs. Depending on the nature of the incident, some differences may exist. For instance, an I/R into a claim involving a possible job-related illness may take more time and require the services of a specialty organization, such as one that samples air or water. Although this type of I/R may occur after the actual event, the I/R officer must focus on the idea that information gained from the inquiry may lead to the discovery of an environmental cause for the illness.

An exposure I/R may occur immediately following the incident or at some point in the future when symptoms of exposure appear. Accurate and thorough documentation of all incidents and exposures is basic to any current or future I/R.

Near Miss

A near miss is an unplanned event that had the potential to cause injury, illness, or damage — but did not. According to the National Safety Council website, the root cause for a near miss is almost always a faulty process or managing system. For that reason, as in the case of accidents, injuries, illnesses, and exposures, near-miss events should require an I/R that seeks to understand what occurred and what may have been contributing factors. This process assumes members understand and can identify a near miss and that they trust the I/R process to maintain confidentiality and integrity of purpose. If firefighters perceive their reputations are at risk or suspect the organization will single them out as undertrained, they will be reluctant to report a near miss.

To reassure members that the near-miss I/R process exists solely to improve organizational performance and reliability, all dates, unit identifiers, addresses, ranks, gender or incident identifiers must be scrubbed from the report. As I/Rs are gathered, the department can use near-miss events as stories to change the attitudes and beliefs of their members and, eventually, to change their practices.

Chapter Review

1. What safety and health clauses may be included in fire service labor/management agreements?
2. What are the job performance requirements (JPRs) of a health and safety officer (HSO) in a safety and health program??
3. How do the elements of the classic risk management model inform a safety and health program?
4. How do the HSO and the occupational safety and health committee identify hazards and apply corrective measures to these hazards?
5. What could dictate a revision of a safety program?

6. What are the hazard control plans for medical, exposure, and hazardous material incidents?
7. What are the objectives for the investigation/review process?

Discussion Questions

1. Why would gathering information from sources inside and outside the fire service be beneficial to HSOs or committees creating an occupational safety and health program?

Appendices

Appendix A
Chapter and Page Correlation to FESHE Learning Objectives

FESHE Learning Objectives	Chapter References	Page References
1. Describe the history of occupational health and safety.	2	54, 55
2. Identify occupational health and safety programs for industry and emergency services today.	2,4	56, 57, 115-118, 136-155
3. Compare the difference between standards and regulations.	11	360
4. List and describe the components of risk identification, risk evaluation, and incident management.	1, 2, 3, 4, 7, 9, 11	21-27, 57-58, 64, 83, 230, 312, 360, 365, 367-368, 370-375
5. Describe the relevance for safety in the workplace, including the importance of PPE.	5, 7, 8, 10	167, 221-222, 225, 229, 237, 240, 243-246, 251-280, 341, 346
6. Apply the knowledge of an effective safety plan to preincident planning, response, and training activities.	1, 4, 6, 8, 9	13, 36, 89, 103, 201, 206, 214, 278, 311, 316, 318-320, 331
7. Explain the components of an accountability system in emergency service operations.	11	288, 290, 291, 317
8. Discuss the need for and the process used for post-incident analysis.	9	330-333
9. Describe the components and value of critical incident management programs.	9	330
10. Describe the responsibilities of individual responders, supervisors, safety officers, and incident commanders, safety program managers, safety committees and fire department managers as they relate to health and safety programs.	3, 10	86, 99, 340
11. Describe the components of a wellness/fitness plan.	3, 4	86, 136-156
12. Identify and analyze the major causes involved in line-of-duty firefighter deaths related to health, wellness, fitness and vehicle operations.	1, 2, 4, 10	45, 59, 66, 75, 115, 122, 129, 136, 143, 156, 180, 204, 285-286

Appendix B
EVERYONE GOES HOME: Firefighter Life Safety Initiatives Program

1. Define and advocate the need for a cultural change within the fire service relating to safety; incorporating leadership, management, supervision, accountability and personal responsibility.
2. Enhance the personal and organizational accountability for health and safety throughout the fire service.
3. Focus greater attention on the integration of risk management with incident management at all levels, including strategic, tactical, and planning responsibilities.
4. All firefighters must be empowered to stop unsafe practices.
5. Develop and implement national standards for training, qualifications, and certification (including regular recertification) that are equally applicable to all firefighters based on the duties they are expected to perform.
6. Develop and implement national medical and physical fitness standards that are equally applicable to all firefighters, based on the duties they are expected to perform.
7. Create a national research agenda and data collection system that relates to the initiatives.
8. Utilize available technology wherever it can produce higher levels of health and safety.
9. Thoroughly investigate all firefighter fatalities, injuries, and near misses.
10. Grant programs should support the implementation of safe practices and/or mandate safe practices as an eligibility requirement.
11. National standards for emergency response policies and procedures should be developed and championed.
12. National protocols for response to violent incidents should be developed and championed.
13. Firefighters and their families must have access to counseling and psychological support.
14. Public education must receive more resources and be championed as a critical fire and life safety program.
15. Advocacy must be strengthened for the enforcement of codes and the installation of home fire sprinklers.
16. Safety must be a primary consideration in the design of apparatus and equipment.

Appendix C
The IAFF/IAFC Fire Service Joint Labor Management Candidate Physical Ability Test Program Summary

The IAFF/IAFC Wellness-Fitness Task Force

The IAFF and IAFC have teamed up with 10 of North America's leading fire departments and unions through the Fire Service Joint Labor Management Wellness/Fitness Initiative to develop the IAFF/IAFC Candidate Physical Ability Test (CPAT). The participating departments on the Task Force are:

- Austin, TX / IAFF Local 975
- Los Angeles Co., CA / IAFF Local 1014
- Calgary, ALB / IAFF Local 255
- Metro Dade Co., FL / IAFF Local 1403
- Charlotte, NC / IAFF Local 660
- New York City, NY / IAFF Local 94/854
- Fairfax Co., VA / IAFF Local 2068
- Phoenix, AZ / IAFF Local 493
- Indianapolis, IN / IAFF Local 416
- Seattle, WA / IAFF Local 27

The Task Force successfully developed the Fire Service Joint Labor-Management Wellness-Fitness Initiative in 1997 to address the need for a holistic and non-punitive approach to wellness and fitness in the fire service. The Task Force then discovered that municipalities were hiring people who would not be physically capable of a successful career in the fire service. The Task Force, therefore, unanimously agreed to develop a physical ability test for pre-employment testing of candidates. The Task Force has developed and validated the Candidate Physical Ability Test (CPAT), resulting in a consistent test for hiring of candidate fire fighters.

Developing the CPAT

The Task Force directed the Technical Committee to develop a performance test for the ten jurisdictions that measures the critical skills of fire fighter candidates. Committee members reviewed six of the ten jurisdictions job analysis and job task surveys. They also reviewed each of the ten jurisdiction's current candidate performance tests and job descriptions.

Additionally, a complete equipment and demographics survey was also completed by the ten jurisdictions. The equipment survey provided types and weights of all fire fighter protective clothing, protective equipment, fire department equipment and fire department tools. It also assessed the average weight of fire fighters in each jurisdiction as well as the average weight of emergency room and hospital admitted patients. Demographic profiles of each jurisdiction were also developed.

Using the data collected, the Committee then derived a list of 31 tasks to investigate. They developed survey questions directly related to these 31 tasks. The critical and physical task skills that all fire fighters should possess were validated (through questionnaires) by 1,000 fire fighters from the ten departments. Selection of the 1,000 fire fighters was random and anonymous, but consistent with the race and gender diversity of the specific department. The completed results of the surveys resulted in the development a series of physical ability tests that were selected based on the results of the data indicating the physicality and criticality of the tasks performed by fire fighters.

Props for each event were built and modified to obtain the necessary information regarding candidate's ability. The props were then placed in a sequence that would best simulate their use in a fire scene. The Technical Committee chose to move the candidates along a predetermined path from event to event in a continuous manner requiring a mandatory walk between events.

Because of the test set up, individual event times would be an ineffective measure of a candidate's ability, so an overall test time needs to be established. Numerous individuals were run through the test including: technical staff members, incumbent fire fighters, lay people and a recruit class.

The overall consensus found the test to be a good predictor of an applicant's ability to perform basic fire fighting tasks. The Technical Committee was confident that the ability test would provide the fire service with a physically competent recruit. There was also unanimous agreement that the test equaled or was superior to current tests run by each of the ten jurisdictions.

These tests were filmed at different speeds and presented to supervisors in the 10 jurisdictions (captains and battalion chiefs with experience in fire fighter training). At the completion of this process an entry test was finalized and presented to the Task Force for adoption. The Task Force unanimously adopted the full Candidate Physical Ability Test with a pass/fail time of 10 minutes and 20 seconds. The test may only be administered on a pass/fail basis, municipalities may not rank candidates based upon CPAT completion times.

Administering the CPAT

In developing the CPAT Program, the Task Force recognized that the fire department should reflect the community it serves. In today's society, communities are increasingly diverse and fire fighters are continually challenged to operate in multi-cultural environments. The goal of the CPAT is to test for those individuals physically qualified to perform the job of fire fighter without separating from the department's broader goal of attaining a properly trained and physically capable workforce whose members reflect the diversity of the community. Diversity should never come by lowering validated entry standards. Rather, it should come from actively recruiting qualified men and women candidates from all racial and ethnic backgrounds for careers in the fire service. It is the position of the IAFF/IAFC Joint Labor Management Wellness-Fitness Initiative Task Force that fire departments should increase the diversity of their workforce by actively recruiting candidates from throughout their communities rather than lowering candidate physical ability standards. The CPAT Program details avenues fire departments can pursue to raise awareness of job opportunities and to recruit and mentor qualified candidates within the fire service before the test is administered.

The CPAT goal was to develop a fair and valid evaluation tool in the selection of fire fighters to ensure that all fire fighter candidates possess the physical ability to complete critical tasks effectively and safely. This CPAT Program covers every aspect of administering the CPAT: from recruiting and mentoring programs, to providing recruits with fitness guidance to help prepare them for the CPAT, to setting up and administering the test. The entire validation process is discussed in detail, as well as the legal issues that departments might face when implementing the actual program.

After a municipality has completed all aspects of recruiting and mentoring candidates they may administer the actual eight-event test. During the entire test the candidate must wear a 50 lb. weighted vest (simulating the weight of a fire fighters protective clothing and equipment). The eight events are:

- Stair Climb (climbing stairs while carrying an additional 25 lb. simulated hose pack)
- Ladder Raise and Extension (placing a ground ladder at the fire scene and extending the ladder to the roof or a window)
- Hose Drag (stretching uncharged hoselines, advancing lines)
- Equipment Carry (removing and carrying equipment from fire apparatus to fireground)
- Forcible Entry (penetrating a locked door, breaching a wall)
- Search (crawling through dark unpredictable areas to search for victims)
- Rescue Drag (removing victim or partner from a fire building)
- Ceiling Pull (locating fire and checking for fire extension)

Appendix D

Sample Facility Inspection Checklist

Facility: ______________________ **Time:** ____________ **Date:** ____________

Compliance with each subject MUST be 100% or attach documentation of corrective action taken.

☐ **ITEM OK** ☐ **ITEM HAS PROBLEM**

1. **Walking & Working Surfaces**
   - A. Aisles
   - B. Stairways
   - C. Catwalks
   - D. Guard Railings
   - E. Work Platforms
2. **Compressed Air & Gas**
   - A. Storage & Restraint
   - B. Handling
   - C. Labeling
   - D. Proper Use
3. **Mechanical Safeguards**
   - A. Point of Operation
   - B. Drive Train
   - C. Operating Controls
   - D. Emergency Stops
   - E. Start Up Alarms
4. **General Environment**
   - A. Noise
   - B. Lighting
   - C. Ventilation
5. **Material Handling Operations**
   - A. Conveyors
   - B. Crane
   - C. Fork Truck
   - D. Slings & Rigging
   - E. Attachments
6. **Housekeeping & Sanitation**
   - A. Work Areas
   - B. Rest Rooms
   - C. Locker Rooms
   - D. Break Areas
   - E. Waste Disposal
   - F. Leakage & Spillage
   - G. Proper Storage of Materials
7. **Material Storage & Stacking**
   - A. Damaged Containers
   - B. Improperly Stacked
   - C. Distance from Sprinklers
   - D. Blind Corners
   - E. Blocked Aisles
   - F. Projecting Materials
   - G. Safe Heights
   - H. Floor Load Limits
   - I. Proper Coil Steel Storage
   - J. Proper Blank Steel Storage
8. **Personal Protective Equipment**
   - A. Side Shields
   - B. Face Protection
   - C. Hearing Protection Needed
   - D. Respirators Required
   - E. Hard Hats Required
   - F. Protective Clothing
9. **Fire & Emergency Equipment**
   - A. Alarms
   - B. Sprinklers
   - C. Extinguishers
   - D. Fire Hoses
   - E. Eye Washes & Showers
   - F. Stretchers
   - G. Emergency Lighting
10. **Fall Hazard Control**
    - A. Anchorage Points
    - B. Fall Protection Equipment
    - C. Ladders
    - D. Scaffolds
    - E. Manlifts
11. **Means of Egress**
    - A. Exits Marked
    - B. Clear & Unobstructed
    - C. No Locked Exits
    - D. Posted Evacuation Plan
    - E. Posted Shelter Areas
12. **Electrical Safeguarding**
    - A. Covers for Switches
    - B. Covers for Outlets
    - C. Doors on Panel Boxes
    - D. Disconnects can Lockout
    - E. Proper Labeling
    - F. Posted Lockout Procedure
    - G. Identify High Voltage
    - H. Extension & Fixture Cords
    - I. Temporary Hookups
13. **Hazardous Materials**
    - A. Storage
    - B. Handling
    - C. Personal Protective Equipment
    - D. Labeling
    - E. Warning Signs
    - F. All Chemicals Approved by HMCC
    - G. Safe Use Instructions Reviewed with Employees
    - H. AC240 Completed for New or Transferred Employees
14. **Cranes & Hoisting Equipment**
    - A. Daily Inspection prior to use
    - B. Inspection tag attached to every hoist
    - C. Authorized/Trained user
    - D. Preventive Maintenance Performed
15. **Hand & Portable Power Tools**
    - A. Ergonomics
    - B. Carrying & Handling
    - C. Damaged Tools
    - D. Guards in Place
    - E. Special Tools
    - F. Grounding
16. **Industrial Vehicles**
    - A. Operation of Trucks
    - B. Railroad Cars
    - C. Utility Vehicles
    - D. Manlifts
    - E. Operation of Heavy Equipment
17. **Safe Working Practices**
    - A. Adherence to Safety Rules and Regulations
    - B. Adherence to Lockout Procedure
    - C. Work Habits
    - D. Unsafe Acts
    - E. Corrective Instruction or Training of Employees

Based on form developed by the National Safety Council

Appendix E
Safety Committee Checklist

The following checklist represents the minimum requirements for a Safety Committee in most states:

_____ Form a safety committee.
_____ Committee shall be composed of an equal number of employer and employee representatives (number of employees may be greater).

Committee members will be provided with:
_____ Purpose and Operation of the Committee
_____ Access to Applicable State and Federal Rules
_____ Methods of Conducting Meetings
______Staff to Record and Distribute Agendas and Minutes

Safety Committee Operations:
_____ Elect a chairperson.
_____ Have a written agenda for conducting committee business during meetings.
_____ Hold monthly meetings.
_____ Take minutes at each meeting.
_____ Maintain minutes for three years.
_____ Post copy of minutes of each meeting and send copy to committee members.
_____ Establish reasonable time limit for management response.
_____ Establish hazard reporting system.
_____ Inform employees of their rights to report workplace hazards.
_____ Evaluate employer accountability system.
_____ Establish procedures of investigating accidents/incidents.
_____ Provide access to safety/health codes.

Inspections:
_____ Establish procedures for inspections.
_____ Conduct quarterly inspection of satellite locations
_____ Train representative in hazard identification
_____ Committee shall review inspection reports
______Establish methods for correction and follow-up

Train committee members regarding:
_____Duties and Responsibilities
_____Hazard Identification and Correction.
_____ Principles of Accident Investigation.

Glossary

Glossary

A

Airborne Pathogens — Disease-causing microorganisms (viruses, bacteria, or fungi) that are suspended in the air.

All-Hazard Concept — Provides a coordinated approach to a variety of incidents; all responders use a similar, coordinated approach with a common set of authorities, protections, and resources.

B

Braking Distance — The distance the vehicle travels from the time the brakes are applied until the apparatus comes to a complete stop.

C

Cardiovascular disease (CVD) — A family of diseases that includes hypertension (high blood pressure), atherosclerosis, cardiovascular heart disease, and stroke.

Cheater Bar — Piece of pipe added to a prying tool to lengthen the handle and provide additional leverage.

Chemical Protective Clothing (CPC) — Clothing designed to shield or isolate individuals from the chemical, physical, and biological hazards that may be encountered during operations involving hazardous materials.

Collapse Zone — Area beneath a wall in which the wall is likely to land if it loses structural integrity.

Communication Model —A representation of the process that people use when interacting with others on a daily basis.

Community-Risk Reduction — The process of addressing preventable injury occurring in a community. It involves identifying leading risks and creating mitigation strategies through the use of integrated prevention interventions.

Control Zones — System of barriers surrounding designated areas at emergency scenes intended to limit the number of persons exposed to the hazard, and to facilitate its mitigation. At a major incident there will be three zones — restricted (hot), limited access (warm), and support (cold).

D

Driver Reaction Distance — The distance a vehicle travels while a driver is transferring the foot from the accelerator to the brake pedal after perceiving the need for stopping.

E

Ergonomics — The applied science of workplace equipment design intended to maximize productivity and minimize potential injuries by reducing operator fatigue and discomfort.

Escape Rope - A single-purpose, emergency self-escape (self-rescue) rope; not classified as a life safety rope.

F

Freelance — Operating independently of the Incident Commander's command and control.

H

Hazard — Condition, substance, or device that can directly cause injury or loss; the source of a risk.

Health and Safety Officer (HSO) — Officer authorized to manage the organization's health and safety program; performs the duties, functions, and responsibilities described in NFPA 1521, *Standard for Fire Department Safety Officer Professional Qualifications.* Must meet the qualifications of this standard or an approved equivalent.

Health Program — Any program designed to assist and encourage employees to adopt and attain a healthier lifestyle.

I

Immediately Dangerous to Life and Health (IDLH) — Description of any atmosphere that poses an immediate hazard to life or produces immediate, irreversible, and debilitating effects on health; represents concentrations above which respiratory protection should be required. Expressed in parts per million (ppm) or milligrams per cubic meter (mg/m3); companion measurement to the permissible exposure limit (PEL).

Incident Action Plan (IAP) — Written or unwritten plan for the disposition of an incident; contains the overall strategic goals, tactical objectives, and support requirements for a given operational period during an incident. All incidents require an action plan.

Incident Command System (ICS) — Standardized approach to incident management that facilitates interaction between cooperating agencies; adaptable to incidents of any size or type.

Incident Commander (IC) — Person in charge of the incident command system and responsible for the management of all incident operations during an emergency.

Intrinsically Safe — Describes equipment that is approved for use in flammable atmospheres; must be incapable of releasing enough electrical energy to ignite the flammable atmosphere.

L

Level A Protection — Highest level of skin, respiratory, and eye protection that can be afforded by personal protective equipment. Consists of positive-pressure, self-contained breathing apparatus, totally encapsulating chemical-protective suit, inner and outer gloves, and chemical-resistant boots.

Level B Protection — Personal protective equipment that affords the highest level of respiratory protection but a lesser level of skin protection. Consists of positive-pressure self-contained breathing apparatus, hooded chemical-resistant suit, inner and outer gloves, and chemical-resistant boots.

Level C Protection — Personal protective equipment that affords a lesser level of respiratory and skin protection than Levels A or B. Consists of full-face or half-mask APR, hooded chemical-resistant suit, inner and outer gloves, and chemical-resistant boots.

Level D Protection — Personal protective equipment that affords the lowest level of respiratory and skin protection. Consists of coveralls, gloves, and chemical-resistant boots or shoes.

Life Safety Rope — Rope dedicated to supporting people during rescue, fire fighting, other emergency operations, or training evolutions.

Line-of-Duty Death (LODD) — Firefighter or emergency responder death resulting from the performance of fire department duties.

M

Mayday — Internationally recognized distress signal.

Mitigate —To cause to become less harsh or hostile; to make less severe, intense or painful; to alleviate.

N

Narcotic — Any class of substances or drugs that numbs, paralyzes, or causes loss of feeling in the body. Narcotics may be addictive, alter moods or behavior, and induce sleep.

National Fire Protection Association (NFPA) — U.S. nonprofit educational and technical association devoted to protecting life and property from fire by developing fire protection standards and educating the public.

National Incident Management System — Incident Command System (NIMS-ICS) — The U.S. mandated incident management system that defines the roles, responsibilities, and standard operating procedures used to manage emergency operations; creates a unified incident response structure for federal, state, and local governments.

Normalization of Deviance — State of a safety culture in which acting against SOPs becomes normal behavior.

O

Occupational Safety and Health Administration (OSHA) — U.S. federal agency that develops and enforces standards and regulations for occupational safety in the workplace.

On-the-Job Training — System of training firefighters that makes full use of personal contact between firefighters and their immediate supervisor; trains firefighters physically and psychologically for the position they will perform.

P

Particulates — Very small particle of solid material, such as dust, that is suspended in the atmosphere.

Personal Alert Safety System (PASS) — Electronic lack-of-motion sensor that sounds a loud alarm when a firefighter becomes motionless. It can also be manually activated. Also known as Personal Alert Device.

Personal Protective Equipment (PPE) — General term for the equipment worn by fire and emergency services responders; includes helmets, coats, trousers, boots, eye protection, hearing protection, protective gloves, protective hoods, self-contained breathing apparatus (SCBA), personal alert safety system (PASS) devices, and chemical protective clothing.

Point of No Return — Point at which air in the SCBA will last only long enough to exit a hazardous environment.

Pre-Incident Plan — Document, developed during pre-incident planning that contains the operational plan or set procedures for the safe and efficient handling of emergency situations at a given location, such as a specific building or occupancy.

Pre-Incident Survey — Assessment of a facility or location made before an emergency, in order to prepare the appropriate emergency response.

R

Reaction Distance — The distance a vehicle travels while a driver is transferring the foot from the accelerator to the brake pedal after perceiving the need for stopping.

Regulations — Rules or directives issued by authorized administrative agencies.

Risk — (1) Likelihood of suffering harm from a hazard; exposure to a hazard. The potential for failure or loss. (2) Estimated effect that a hazard would have on people, services, facilities, and structures in a community; likelihood of a hazard event resulting in an adverse condition that causes injury or damage.

Risk Management Plan — Written plan that identifies and analyzes the exposure to hazards, selects appropriate risk management techniques to handle exposures, implements those techniques, and monitors the results.

S

Safety Data Sheet (SDS) - Reference material that provides information on chemicals that are used, produced, or stored at a facility. Form is provided by chemical manufacturers and blenders; contains information about chemical composition, physical and chemical properties, health and safety hazards, emergency response procedures, and waste disposal procedures. Also known as Material Safety Data Sheet (MSDS) or Product Safety Data Sheet (PSDS).

Situational Awareness — The ability to comprehend the implications of surrounding conditions and apply that understanding to help predict how and when the situation may change in order to take appropriate action.

Standards — Criterion documents developed to serve as models or examples of desired performance or behaviors that contain requirements and specifications outlining minimum levels of performance, protection, or construction. No one is required to meet the requirements set forth in standards unless they are legally adopted by the authority having jurisdiction, in which case they become law.

Strategic Safety Plan — Plan that identifies specific goals that support and enhance the safety and risk management activities of the department.

Size-Up — Ongoing evaluation of influential factors at the scene of an incident.

T

Target Hazard — Facility in which there is a great likelihood of life or property loss in the event of a fire, terrorist attack, or natural disaster.

Terrain — The physical features of a tract of land.

Total Stopping Distance — Sum of the driver/operator reaction distance and the vehicle braking distance.

Train-the-trainer — Training program in which key members of the organization train other company personnel in a specific topic, procedure, or policy.

V

Visual Lead Time — The concept that driver/operators should scan far enough ahead of the apparatus, for the speed it is being driven, to assure that evasive action can be taken if it becomes necessary.

W

Weapon of Mass Destruction (WMD) — Any weapon or device that is intended or has the capability to cause death or serious bodily injury to a significant number of people through the release, dissemination, or impact of toxic or poisonous chemicals or their precursors, a disease organism, or radiation or radioactivity; may include chemical, biological, radiological, nuclear, or explosive (CBRNE) type weapons.

Wildland/Urban Interface (WUI) — Line, area, or zone where an undeveloped wildland area meets a human development area.

Workload Management — The Crew Resource Management factor that attempts to reduce stress by reducing a job to its most basic tasks and assigning those tasks to individuals and groups.

CAPTAIN
SKILLY
FIREFIGHTER
DILLON

Index

CAPTAIN
SKILLY
FIREFIGHTER
DILLON

Index

B

E

F

J

K

L

O

P

S

T

U

Y

Z